Investigating Science Communication in the Information Age

Investigating Science Communication in the Information Age

Implications for public engagement
and popular media

Edited by

Richard Holliman, Elizabeth Whitelegg,
Eileen Scanlon, Sam Smidt, and Jeff Thomas

Published by Oxford University Press, Great Clarendon Street, Oxford OX2 6DP
in association with The Open University, Walton Hill, Milton Keynes MK7 6AA.

Oxford University Press is a department of the University of Oxford.
It furthers the University's objective of excellence in research, scholarship,
and education by publishing worldwide in

Oxford New York

Auckland Cape Town Dar es Salaam Hong Kong Karachi
Kuala Lumpur Madrid Melbourne Mexico City Nairobi
New Delhi Shanghai Taipei Toronto

With offices in

Argentina Austria Brazil Chile Czech Republic France Greece
Guatemala Hungary Italy Japan Poland Portugal Singapore
South Korea Switzerland Thailand Turkey Ukraine Vietnam

Oxford is a registered trade mark of Oxford University Press
in the UK and in certain other countries

Published in the United States
by Oxford University Press Inc., New York

First published 2009

© The Open University 2009

British Library Cataloguing in Publication Data

Data available

Library of Congress Cataloging in Publication Data

Data available

Typeset by Graphicraft Limited, Hong Kong
Printed in Great Britain
on acid-free paper by
Ashford Colour Press Ltd, Gosport, Hampshire

This book forms part of the Open University course SH804 *Communicating science in the
information age*. Details of this and other Open University courses can be obtained from
the Student Registration and Enquiry Service, The Open University, PO Box 197, Milton Keynes
MK7 6BJ, United Kingdom: tel. +44 (0)845 300 60 90, email general-enquiries@open.ac.uk

http://www.open.ac.uk

ISBN: 978–0–19–955266–5

1 3 5 7 9 10 8 6 4 2

■ PREFACE

This volume, like its companion (Holliman *et al.* 2009), deals with developments that have occurred in science communication over the past 10 years, in this case in the theory, research and practice thereof. During this time we have continued to teach a previous Open University (OU) course called *Communicating Science* (see Scanlon *et al.* 1999a,b), as part of the MSc in Science, MSc in Science and Society and Postgraduate Diploma in Science and Society.

It goes without saying that the study and practice of science communication has developed quite dramatically during this time. We are therefore eager to acknowledge that this volume is, at least in part, informed by these developments, and also by our experiences of producing (and running for 10 years) *Communicating Science* and our interactions with a wide range of students, many of whom have been practising science communicators, some of whom have gone on to become researchers and practitioners in this burgeoning field of scholarship.

This volume has a particular focus on the investigation of contemporary forms of science communication. Our planning for this volume has been informed by what we have learnt from our experiences of participating in a number of research projects, including research into different forms of contemporary science communication (e.g. see Carr *et al.*, Chapter 6.2 this volume; Holliman and Jensen, Chapter 1.3 this volume; Holliman and Scanlon, Chapter 6.3 this volume; Jensen and Holliman, Chapter 2.1 this volume; Thomas 2009). In addition, collaborative work undertaken by members of the editing team (RH, JT) with colleagues from across the European Union as part of the European Network of Science Communication Teachers (ENSCOT Team 2003), has extended our experience beyond the confines of the UK.

We are grateful to a number of academics who have supported our efforts in the past, including: our two respective external examiners, Roger Hartley and Brian Trench; external assessors on the first course and this one (Robin Millar and Joan Leach, respectively); guest lecturers, including Jane Gregory, Alan Irwin, Joan Leach and Brian Trench; members of the original production course team, including Roger Hill, Jill Tibble (BBC), Carol Johnstone, Kirk Junker, Hilary MacQueen, Cheryl Newport, Shelagh Ross, Rissa de la Paz (BBC) and Simeon Yates; and the support from Peter Taylor and Susan Tresman who each led the postgraduate programme offered by the Science Faculty at the OU over this period.

No course of this nature would be possible without a committed and skilled support team. To this end, we are especially grateful to the ongoing efforts of Christine Marshall and Carol Johnstone, and to James Davies and Martin Chiverton. Furthermore, we are also grateful to Jonathan Crowe from Oxford University Press and Christianne Bailey from the OU's co-publishing department for their support in the development of this project, and to Eric Jensen and Jane Perrone.

We acknowledge the following for permission to use images for the Section and Chapter openings. In Chapter 1.3 the photograph of 'The Magic of Oxygen' was provided by Mike Batham and Rob Janes. (For more information about this activity, see http://www.open.ac.uk/science/outreach/outreach-projects/the-magic-of-oxygen.php.)

In Chapter 6.2 the drawing of a scientist was provided by one of the participants in the (In)visible Witnesses research project.

Finally, we would also like to acknowledge those who have worked at the 'coal-face' of OU teaching, the course tutors; the course would not have been a success without them. Over the years we have worked with a number of academics who have worked in this capacity. In alphabetical order they are: Kim Alderson, Susan Barker, Bruce Etherington, John Forrester, Richard Holliman, Vic Pearson, Charlotte Schulze, Zbig Sobiersierski, Rachel Souhami, Anna Tilley and Simeon Yates.

Richard Holliman,
Elizabeth Whitelegg,
Eileen Scanlon,
Sam Smidt and
Jeff Thomas

■ **REFERENCES**

ENSCOT Team (2003). ENSCOT: the European Network of Science Communication Teachers. *Public Understanding of Science*, **12**(2), 167–81.

Holliman, R., Thomas, J., Smidt, S., Scanlon, E. and Whitelegg, E. (eds) (2009). *Practising Science Communication in the Information Age: Theorizing Professional Practices*. Oxford University Press, Oxford.

Scanlon, E., Hill, R. and Junker, K. (eds) (1999a). *Communicating Science: Professional Contexts*. Routledge, London.

Scanlon, E., Whitelegg, E. and Yates, S. (eds) (1999b). *Communicating Science: Contexts and Channels*. Routledge, London.

Thomas, J. (2009). Controversy and consensus. In: *Practising Science Communication in the Information Age: Theorizing Professional Practices* (ed. R. Holliman, J. Thomas, S. Smidt, E. Scanlon and E. Whitelegg). Oxford University Press, Oxford.

■ CONTENTS

■ ABBREVIATIONS AND ACRONYMS

AAAS	American Association for the Advancement of Science	**NESTA**	National Endowment for Science, Technology and the Arts
ABSW	Association of British Science Writers	**NICE**	National Institute for Health and Clinical Excellence
BA	British Association (for the Advancement of Science)	**NGOs**	Non-governmental organisations
BBC	British Broadcasting Corporation	**Ofcom**	Office of Communications
BSE	Bovine spongiform encephalopathy	**PEST**	Public engagement with science and technology
CAQDAS	Computer-aided qualitative data analysis software	**PRP**	Public relations professional
CGI	Computer-generated imagery	**PUS**	Public understanding of science
COPUS	Committee on the Public Understanding of Science	**RSS**	Really simple syndication
DEFRA	Department of the Environment, Food and Rural Affairs	**SciDev.Net**	Science Development Network
		SPSS	Statistical Package for Social Sciences
DfES	Department for Education and Skills	**STAGE**	Science, Technology and Governance Network
DOGs	Digital on-screen graphics	**STEM**	Science, technology, engineering and mathematics
EC	European Commission		
EU	European Union	**STEMPRA**	Science, Technology, Engineering and Medicine Public Relations Association
eTOC	Electronic table of contents		
GM	Genetically modified	**vCJD**	variant Creutzfeld–Jakob disease (previously new variant)
ICT	Information and communication technology	**UKRC**	United Kingdom Resource Centre for Women in Science, Engineering and Technology
MAFF	Ministry of Agriculture, Food and Fisheries		
NASW	National Association of Science Writers	**URL**	Uniform resource locator

■ BIOGRAPHIES OF CONTRIBUTORS

Stuart Allan is Professor of Journalism in the Media School, Bournemouth University, UK. He is the author of *News Culture* (1999, 2nd edn 2004), *Media, Risk and Science* (2002) and *Online News: Journalism and the Internet* (2006). His edited collections include *Environmental Risks and the Media* (2000, with B. Adam and C. Carter). Currently, his research focuses on newspaper and online news reporting of nanotechnology, findings from which have been published in journals such as *Science Communication, Health, Risk and Society* and *Public Understanding of Science*, as well as in the book *Nanotechnology, Risk and Communication* (with A. Anderson, A. Petersen and C. Wilkinson). He is a book series editor for the Open University Press.

James Bennett is Senior Lecturer in Media Studies at London Metropolitan University and the author of a number of articles on digital, interactive television in the UK. His work has been published in *Convergence, Screen, New Review of Film and Television Studies* and a number of edited collections. He has been a member of the Midlands Television Research Group (2002–07), chairing the group in 2004–05. He has presented papers at Screen (2003, 2005 and 2006), SCMS (2005 and 2007), Bradford Museum of Film, Photography and Television (2003), Roehampton University and the University of Madison-Wisconsin (2005). He organized the international conferences *'What is a DVD?'* . . . *Some people are disappointed to only get the film* at the University of Warwick (2005) and *Television (Studies Goes Digital)* at London Metropolitan University. He is currently editing the book *Film and Television after DVD* (Routledge, 2008).

Jenni Carr is a Postdoctoral Research Fellow in the Faculty of Science at the Open University (OU), currently working on the *(In)visible Witnesses* project. She has also been an Associate Lecturer at the OU for a number of years, working on both undergraduate social policy courses and the postgraduate social science programme. Her research interests focus on policies and initiatives designed to widen participation in learning and the impact that these policies may have on both learners and practitioners. As such, her PhD project explored the ways in which informal learning in non-traditional settings could play a positive role in overcoming barriers to participation in more formal learning programmes.

Sarah Davies has a BSc in Biochemistry, a MSc in Science Communication, and a PhD examining 'the public' of scientists' talk and public engagement events. She has worked in exhibition development at London's Science Museum and taught science undergraduates about science communication at Imperial College London. Her publications include theoretical and empirical analyses of 'informal' public dialogue events and descriptions of the complexity of scientists' talk about science and the public. Based at Durham University's Institute of Hazard and Risk Research, she is currently working on a project looking at 'lay' ethics of nanotechnology.

Anders Hansen is Deputy Director of the Centre for Mass Communication Research and Lecturer in Mass Communications in the Department of Media and Communication, University of Leicester, UK. He is Head of Teaching for the department's programme of undergraduate and postgraduate courses in media and communication, and Course Director for the department's popular MA in Mass Communications. His main areas of research include environmental issues and science communication, health communication, news management and journalistic practices, and media roles in the construction and careers of social/political problems. He is Chair of the Working Group on Environmental Issues, Science and Risk Communication under the International Association for Media and Communication Research (IAMCR). His published work includes articles in *Media, Culture and Society*, *Public Understanding of Science* and *Communications*, as well as the books *The Mass Media and Environmental Issues* (Leicester University Press, 1993) and, as lead-author, *Mass Communication Research Methods* (Palgrave/Macmillan, 1998).

Barbara Hodgson is Senior Lecturer in Educational Technology at the Open University (OU) and currently chairs a course on the OU MA in Online and Distance Education. Trained as a physicist she researched and taught physics in higher education before moving into science education and educational technology. She has worked on the development of a wide range of science and science education courses and continued to teach physics at the OU. Until recently she directed the OU 'Postgraduate Certificate in Teaching and Learning in Higher Education' programme. Throughout her career she has had a research and implementation/intervention interest in gender and science at all levels of education. Most recently her research has been concerned with women's careers in STEM and, as a partner in the UKRC, she has been working to help women return after career breaks. She is a member of the OU's Centre for Research in Education and Educational Technology and is part of the *(In)visible Witnesses* research project team.

Richard Holliman is Senior Lecturer in Science Communication at the Open University (OU), UK and production course team chair of *Communicating Science in the Information Age*. After completing a PhD investigating the representation of contemporary scientific research in television and newspapers in the Department of Sociology at the OU, he moved across the campus to the Faculty of Science. Since that time he has worked on a number of undergraduate and postgraduate course teams, and as part of the European Network of Science Communication Teachers (ENSCOT), producing mixed-media materials that address the interface between science and society. He edited (with Eileen Scanlon) *Mediating Science Learning Through ICT* (2004, Routledge) and, more recently (with Jeff Thomas) a special issue of the *Curriculum Journal* (**17**: 3) on science learning and citizenship. He is a member of the OU's Centre for Research in Education and Educational Technology and is leading (with colleagues) the ISOTOPE (Informing Science Outreach and Public Engagement) and *(In)visible Witnesses* research project teams.

Susanna Hornig Priest has written and taught about science communication and its impact on audiences for close to two decades, and she has written many dozens of articles and book chapters on this topic. She is also interested in the impact of new media on public access to information about science. She is presently Professor of Journalism and Media Studies at the Hank Greenspun School, University of Nevada, Las Vegas, from where she edits the journal *Science Communication*.

Alan Irwin is Dean of Research at Copenhagen Business School. His PhD is from the University of Manchester and he has held previous appointments at Manchester, Liverpool and at Brunel University. Alan currently chairs the UK BBSRC (Biotechnology and Biological Sciences Research Council) Strategy Panel on 'Bioscience for Society'. Alan Irwin has published widely on issues of science and technology policy, risk and science–public relations. His books include *Risk and the Control of Technology* (Manchester University Press, 1985), *Citizen Science* (Routledge, 1995), *Sociology and the Environment* (Polity, 2001), and (with Mike Michael) *Science, Social Theory and Public Knowledge* (Open University Press, 2003). He was also co-editor (with Brian Wynne) of *Misunderstanding Science?* (Cambridge University Press, 1996), and co-author of *The Received Wisdom* (Demos, 2006). His most recent research has been on the governance of science—including work with the UK Department of Environment, Food and Rural Affairs on 'lay' advice in the policy process.

Eric Jensen is the ISOTOPE Project Post-Doctoral Research Fellow in the Department of Chemistry, Open University. He studied at the University of Cambridge as a Gates-Cambridge Scholar for his MPhil and PhD in sociology. He has supervised Cambridge undergraduates on qualitative methodology, social theory and the media, and has lectured on the human cloning debate, culture, media and the role of religion in politics. His PhD research examined news production and content in Anglo-American press coverage of therapeutic cloning. Recent publications include: 'The Dao of human cloning: hope, fear and hype in the UK press and popular films' in *Public Understanding of Science*; 'Scientific controversies and the struggle for symbolic power', in B. Wagoner (ed.), *Symbolic Transformations: the Mind in Movement Through Culture and Society* (Routledge); and 'Abortion rhetoric in American news coverage of human cloning', *New Genetics and Society*.

Joan Leach convenes the Science Communication Programme at the University of Queensland in Australia. She edits the quarterly journal *Social Epistemology*, and has published on rhetorics of science, popular science and science mediation. She has taught at Imperial College, London (as part of the Science Communication Group), the Open University (contributing as an author and associate lecturer to the courses *Science and the Public* and *Communicating Science*, respectively) and the University of Pittsburgh (Rhetoric of Science Program) before moving to Australia. She has been named a Science Communication Laureate by Purdue University for her work in science communication and is the author of a forthcoming book, *Valuing Communication in Science*.

Robin Meisner is a recent graduate of the Center for Informal Learning and Schools (CILS) at King's College London where she earned her doctorate in education research. Her work focuses on investigating children's activity at interactive science exhibits and looking at ways to improve the design and deployment of such exhibits. She has been involved in a range of research and evaluation projects concerned with how people use and make sense of novel science exhibits. Prior to focusing on research, she was a science exhibit and programme developer at Providence Children's Museum and a course instructor at Boston's Museum of Science, both in the USA.

Felicity Mellor is Lecturer in Science Communication at Imperial College London, where she runs the MSc in Science Communication and teaches media theory. She holds a PhD in theoretical physics from Newcastle University and was lecturer in astronomy at Sussex University before switching her focus to critical analyses of science. Her current research interests include the popularization of physics, representation of science in the media and the role of narrative within science. Her publications include papers in *Social Studies of Science* and *Public Understanding of Science*.

Jonathan Osborne is Head of the Department of Education and Professional Studies and the Chair of Science Education at King's College London where he has been since 1985. Prior to that he taught physics in Inner London for 9 years. He has an extensive record of publications and research grants in science education in the field of primary science (the SPACE project), science education policy (*Beyond 2000*), the teaching of the history of science, argumentation (the IDEAS project) and informal science education. He was an advisor to the House of Commons Science and Technology Committee for their report on science education. He was also President of the US National Association for Research in Science Teaching (NARST) (2006–07) and has won this association's award for the best research publication in both 2003 and 2004 in the *Journal of Research in Science Teaching*.

Eileen Scanlon is Professor of Educational Technology at the Open University and a Visiting Professor at the Moray House School of Education, University of Edinburgh. She co-directs the Centre for Research in Education and Educational Technology at the Open University. Her books include *Communicating Science: Professional Contexts* (Routledge, 1999), edited with Roger Hill and Kirk Junker; *Communicating Science: Contexts and Channels* (Routledge, 1999) edited with Elizabeth Whitelegg and Simeon Yates; *Reconsidering Science Learning* (Routledge Falmer, 2004) edited with Patricia Murphy, Jeff Thomas and Elizabeth Whitelegg; and *Mediating Science Learning with ICT* (Routledge Falmer, 2004) edited with Richard Holliman.

Sam Smidt is a senior lecturer based in the Department of Physics and Astronomy at the Open University, and Award Director of the MSc in Science. She has contributed to a number of courses in physics and also to courses concerned with contemporary science and society issues at both undergraduate and Masters level, including *Science in Context*, *Contemporary Issues in Science Learning* and *The Science Project Course: Science and Society Project*. She has interests in physics education and outreach work that promotes science to the public.

Jack Stilgoe is a senior researcher at the think tank Demos. He works on science and technology projects and specializes in issues of science, expertise and public engagement. Previously, he was a research fellow in the Science and Technology Studies department at University College, London, where he looked at debates between scientists and the public about the possible health risks of mobile phones. He has a degree in economics, an

MSc in science policy and a PhD in the sociology of science. He has recently published academic papers in the journals *Science and Public Policy* and *Public Understanding of Science*, and is the co-author of a number of Demos publications, including: *The Public Value of Science* (2005, with B. Wynne and J. Wilsdon); *The Received Wisdom* (2006 with A. Irwin and K. Jones) and *Nanodialogues* (2007).

Brian Trench is senior lecturer and former head of school at the School of Communications, Dublin City University, where he coordinates the Masters in Science Communication and chairs the university's research ethics committee. He teaches courses in Science and Society and Science in the Media, as well as other courses in communication and journalism. His science communication-related research interests include: uses of the internet in science communication; representations of science in discourses of the knowledge society; and models and strategies in science communication. He is co-editor, with Massimiano Bucchi, of the *Handbook on Public Communication of Science and Technology* (Routledge, 2008), a member of the scientific committee of the Public Communication of Science and Technology (PCST) international network, and a former member of the Irish government's Council for Science, Technology and Innovation.

Jeff Thomas is a senior lecturer within the Department of Life Sciences at the Open University (OU). He has worked at the OU all his professional life, contributing to a wide range of teaching initiatives in biology and in health sciences, and more recently to a range of projects concerned with contemporary science issues and on the relationships between science and different publics, at both undergraduate and Masters level, and as part of the European Network of Science Communication Teachers (ENSCOT). He co-edited *Science Today; Problem or Crisis* (with Ralph Levinson; Routledge, 1997) and *The Sciences Good Study Guide* (with Andrew Northedge, Andrew Lane and Alice Peasgood; Open University, 1997), and (with Richard Holliman) co-edited a special issue of the *Curriculum Journal* (**17**: 3). His research interests are concerned with the influence of contemporary science controversies on public attitudes, on conceptual problems of learning biological science and in public involvement in science-based policy-making. He also teaches part-time for Birkbeck College, University of London on its Diploma in Science Communication.

Elizabeth Whitelegg is Senior Lecturer in Science Education working in the Science Faculty at the Open University, and inaugural Award Director for the Science Short Course Programme which she continues to lead. She has worked on a range of science and society-related courses, including *Communicating Science* and *Contemporary Issues in Science Learning*. As a member of the Centre for Research in Education and Educational Technology her main research interest is in girls' and women's participation in science and in learning science (particularly physics) at all levels and she recently produced (with Professor Patricia Murphy) a review of the research literature on the participation of girls in physics for the Institute of Physics. She is currently leading (with Richard Holliman) the *(In)visible Witnesses* project. In 2003 she was invited to become a Fellow of the Institute of Physics.

James Wilsdon is head of science and innovation at Demos and a senior research fellow at the Institute for Advanced Studies at Lancaster University. His research interests include science and innovation policy, emerging technologies, sustainability and globalization. He regularly advises government agencies, companies and NGOs on these topics. Together with Charles Leadbeater, he is the coordinator of *The Atlas of Ideas* project. His recent publications include: *The Atlas of Ideas* (2007); *Governing at the Nanoscale* (with M. Kearnes and P. Macnaghten, Demos, 2006); *The Public Value of Science* (with B. Wynne and J. Stilgoe, Demos, 2005); and *See-through Science* (with R. Willis, Demos, 2004).

Simeon Yates is the Director of the Culture, Communication and Computing Research Institute (C3RI) at Sheffield Hallam University. He has a background in social science (communication studies) as well as an interest and training in science (geology). His current broad research interests include: new media, language, culture and interpersonal interaction; scientific and technical communication; discourse analysis—theory and methods. He is joint editor of the web-based journal *Discourse Analysis Online* and a member of the editorial board of the *Journal of Computer Mediated Communication*. He previously worked at the Open University, contributing as an author and associate lecturer to a number of courses, including *Communicating Science*, and at the University of Leeds.

■ INTRODUCTION TO THE VOLUME

In this introduction we offer some advice on how the chapters in this volume could be used by students, researchers and teachers of science communication, and provide an overview and rationale for the selection of themes, sections and chapters by which the volume is organized.

How to use this book

The field of science communication, as an active field of systematic research, is still relatively young (see Hansen, Chapter 3.1 this volume). Social researchers from a number of disciplinary backgrounds continue to conduct science communication research, contributing to what is a diffuse and multi-disciplinary corpus of literature. As such, there is currently no single overarching theoretical perspective that informs these studies. As a multi-disciplinary editing team we welcome this diversity and the opportunities for collaborative working that it has brought to us. This edited collection reflects some of this diversity. Indeed, we see great strengths in not confining students (researchers and lecturers) of science communication to what would inevitably be an artificially enforced single perspective. Rather, we wish to engender critical engagement with what we (and many others) consider to be important and often complex issues. Neither do we claim to have addressed everything that a reader might want to know about the study of contemporary science communication. Instead, our aim has been to commission and edit what we consider to be stimulating and thought-provoking chapters from leading academics, addressing issues that we hope students, researchers and lecturers will explore further.

Having said this, we note that many of the authors draw on sociologically informed theories to inform their contributions. Such choices, in all forms of communication, are rarely accidents. Rather they reflect active, if sometimes unconscious, decisions; in this case our conscious decision to commission authors whom we knew to be broadly sympathetic to sociologically informed approaches to the study of science communication. In a large part, this reflects our prior knowledge and experience of science communication research. It should therefore come as no surprise that we have commissioned authors who we have found to be persuasive, engaging and thought-provoking to us in the past.

Researching science communication

One of the main focuses of science communication research is studying the choices that producers and receivers—concepts that are becoming increasingly blurred with the emergence of user-generated forms of communication—make when participating in acts of science communication; to make more visible the process of mediation and agency that are fundamental to all forms of (science) communication. Concomitant to these studies are those that aim to provide informed accounts of various forms of media content, where media is defined in the plural sense to cover different forms (newspapers, radio, television, blogs, etc.) and genres (news and current affairs, science fiction, comedy, docudrama, etc.).

Such investigations centre on the 'why', 'when', 'where', 'who', 'how', 'what for' and 'to what effect' of science communication acts. In the 'information age', where the proliferation of alternative accounts of (scientific) knowledge equates not only to power, assuming one has access to such information, but also to unequally distributed economic prosperity in a competitive globalized economy, these questions place additional requirements on the science communication researcher. Challenging as the ever-changing context may be, we argue that by attempting to answer these sorts of important questions from a critical perspective, science communication researchers can shed light on the often-overlooked processes of mediation and agency, documenting some of the motivations and constraints of communicating science in the information age.

The structure of the book

This collection is based on two broad themes: public engagement and popular media. Necessarily this has been at the expense of other themes, so it is worth briefly explaining our reasoning for these choices.

Much has been written about the so-called 'dialogic turn' from public understanding of science towards public engagement with science; notably whether such a shift is evident in the practices of those who choose to engage. Increasingly, critical authors have raised important questions that have begun to address apparently 'taken-for-granted' assumptions about the nature and purpose of public engagement with science, and the methods enacted under this ascendant branding. In choosing this theme for the first two sections of the book, we hope to focus an additional critical lens on the current context for science–society relations, inviting students, researchers and lecturers to challenge how they (re)conceptualize and investigate this important area for study.

In conceiving the second theme of the book—popular media, Sections 3 to 6 of the volume—three issues are worthy of note. First, we have defined popular media in the plural. This may seem an obvious assumption to state, but we note the number of occasions when the term 'media' has been used as an umbrella term, somewhat unfortunately, for a whole range of different communicative forms and genres. Of course, we have been as guilty in this respect as many others, but we hope to begin to redress this situation in this volume. Second, we have defined popular media in a broad sense of the term to include a range of communicative forms and genres. These include, but are not confined to, news and current affairs, as represented in newspapers and on television, the combination of which has often been the focus of science communication research. Rather, we have attempted to broaden the discussion of 'sciences-in-media' in this volume to include other genres and forms of digital media. Third, we have defined the communication process, involving production, content and reception, symmetrically, challenging what we perceive to be a previous emphasis on content and production. Such a shift in focus, we argue, not only helps to illuminate the equally important element of reception within communicative acts, but is a necessary requirement given the increasingly blurred relationship between producers, content and reception that the information age affords.

Section 1: Engaging with public engagement

The three chapters comprising this section are rich in notions of 'challenge', 'potential' and 'opportunity'; they are less taken up with ideas about 'achievement', 'success' and 'established good practice'.

A mere 40 years ago, a key work on the politics of science (Rose and Rose 1969) made no mention of 'public engagement' and neither was it hopeful about the potential for greater democratization of science. Now hopes for 'dialogic' and 'contextual' science communication are unapologetically ambitious—the instrument of a reinvigorated social contract for science. Science communication is now indeed at a crossroads (Miller 2001), but frustratingly in need of a reliable map.

Alan Irwin's chapter (1.1) sets the broader scene with characteristic vision, but is more guarded about future directions. He too hopes that experiences from the past can better inform practice in future; BSE is his starting point—an episode now burnt deep into the consciousness of those who study science communicators, since it embodies so many of the unlamented practices of the past. But putting 'lessons learnt' into practice is far from straightforward and the challenges ahead carry such optimism and enthusiasm in part because they remain largely untested. As Irwin points out, if moving forward is to be achieved, the new territory will ally science communication with far-reaching issues of 'direction, sponsorship and control' of science—truly then a coming of age for what has too long been a fledgling discipline.

Stilgoe and Wilsdon's chapter (1.2) looks critically at recent progress towards 'upstream public engagement', which they see as a hallmark of a much healthier and productive interactivity between science and society. Such a process opens science up to public scrutiny and influence much more than do routine debates about levels of risk and uses of the products of science; it is concerned with questions of 'why?', of 'who benefits?', of 'who recognizes my worries?'—questions about what type of society should science help shape. Theirs is an ambitious recipe for encouraging and enabling scientists to reflect on 'the public within themselves'—by considering the ethical and social dimensions of their work. They offer some credible and tangible examples of their take on 'upstream public engagement'—public inputs helping to 'shape research priorities, review proposals from scientists, assess researchers and monitor research'—and contributing to a new social contract for science 'based on a deeper appreciation of the politics of science'.

The final chapter in Section 1, by Holliman and Jensen (1.3), also has interesting things to say about the relationship between present-day practice and the better territory that may lie around the corner. They looked for evidence in their research interviews with scientists for expressions of genuine commitment to the deliberative and contextual ambitions of authentic public engagement. Their findings suggest that such ambitious agendas are only rarely articulated by scientists, who are generally keen to contribute in ways they understand best. It is fitting to finish the section with a reminder of the difficult navigational tasks required for changing established practice and for reaching those promised new pastures—hopefully greener but undoubtedly different.

Section 2: Researching public engagement

Theoretically informed research investigating whether and how citizens—scientists and non-scientists—are engaging with science and what the outcomes of their engagement are should form a core part of public engagement activity if it is to grow and develop successfully. The three chapters in Section 2 address this issue to illustrate some of the ways this research is conducted.

The scene is set by Jensen and Holliman in Chapter 2.1, which provides a critically informed rationale for some of the various analytical techniques that can be employed by researchers investigating scientists' (and that of other practitioners) participation in science outreach and public engagement activities. This chapter is enriched by setting these techniques within the context of a current action research project, ensuring the outputs of the research inform future practice (see also Chapter 1.3).

The second chapter in this section, by Davies (2.2), puts one specific methodology into practice by adopting a case study approach to investigate practices of 'informal dialogue'. The case study, situated within the Dana Centre at London's Science Museum, examines the multiple 'framings' made by the organizers of panel debates, and the various ways that the interaction between the participants—organizers, speakers and audience members— raises questions about the role and function of informal dialogue events.

The final chapter in this section, by Meisner and Osborne (2.3), is concerned with children's interaction with interactive exhibits at a science museum. Using quantitative and qualitative data collection and analysis, the research examines children's behaviours when they are engaged in 'playing' with exhibits and investigates whether any signifiers of learning can be identified from a close examination of these behaviours.

Section 3: Studying science in popular media

Two chapters are included in this section. One by Anders Hansen (3.1) is a review of research on the topic of science, communication and media. The review is an eclectic one, guided by a perspective of media and communications work with a sociological inclination. The review is subdivided into three sections. First a section on the production of science in media, second on studies of media and content, and third studies of media audiences. Hansen concludes with a discussion of the challenges for science communication research of the fragmentation of media forms and the availability of new technologies and the resulting easy public access to resources of information, opinion and debate.

The paper by Joan Leach and colleagues (3.2) presents a discussion of various approaches to the development of models of communication that can be applied to the different settings in which science communication occurs. Three perspectives are presented as example models: the transmission model, the ritual model and the media studies model. The paper discusses the usefulness of models of science communication and describes their value as analytic frameworks that are of use in the study of science communication.

Section 4: Mediating science news

Section 4 examines the mediation of science news. The two chapters, from Stuart Allan (4.1) and Brian Trench (4.2) respectively, illustrate some of the challenges and opportunities facing contemporary science journalists, and those researching these issues.

In Chapter 4.1 Allan describes recent and historical examples in which the accuracy of science-based news reports was called into question. Importantly, he notes that the more recent example that he describes could be attributed, at least in part, to the internet; the result of overworked journalists resorting to single web-based sources of questionable provenance, with the resulting reports being challenged by 'bloggers'; the latter having been described elsewhere as an emerging 'fifth estate' (see Holliman 2008 for discussion). Having considered these examples and discussed the professionalization of science journalism, Allan then considers 'how science news is made', paying particular attention to the concepts of news value(s) and framing in relation to the selection and subsequent construction of news reports. He concludes with a discussion of the role of the internet as a source of scientific information for journalists, scientists and citizens, noting how ever-greater convergence requires that *users* adopt a 'digital mindset'.

In Chapter 4.2 Trench examines the management of science news sources, mediated increasingly through a range of electronic forms. In particular, he examines the proactive (often global) promotion of science through a range of press services, for example through the routine use of information subsidies by high-profile scientific journals, and other scientific, higher education and privately funded institutions. He makes a convincing argument for one of the outcomes of the success of this strategically managed flow of electronically disseminated scientific information: the increasingly deskbound nature of research conducted by science journalists. He argues that this situation may result in a reduction in investigative newsgathering as opposed to reactive news processing.

Section 5: Communicating science in popular media

Section 5 examines the communication of science in two popular media; television and newspapers. A further distinction can be made in terms of the genres discussed in the chapters that form Section 5, which examine examples of news and current affairs, documentaries and 'un-natural history'.

In Chapter 5.1 Bennett makes a critical examination of recent science programming on television, focusing on examples of 'un-natural history', and relating his arguments to current discussions in the UK about the shifting role of public sector broadcasting in the emerging digital television landscape. Drawing on Raymond Williams' notion of 'flows', he argues for extending this analytical concept to address 'user flows'. In so doing, he both acknowledges and critiques the increasing opportunities for audience members—viewers, viewsers, users—to engage selectively with digital output, considering the implications of this shift both in relation to how public sector broadcasting remits are (not) being fulfilled, and with respect to ongoing concerns about dumbing down and (im)partiality.

In Chapter 5.2 Mellor describes a detailed qualitative analysis of media content through two examples of science communication in popular media: a newspaper article and the opening sequence from a television documentary. In so doing she draws selectively on analytical concepts from semiotics to examine how images, including photographs and moving images, sounds, including music, and texts, in both written and spoken form, play an important role in the construction of preferred polysemic meaning(s) of popular science.

Section 6: Examining audiences for popular science

Section 6, which includes three chapters, makes a critical examination of contemporary audiences for science in popular media.

In Chapter 6.1 Susanna Hornig Priest reconsiders the concept of audiencehood in relation to popular science. She argues that issues of access, fragmentation, social networking and participation require a broader conceptualization of what it means to communicate science. In arguing that audiences have contingent motivations and constraints that encourage active consumption (or avoidance) of science in popular media, she discusses the concept of democratic citizenship in relation to science, in particular in relation to the proliferation of alternative sources of scientific information in digital media. At least some of these sources are produced and consumed by multiple types of (non-science) users who draw on different forms of expertise, communicating for particular purposes. She argues that this proliferation of scientific information requires continued reskilling among those who communicate science.

Jenni Carr and colleagues (6.2) provide a critically informed rationale for some of the qualitative and quantitative methods that can be used to investigate how audiences make sense of science in popular media, in particular in relation to the development of self-concept and science. These multi-disciplinary authors describe the data collection and analytical methods used in a recent study of children's reception of science on television. The authors document how they adapted and combined a range of previously used analytical methods to provide a triangulated approach that could take account of the gender and age differences of participants, whilst ensuring that they were an integral part of the research process.

Richard Holliman and Eileen Scanlon (6.3) document the findings from a study of newspaper reception of 'contested science', where participants were encouraged to discuss how they actively interpreted and contextualized (or avoided) this reporting. Drawing on two examples—finger length and sexuality, and genetics and intelligence—the chapter compares the main findings from a study of newspaper content with that of a reception study involving 14 focus groups. The authors argue that controversial topics in science provide scientific citizens with interesting opportunities for discussion and action that can have consequences for their engagement with science.

Further reading and useful web sites

Each of the chapters features a selection of briefly annotated 'further reading' and 'useful web sites'. As stated above, in producing this collection we started from the premise that the enclosed readings could be considered as a starting point for students (researchers and lecturers), but that we would also encourage further exploration of the corpus of science communication research and associated mixed media resources. These annotated selections can be seen as initial guidance for these further explorations. Some of these have been chosen and written by the authors, others by the editing team, others still by Eric Jensen; we would argue that they can and should be considered worthy of further study.

It should come as no surprise that several authors referred to the same sources in preparing these annotated sources. Rather than duplicate what are considered to be a small number of generative texts—House of Lords (2000), Irwin and Wynne (1996) and Nelkin (1995) being the most obvious—we have attempted to provide a selection from which students (and lecturers) can choose. It follows that we have made the decision not to mention a given 'further reading' or 'useful web site' in more than one of the chapters.

Richard Holliman,
Elizabeth Whitelegg,
Eileen Scanlon,
Sam Smidt and
Jeff Thomas

■ **REFERENCES**

Holliman, R. (2008). Communicating science in the digital age—issues and prospects for public engagement. In: *Readings for Technical Writers* (ed. J. MacLennan), pp. 68–76. Oxford University Press, Toronto.

House of Lords, Select Committee on Science and Technology (2000). *Science and Society*, Third Report. HMSO, London.

Irwin, A. and Wynne, B. (eds) (1996). *Misunderstanding Science? The Public Reconstruction of Science and Technology*. Cambridge University Press, Cambridge.

Miller, S. (2001). Public understanding of science at the crossroads. *Public Understanding of Science*, **10**(1), 115–20.

Nelkin, D. (1995). *Selling Science: How the Press Covers Science and Technology*, 2nd revised edn. W.H. Freeman, New York.

Rose, H. and Rose, S. (1969). *Science and Society*. Penguin, London.

SECTION 1

Engaging with public engagement

Society's relationship with science is in a critical phase. Science today is exciting, and full of opportunities. Yet public confidence in scientific advice has been rocked by BSE; and many people are uneasy about rapid advances in areas such as biotechnology and IT—even though for everyday purposes they take science and technology for granted. This crisis of confidence is of great importance both to British society and to British science.

House of Lords Select Committee for Science and Technology (2000)
Science and Society (Third Report)

1.1

Moving forwards or in circles? Science communication and scientific governance in an age of innovation

Alan Irwin

Just how complicated can science communication be? Surely all it requires is that we choose the relevant facts, select the right audience (sometimes eager, sometimes more difficult to entice) and put the information across as persuasively as possible? And we hardly need to justify science communication in a world where technical change is taking place all around us. As the Royal Society (the UK's elite scientific organization) expressed it back in 1985:

. . . better public understanding of science can be a major element in promoting national prosperity, in raising the quality of public and private decision-making and in enriching the life of the individual. . . . Improving the public understanding of science is an investment in the future, not a luxury to be indulged in if and when resources allow.

Royal Society (1985, p. 9)

Partly inspired by the Royal Society report (but drawing also on deeper roots), many practical initiatives, academic courses and conferences have taken place promoting what in the late 1980s and 1990s became known as the 'public understanding of science'. Writing at a time when nanotechnology, stem cell research and energy policy are all matters of political as well as scientific concern, it is hard to resist the blunt argument that the more science communication takes place the better. As J. B. S. Haldane, one of the great science communicators of the 1930s and 1940s, put it:

I am convinced that it is the duty of those scientists who have a gift for writing to make their subject intelligible to the ordinary man and woman. Without a much broader knowledge of science, democracy cannot be effective in an age when science affects all our lives continually.

Haldane (1939, p. 8)

Of course, science communication now extends beyond Haldane's imaginative newspaper articles on such topics as 'Why bananas have no pips' and into any number of televisual, multimedia and information and communication technology (ICT)-connected conduits. However, the basic argument remains the same. Science is central to the modern world. Public knowledge of science is a democratic, personal and cultural asset. Enhanced science communication is therefore vital. Occasional controversies over nanotechnology, nuclear power and genetic research only increase the importance of science communication. What more is there to say?

This chapter is written in the belief that actually there is an awful lot more to be said about the theory and practice of science communication. In part, this reflects the substantial experience in science communication that has built up since the 1980s—and especially the experience of trying to 'engage' with the wider publics about science-related issues. Whether this experience suggests that science communication has moved forward or in circles is a question we will consider later. Certainly, a frustration has developed—especially among policy-makers and practitioners—that one-way communication may not be enough but that two-way communication is much more challenging than it might at first appear. It should also be stressed from the start that the point of considering science communication more deeply is not simply to make life complicated—or turn a lively endeavour into a dusty academic exercise—but rather to understand how scientific and social communications can be *improved*. As I hope to suggest, the intellectual and practical challenges of communicating science are profoundly interlinked.

Mad cows and cognitive deficits (and zombies too)

Perhaps the best way of opening up these issues is through a quick journey back in time—a journey which will take special account of British experience but without forgetting that these are international issues and concerns. We can start with one of the most-discussed episodes in UK science–public relations: the case of BSE (bovine spongiform encephalopathy—or 'mad cow disease' as it was generally known).

At the beginning of the 1990s, BSE was seen as an animal problem with an uncertain connection to human health and illness (Irwin 1995). No specific infectious agent had then been identified (talk of prions was to come later), but equally the existence of such a mechanism could not be ruled out. However, and as should have been clear at the time, 'absence of evidence' and 'evidence of absence' are certainly not synonymous. In demonstration of that point, by the end of the 1990s, a growing number of human deaths—in the specific form of new variant Creutzfeldt–Jakob disease (nvCJD)—were being linked to BSE. Over 4 million cattle were slaughtered as a precaution. Sales of British beef plummeted both nationally and internationally. The government department responsible found itself under great public and political pressure with many arguing that the dual roles of industry promoter and risk regulator were fundamentally incompatible. This pressure eventually led to the department being split apart. Reflecting widespread criticism of the government's handling of the whole episode, an influential report was later to declare: 'Confidence in government pronouncements about risk was a further casualty of BSE' (Lord Phillips *et al.* 2000, Vol. 1, Section 1)

As can already be gathered, communicating science in such a situation presents considerable challenges. What form should 'the science' to be communicated take when 'the facts' do not simply speak for themselves? When there are many different, albeit overlapping, publics (consumers, activists, farming communities, elected representatives, households, audiences at home and abroad), which are the publics with whom one should be communicating and in what way? One question would prove especially important within this case: *what assumptions about uncertainty should be built into public communications*? Is it better to acknowledge doubt (and risk appearing ignorant and out of control) or to attempt reassurance (and risk appearing at best complacent and at worst deceitful)? On that basis, the handling of the BSE crisis in the UK is generally viewed as a classic example of how science communication (and science governance) can go badly wrong.

There are many possible ways of telling the story of BSE. It could be recounted as a sombre tale of experts failing to identify (or at least to identify with sufficient alacrity) a causal link between an animal health problem and human morbidity/mortality. Equally, it could be told as a heroic account of a small group of scientists fighting to convince sceptical civil servants and politicians about the underlying threat. The story could also be presented as a manifestation of the commercial pressures which often surround (and indeed permeate) apparently 'technical' issues. Certainly, the British beef industry was very concerned in the early 1990s about the impact on sales of yet another food and health scare (for this was not the first). International politics also entered the picture, as British officials attempted to prevent their European counterparts from imposing a ban on beef exports (with much muttering in the British corridors of power about the Continental cousins imposing unfair barriers to trade).

All of these versions of the 'mad cow' crisis could be offered. However, two very significant quotations, separated by 10 years of bitter experience with BSE, suggest the most common lesson to be drawn from this case:

As the Chief Medical Officer has confirmed, British beef can continue to be eaten safely by everyone, adults and children.

John Gummer, Minister of Agriculture, Fisheries and Food, May 1990

The Government did not lie to the public about BSE. It believed that the risks posed by BSE to humans were remote. The Government was preoccupied with preventing an alarmist over-reaction to BSE because it believed that the risk was remote. It is now clear that this campaign of reassurance was a mistake. When on 20 March 1996 the Government announced that BSE had probably been transmitted to humans, the public felt that they had been betrayed.

Lord Phillips *et al.* (2000, Vol. 1, Section 1)

The story that emerges here is of a closed group of civil servants and advisors convincing one another that the risks were low and, having done that, embarking on a campaign of public reassurance which played down any element of doubt. In these circumstances, spokespeople tended to dismiss other, more critical, viewpoints as uninformed and irrational. As my own experience at the time suggested, attempts to engage the Ministry in dialogue about issues of risk and public communication were met with a defensiveness which often employed science as a rhetorical weapon aimed at closing down discussion (Did I have better expertise in the subject than the government's scientists? Was I aware of specific empirical evidence that had not been located by the Ministry?). Of course, this

Figure 1 The then Secretary of State for Agriculture, Fisheries and Food, John Selwyn Gummer, and his daughter, Cordelia, eating beefburgers. (First broadcast on BBC2, *Newsnight*, 16 May 1990. Picture from Empics.com.)

in itself represents a form of science communication—although hardly what the Royal Society (1985) and Haldane (1939) had in mind.

In this climate, even the most gentle of inquiries was robustly dismissed as mischievous and politically motivated. Meanwhile, government statements on BSE risk adopted a categorical and confident tone which jarred badly with the scientific uncertainties. This was typified (and captured for all time) by the Minister's feeding of a beefburger to his young daughter before a cluster of eager photographers (Figure 1). As the minister in question subsequently explained it, this was intended as a sincere demonstration of confidence in the safety of British beef. Subsequent commentators have tended to be less generous of what they saw as an attempted exploitation of both the media and (even worse) the minister's own offspring.

It is unsurprising in these circumstances that the official report on the government's handling of BSE was so deeply critical. Terms such as 'unwarranted reassurance' (Lord Phillips *et al.* 2000, paragraph 1150) and 'culture of secrecy' (ibid., paragraph 1258) recur throughout the Phillips report. Indeed, this goes so far as to suggest that the main object of MAFF's communication strategy was 'sedation' of the public (ibid., paragraph 1179). The report focuses especially on the perceived need by civil servants and others to counteract what they saw as potentially alarmist public and media reactions to the existence of risk: 'the approach to communication of risk was shaped by a consuming fear of provoking an irrational public scare' (ibid., paragraph 1294). More broadly, the official report stressed several points that, in the wake of BSE, have become central to science communication and scientific governance more generally:

- Trust can only be generated by openness.
- Openness requires recognition of uncertainty, where it exists.
- The public should be trusted to respond rationally to openness.
- Scientific investigation of risk should be open and transparent.
- The advice and reasoning of advisory committees should be made public.

Before we go on to discuss this new mantra of openness, transparency, trust and dialogue (and, especially, what it means for science communication), it is worth pausing to consider some of the wider lessons from the BSE case. The governmental handling of the issue can certainly be presented as a textbook example of what has become known as the 'deficit' approach to science communication. This is usually defined as the assumption on the part of institutions and their science communicators that the public is ignorant about science—but that it (for this is a singular presentation of 'the public') would accept science readily if it only knew more (with 'science' similarly being singular rather than plural or heterogeneous). The deficit perspective suggests one-way communication with a passive audience soaking up 'the facts'. It also suggests that one approved model of 'science' should be central to the communication rather than (in this case) information about why usually herbivorous animals were being fed on meat and bone meal, what farmers and abattoir workers had to say about the attempted controls or else what minority views existed within the scientific community.

This model of the public as operating within a cognitive deficit is usually contrasted with the approach offered by the Phillips report which emphasizes a more open and two-way relationship between what we can term (rather awkwardly) 'the sciences' and 'the publics'. Going further, the alternative 'contextual' model of science communication emphasizes that lay people can also be informed and knowledgeable within the conditions of everyday life and, indeed, that science might have as much to learn as to communicate when it comes to understanding the social realities of farming, abattoir operation and feeding a family (Irwin and Wynne 1996).

This is not to say that a farmer's understanding of structural biology is likely to be as good as a qualified scientist's. But it is to suggest that scientists (and scientific communicators) must, on the one hand, be careful about generalizing outside the closed conditions of the laboratory and, on the other, appreciate that issues such as BSE are not solely scientific in character but incorporate wider questions of government competence and credibility, economic and political priorities, ethical concerns and understandings built on direct experience (including that of farmers and those responsible for domestic food preparation). Once we move away from a crude deficit model, science communication becomes rather more complicated—and certainly it involves listening as well as speaking. It also becomes more interesting as we start to consider the role of science within particular social contexts —and also the limits as well as the strengths of scientific analysis.

Now despite all of the post-BSE talk about the 'death of the deficit model', it has to be said that 'deficit talk' still possesses a zombie-like longevity, refusing to lie down in its grave but instead lurking within many attempts at science communication. Just when we all agree that the model of science communication employed in the early 1990s during the BSE crisis has no place in modern science communication, then it can again be spotted within the statements of government and other organizations (including environmental and campaigning groups). Science communication often takes the form of imploring (or shaming) us to be more rational about 'the facts', explaining to us what we evidently have not understood (or else why would we disagree with the communicator?), and persuading us that science and technology have the answers to our (usually unformulated and unasked) questions. On such occasions, we can feel that—far from making progress in science communication—we are actually just circling the same old issues of how to overcome what is presented as societal resistance to technical change.

Partly this is a structural and institutional problem: it can be hard for 'scientific' organizations to be 'two-way and contextual' in areas which they see as firmly within their professional competence. Partly also, it reflects the widespread emphasis within modern societies on science as *the* form of knowledge and understanding. However, the point must also be acknowledged that 'deficits' are inevitably and unavoidably embedded in our attempts at communication (Why else write this chapter? Why else read it?) and are not always to be avoided. In that spirit, I am willing to declare my cognitive deficit concerning large areas of mathematics, many world languages, the operation of numerous forms of domestic technology and (especially) the music of Barbra Streisand. 'Deficits' are not necessarily a bad thing—and are not necessarily to be avoided or denied. Taking that point further, it appears an inevitable characteristic of everyday life that we make choices (not always explicitly) about the kinds of knowledge and information we wish to acquire. Equally, and as those working with medical patients are often aware, so-called 'ignorance' of a topic may not simply represent the absence of information but a deliberate decision not to engage ('I don't know, and I don't want to know'). As one resident living close to a hazardous industrial installation put it to me: 'I could have a PhD in chemistry, but I'd still be living here'.

In challenging the 'deficit' approach to science communication, the point is not to deny the existence of knowledge deficits, but instead to be sensitive to underlying assumptions about science–public relations (especially what is being communicated and to whom), to consider alternative forms of knowledge and understanding and to take full account of the communication setting. It is not deficits themselves that are the problem but rather how we construct, respond to and make sense of these within specific contexts. One crucial element within this will be how we communicate scientific and social uncertainty in situations where there may be many understandings in operation. Very importantly also, the case of BSE suggests the dangers of using deficit talk as a means of avoiding institutional self-analysis and the active pursuit of alternative understandings and perspectives. Viewing 'the public' as the problem—and better communication as the answer—can be very comforting for an organization under scrutiny (at least in the short term). It can also mean that the all-important 'what if we are wrong?' question never gets asked.

Trust as the new deficit?

So what then of the new approach to openness and trust-building signalled by the UK Phillips report? Certainly, Phillips has not been alone in advocating a fresh perspective on science–public relations. Thus, 2000 saw publication not only of the Phillips inquiry's findings on BSE but also of an influential report from the House of Lords Select Committee on Science and Technology. *Science and Society* identified a 'crisis of trust' and argued boldly for a much greater degree of public engagement and dialogue around science. As the Lords report began:

Society's relationship with science is in a critical phase. . . . On the one hand, there has never been a time when the issues involving science were more exciting, the public more interested, or the opportunities more apparent. On the other hand, public confidence in scientific advice to Government has been rocked by a series of events, culminating in the BSE fiasco . . . public unease, mistrust and occasional outright hostility are breeding a climate of deep anxiety among scientists themselves.

House of Lords (2000, p. 11)

At the core of the Lords' report was the question of how to improve public confidence in science and technology. For the Select Committee, this issue is central to any future strategy for science and innovation; how can the UK (or any other nation) be successful in science and technology unless the wider publics are broadly supportive? In particular, the governmental agenda of innovation, economic competitiveness and strategic invest-ment has the potential to be in conflict with citizen assessments of science, technology and the direction of change. As the then UK Prime Minister Tony Blair put the same point (in decidedly loaded terms):

I want Britain and Europe to be at the forefront of scientific advance. But its [*sic*] no exaggeration to say that in some areas we're at a crossroads. We could choose a path of timidity in the face of the unknown.

Or we could choose to be a nation at ease with radical knowledge, not fearful of the future, a culture that values a pragmatic, evidence-based approach to new opportunities. The choice is clear. We should make it confidently.

Blair (2002)

For the Lords Committee, the means of putting the nation 'at ease with radical knowledge' was enhanced *dialogue* around science and technology. This 'direct dialogue' should not simply be an 'optional add-on' to science-based policy-making and to the activities of research organizations but should instead be a 'normal and integral part of the process.' (House of Lords 2000, paragraph 5.48) Similar talk can be heard at the European level. Thus, the European Commissioner for Research called in the EC's Action Plan for Science and Society for a pooling of 'efforts at European level to develop stronger and more harmonious relations between science and society' (Busquin 2002, p. 3). As the EC Action Plan puts this:

The relationship between science, technology and innovation, on the one hand, and society, on the other, must be reconsidered. Science activities need to centre around the needs and aspira-tions of Europe's citizens to a greater degree than at present . . . A true dialogue must therefore be instituted between science and society.

CEC (2002, pp. 7, 14)

Here we have an approach to the theory and practice of science communication which explicitly attempts to learn lessons from BSE and other such cases. Whereas the 'old' perspective on science communication presented knowledge and information as the barriers to public acceptance of science, the 'newer' approach (bearing in mind that such 'old/new' formulations are always open to challenge) puts trust at the core of science–public interactions and presents openness, dialogue and mutual engagement as the means of (re)building this. The treatment of socio-technical uncertainty looms large within this change of perspective. Uncertainties should be acknowledged not denied. Rather than being feared by experts and civil servants (as seemed apparent during the BSE crisis), the publics should be trusted to deal with uncertainties in an appropriate fashion. Characteristically, also, the 'new' perspective on science communication involves an acknowledgement that matters of scientific and technological change are not about science and technology alone but involve social, political and (especially) ethical questions which must be opened up to public scrutiny and debate. Rather than treating the publics as wallowing in a miasma of misunderstanding, the challenge instead is to engage and consult over matters of scientific and societal concern.

There are a number of possible responses to this new emphasis within the language of science communication and science governance. One commonly held position is to view all this as merely lip service—a rhetorical move to silence critics without actually changing very much. Is 'dialogue' around controversial topics such as nuclear power, genetically modified food and nanotechnology to be taken seriously when so many economic and commercial forces are so firmly in their favour? Is 'engagement' at all compatible with the demands of international competitiveness, substantial levels of corporate investment (accompanied by the need for an economic return), commercial secrecy and intellectual property protection?

A more pragmatic form of this criticism would question the practicality of dialogue as a 'normal and integral part of the process' when it comes to the many decisions (large and small) which need to be made about science-based policy (from standard-setting in consumer products to the allocation of research council funds). However, if engagement is not to be 'normal and integral' then who should decide (and how) about which issues to open up to discussion and review? While exhortations about dialogue, openness and transparency seem worthy in themselves, there are many questions to be asked about putting these into practice within administrative systems designed according to very different principles. Of course, these questions only become more challenging when we consider the operation of dialogue at a level beyond the nation-state (would a 'European' citizen dialogue over stem cell research or animal experimentation have any real meaning?).

As we will see, these practical challenges inevitably take us into broader questions about the relationship between science communication and the *direction and governance of science and technology* (and of scientific and technological *institutions*). Once again, we find that—although science communication can appear to be a benignly simple process of 'getting the message' across—it is unavoidably implicated in wider issues of the relationship between science and society. Turning this statement around, we also begin to see that 'communicating science' lies at the core of science–society interactions.

If we start with matters of *practice*, then a number of questions emerge about how the broad rhetoric about dialogue and engagement can translate (and has been translated) into specific initiatives. This will inevitably involve some discussion concerning the form and timing of dialogue exercises. Is it better to engage 'upstream' (i.e. early in the innovation process) when the socio-technical options are still open or to wait 'downstream' until the options are clear? In what (and whose) terms should dialogue be framed? Particular criticism has surrounded 'toothless' exercises where public views do not seem to have 'made a difference' in terms of practical outcomes. It has often been claimed that dialogue can help create social consensus and agreement—the apparent assumption being that 'consensus' is both possible and desirable (a notion that appears profoundly open to challenge in this area). Practical experience meanwhile suggests that dialogue has often led to calls for more dialogue—and that 'democratic' initiatives in consultation have generated angry charges that the activity was insufficiently democratic.

Engaging with GM—and anticipating nanotechnology

One important example of 'putting dialogue into practice' concerns the British *GM Nation?* debate over the commercialization of genetically modified (GM) crops (for a comparison with the equivalent Dutch debate see Hagendijk and Irwin 2006). Taking place in the summer of 2003, the debate was designed to be 'innovative, effective and deliberative' but also 'framed by the public'. Reflecting a generally cautious response, the final report on the debate characterized public opinion over the growth of GM crops in Britain as 'not yet—if ever' (*GM Nation?* 2003).

Here then we have what was, in terms of UK practice, a very well-developed exercise in 'public dialogue' around science and technology (Irwin 2006a). The steering group conducting the exercise received some 37,000 feedback forms from members of the public. There were 2.9 million hits on the debate's website, and over 600 local, regional and 'top tier' (i.e. centrally organized) meetings. Despite this, the exercise was roundly criticized for its limitations. Discussion after the event suggested both that the exercise had been 'hijacked' by activist groups and that it was far too restricted in participation, depth and coverage (House of Commons, Environment, Food, and Rural Affairs Committee, 2003). What we can especially identify in this case is a tendency for 'public communication and dialogue' to be seen as a discrete phase within the policy process: an activity to be fed into decision-making at the appropriate time, alongside other forms of evidence, before business as usual can return. Such an approach imposes fundamental constraints on public dialogue with science. To offer three examples of these constraints:

1. The 'public' strand of the debate ran in parallel with a separate review of the available science and an economic assessment of the costs and benefits of GM crops. It would appear that the construction of public debate, economic and scientific reviews as three separate strands inhibited the possibility of transparent public engagement in 'technical' analysis or of public discussion openly reflecting upon the issues raised by the other streams (Irwin 2006b).

2. The UK government's eventual decision to proceed with GM technology on an 'individual case by case basis' fitted more easily with the economics and science strands

than it did with the public debate. Whilst the economics and science strands appeared to feed directly into the government's decision, the 'public' strand was presented as a viewpoint for government to bear in mind rather than a body of evidence and opinion on which it must act.

3. It was also very apparent in the debate that members of the public typically 'framed' the underlying issues much more broadly than governmental and industrial officials. Whilst for the concerned civil servants this was a matter of deciding about a particular technical and administrative issue, for many members of the public the debate was connected to a much larger set of questions about the power of transnational companies, globalization, the future of UK agriculture and the benefits of current GM technologies to British consumers. Whilst policy-makers tended to frame the issue as a matter of 'risk' (to humans and the environment) this by no means captured the full spectrum of public assessments (a point made more generally by Wynne 2002).

None of the above represents a fundamental critique of the *GM Nation?* exercise. However, it does draw attention to the practical challenges involved in achieving 'dialogue' around science and technology. Openness, transparency and engagement are beguiling concepts but they also provoke (or rather *should* provoke) profound questions about their meaning, formulation and practice (especially when applied to specific contexts and situations). The point should also be made that this particular form of dialogue represents a very 'top down' (i.e. government-led) approach to what is often seen as the 'bottom up' expression of citizen views and concerns. This also makes us aware that dialogue initiatives often draw on consultation methods (citizen juries, people's panels, consensus conferences) which are still relatively unfamiliar and need to be 'learnt' (reminding us too that all such methods represent a specific 'framing' of the scientific citizen; Irwin 2001).

One particular criticism of the 2003 GM debate was that it came too late in the development process—when technologies were already close to market and international economic pressures high. Might these issues of dialogue and communication become easier if they were tackled earlier? In 2004, two of the UK's leading scientific organizations—the Royal Society (RS) and the Royal Academy of Engineering (RAE)—produced a report on the nanosciences and nanotechnologies which considered this very point.

On the one hand, the 2004 report was certainly in sympathy with 'the new mood for dialogue' (as the 2000 Lords report put it)—recommending in particular that:

. . . the government communicate with, and involve as far as possible, the public in the decision-making process in the area of nanotechnologies.

RS/RAE (2004, p. 62)

On the other, the report's authors found themselves confronted with a difficult set of issues. How to establish dialogue when less than a third of the population even recognized the term 'nanotechnology'? How to decide who might be involved in any dialogue over these issues? How to specify the precise form of dialogue when the objectives and terms of discussion are likely to evolve as the issues themselves evolve?

The RS/RAE report presents the nanotechnologies as an 'upstream' issue in terms of the development decisions yet to be made, the social and ethical impacts yet to be envisioned, and the public acceptance (or otherwise) yet to be formulated. Noting that the precise form of dialogue will be 'no simple matter' (RS/RAE, p. 64), the report advocated early dialogue and communication:

The upstream nature of most nanotechnologies means that there is an opportunity to generate a constructive and proactive debate about the future of the technology now, before deeply entrenched or polarised positions appear.

RS/RAE (2004, p. 67)

Although there is much sense in this proposal, 'upstream' dialogue should not be seen as a solution in itself but rather as one element in a continuing process of engagement and communication. It is also the case that engaging earlier does not automatically ensure that engagement will be better—instead all the questions about the form and content of communication presented here still apply. Equally, upstream engagement should not be considered primarily as a means of predicting (and subsequently managing) public responses (Wynne 2006). Instead, the potential for early engagement is precisely that it can shape patterns of technological development in a fashion which takes account of public concerns, aspirations and preferences—and can put the assumptions of the developers of technology into dialogue with those of the wider publics (Wynne 2006).

Far from making science communication an easier task, upstream dialogue needs to consider that even the definition of the 'nanosciences and nanotechnologies' is as yet unstable and unfixed, that expectations of science and technology cannot be separated from social and personal expectations of the future, and that societal responses are likely to be negative as well as positive. Certainly, 'dialogue' undertaken with the aim of 'winning over' the publics to the technological and economic potential of the nano-technologies is not dialogue at all but simply a more sophisticated form of the old (and zombie-like) deficit model.

Inventing the futures

Our discussion seems to have taken us a long way from the 1985 Royal Society report with which we began. Since that time, there have been many practical initiatives in science communication and many analyses of the relationship between 'the sciences' and 'the publics'. Of course, this flurry of activity partly reflects the inherent fascination and chal-lenge of communicating science and technology in an age where the issues appear so pressing and so many channels are now open and available (in the case of the GM debate, from 'village hall' discussions to on-line consultation, but the possibilities are clearly much greater). It is also not hard to identify beneath all of the positive talk about dialogue and open communication an institutional anxiety that public scepticism about science and technology might lead to societal resistance and lost possibilities for innovation.

Tellingly, the 'crossroads' speech by Tony Blair (2002) quoted above was prompted by the former PM's experience not in London or Brussels but Bangalore. As Blair told the story, he was challenged there by a group of academics (who, significantly, were also 'in business') telling him that 'Europe has gone soft on science, we are going to leapfrog you and you will miss out'. This group professed itself to be astonished by the European debate over GM and saw emotion driving out reason in such discussions. Blair's expressed fear was that the Indian business-academics might be proved right 'if we don't get a better understanding of science and its role' (ibid.). It is hard not to see the hand of the zombie behind such words, reminding us that—despite all the claimed progress in science communication—we may not (or at least not consistently) have moved so far forwards.

Similar statements abound as dialogue and communication are put firmly into the context of the need for successful innovation. As the Independent Expert Group on Research and Development (R&D) and Innovation expressed it in a 2006 European Commission report:

> Europe and its citizens should realise that their way of life is under threat but also that the path to prosperity through research and innovation is open if large scale action is taken now by their leaders *before it is too late.*
>
> CEC (2006, p. vii, emphasis in original)

Meanwhile, official statements about the European Research Area make reference to effective sharing of knowledge between 'public research and industry, as well as with the public at large' but do so firmly in the service of developing European economic competitiveness and especially a leading edge in knowledge and innovation (CEC 2007). The Commission might use the strapline of 'inventing our future together' but institutional expectations of that future are already firmly in place and the scope for expressing other inventions (and other futures) appears limited.

At this point it is very clear that the governance of science (including its direction, sponsorship and control) and the communication of science are not separate activities but instead tightly bound up with one another. Discussion of scientific communication cannot therefore stand apart from discussion of research investment, economic development and the agenda for scientific and industrial change. The challenge for science communication now is whether it should simply serve to (to put it crudely) soften up public opinion in the face of pending innovation or else provoke richer and more meaningful discussions over the alternative futures for science, innovation and society. Put differently, should science communication continue to circle the notion that the wider publics are a problem for change—or should it instead view public expressions, cultural understandings and expectations of the future as a valuable resource (see also Stilgoe *et al.* 2006)?

We can conclude this discussion of deficits and dialogues, and moving forwards and in circles (not to mention the occasional zombie), with a final look at the nanosciences and nanotechnologies. Certainly, communication in this area can easily trade in the language of public ignorance and emotion, of potential resistance and fear of the unknown. The alternative approach of dialogue and engagement offers no easy route and raises as many questions as answers. Equally, science communication cannot assume that knowledge and information are readily detachable from the institutions that provide such evidence—which must themselves be open to external scrutiny. The challenges of science communication around the nanosciences are considerable. Science communication alone may not have all the answers to the questions being raised. But without the practice of vigorous, critical, imaginative, multi-level and provocative science communication, our socio-technical futures will be severely constrained.

■ REFERENCES

Blair, T. (10 April 2002). *Science Matters*. Available at **http://www.number-10.gov.uk/output/Page1715.asp**.

Busquin, P. (2002). Foreword. In: Commission of the European Communities (CEC) *Science and Society: Action Plan*. European Communities, Luxembourg, p. 3.

CEC (Commission of the European Communities) (2002). *Science and Society: Action Plan*. European Communities, Luxembourg.

CEC (Commission of the European Communities) (2006). *Creating an Innovative Europe: Report of the Independent Expert Group on R&D and Innovation appointed following the Hampton Court Summit and chaired by Mr. Esko Aho*, EUR 22005. European Communities, Luxembourg.

CEC (Commission of the European Communities)/Directorate-General for Research (April 2007). *The European Research Area: New Perspectives*, EUR 22840 EN. European Communities, Luxembourg.

GM Nation? (24 September 2003). *The Findings of the Public Debate*. Previously available online at: **http://www.gmpublicdebate.org.uk/ut_09_9_6.htm#summary**.

Hagendijk, R. and Irwin, A. (2006). Public deliberation and governance: engaging with science and technology in contemporary Europe. *Minerva*, **44**(2), 167–84.

Haldane, J.B.S. (1939, reprinted 1943). *Science and Everyday Life*. Pelican, Harmondsworth.

House of Commons, Environment, Food and Rural Affairs Committee (2003). *Conduct of the GM Public Debate*, Eighteenth Report of Session 2002–2003. HMSO, London.

House of Lords, Select Committee on Science and Technology (2000). *Science and Society*, Third Report. HMSO, London.

Irwin, A. (1995). *Citizen Science*. Routledge, London.

Irwin, A. (2001). Constructing the scientific citizen: science and democracy in the biosciences. *Public Understanding of Science*, **10**(1), 1–18.

Irwin, A. (2006a). The politics of talk: coming to terms with the 'new' scientific governance. *Social Studies of Science*, **36**(2), 299–320.

Irwin, A. (2006b). The global context for risk governance: national regulatory policy in an international framework. In: *Globalization and Health: Challenges for Health Law and Bioethics* (ed. B. Bennett and G. Tomossy), pp. 71–85. Springer, Amsterdam.

Irwin, A. and Wynne, B. (eds) (1996). *Misunderstanding Science?* Cambridge University Press, Cambridge.

Phillips, Lord, Bridgeman, J. and Ferguson-Smith, M. (2000). *The BSE Inquiry: the Report*. HMSO, London.

Royal Society (1985). *The Public Understanding of Science*. Royal Society, London.

RS/RAE (Royal Society/Royal Academy of Engineering) (July 2004). *Nanoscience and Nanotechnologies: Opportunities and Uncertainties*, Royal Society Policy Document 19/04. Royal Society, London.

Stilgoe, J., Irwin, A. and Jones, K. (2006). *The Received Wisdom: Opening up Expert Advice*. Demos, London. Available online at: **http://www.demos.co.uk/publications/receivedwisdom**.

Wynne, B. (2002). Risk and environment as legitimatory discourses of technology: reflexivity inside out? *Current Sociology*, **50**(3), 459–77.

Wynne, B. (2006). Afterword. In: *Governing at the Nanoscale: People, Policies and Emerging Technologies* (ed. M. Kearnes, P. Macnaghten and J. Wilsdon), pp70–8. Demos, London. Available online at: **http://www.demos.co.uk/publications/governingatthenanoscale**.

■ **FURTHER READING**

- Hagendijk, R. and Irwin, A. (2006). Public deliberation and governance: engaging with science and technology in contemporary Europe. *Minerva*, **44**(2), 167–84. This article draws upon key findings derived from the results of 26 qualitative case studies conducted in eight EU member states as part of the 'Science, Technology and Governance in Europe' (STAGE) Network. The results are interpreted in the light of theories of deliberative democracy, as well as concerns over the level and quality of public participation in science governance.

- House of Lords, Select Committee on Science and Technology (2000). *Science and Society*, Third Report. HMSO, London. This report represents 'official' recognition of some of the key findings of at least two decades of scholarship in science and technology studies. Essentially this report acknowledges a 'crisis of trust' with science. It argues that this problem should not be simply viewed as a public relations issue, but rather it should be answered by a long-term and routine commitment to dialogue and public engagement with science.

- Jasanoff, S. (2005). *Designs on Nature: Science and Democracy in Europe and the United States*. Princeton University Press, Princeton, NJ. This book addresses the public and political issues arising from biotechnology in Europe and the United States. Situating these issues with the wider context of science–society relations, the author makes a convincing case for a form of scholarly analysis that is theoretically informed and empirically aware.

- Leach, M., Scoones, I. and Wynne, B. (eds) (2005). *Science and Citizens: Globalization and the Challenge of Engagement*. Zed Books, London and New York. This edited collection addresses issues of citizenship and engagement from the perspectives of science and technology and development studies, and in relation to a series of pressing science-based issues, including HIV/AIDS and agricultural biotechnology.

- Royal Society/Royal Academy of Engineering (2004). *Nanoscience and Nanotechnologies: Opportunities and Uncertainties*, Royal Society Policy Document 19/04. Royal Society, London. This report represents a consultation process that is widely considered one of the best examples to date of upstream public engagement in UK science governance. The qualitative workshops revealed some key areas of public concern, including financial implications, societal impacts, efficacy, side-effects and future controllability of the technology. However one of the most interesting results is that only 29% of respondents to the quantitative survey of UK adults had ever heard of nanotechnology. This and other aspects of the consultation exercise have raised questions about how effective such early upstream engagement can be, and whether publics may change their orientation towards a technology as it becomes more widely known.

■ **USEFUL WEB SITES**

- **research*eu**: http://ec.europa.eu/research/research-eu. *research*eu* is a European Union magazine aiming to 'broaden the democratic debate between science and society'.

- **The Stem Cell Network**: http://www.stamcellenetvaerket.dk/. The Stem Cell Network is an experiment in 'spatial research communication' which seeks to create dialogue and interaction around the social and political aspects of stem cell research.

- **The 'Science, Technology and Governance in Europe' (STAGE) Network**: http://www.stage-research.net/STAGE/index.html. Papers from the Science, Technology and Governance in Europe (STAGE) Network—a European project covering eight nations.

- **The Danish Board of Technology: http://www.tekno.dk/.** The Danish Board of Technology has an international reputation for creating discussion and dialogue around technological development.

- **The Nanoscale Informal Science Education (NISE) Network: http://www.nisenet.org/.** The US-based Nanoscale Informal Science Education (NISE) network draws together scientists and museum professionals in 'bringing nanoscale science to the public'.

1.2

The new politics of public engagement with science

Jack Stilgoe and James Wilsdon

Introduction

In common with other groups whose work rests on a contract of implicit social support, scientists have long agonized about their relationship with the public. In the last few decades, we have seen a slow recognition, in response to political debates involving science, that we need new ways to understand the relationship between science and society. What was a monologue has become a conversation.

With a move from one-way communication to dialogue, we have seen science's understanding of the public become at least as important as the public's understanding of science. In this chapter, we chart developments in the theory and practice of what has become known as *upstream public engagement with science*. We explain some of the changes that made upstream engagement necessary and learn some lessons from early experiments in this form of dialogue.

A brief history of science's relationship with society

Since the technologically fuelled end of World War II (1939–1945), we have seen six decades of exponential growth in the potential of science and technology. The first atomic bomb reminded the world of science's power, and began a series of discussions about how it should be controlled. As science plays an increasingly significant part in our everyday lives, we have seen various attempts to systematically place it within society. As the controversies over BSE, genetically modified (GM) crops and foods and other technologies have demonstrated, people are questioning scientists more and trusting them less. Drawing on extensive polling data, Ben Page of the opinion pollster MORI recently summed up the current state of public opinion: 'Blind faith in the men in white coats has gone and isn't coming back' (Page 2004, p. 31). This is not a huge surprise. As we move towards knowledge societies that rely on innovation to drive economic growth, science and technology are likely to become increasingly contested sites of public debate (see also Irwin in Chapter 1.1).

The response of the science establishment to the changing public mood has been to reach out and experiment with different models of public engagement. Over the past few decades, in an attempt to claw back public trust, the institutions of science have begun to recognize the rationale for involving the public more intimately in their work. The history of this movement can be told in three phases.

Phase 1: Public understanding of science (PUS)

The instinctive response of scientists to growing levels of public detachment and mistrust was to embark on a mission to inform. Scientists needed to get better at spreading the word, persuading people of the value of science. The diagnosis was that people did not know enough science, or know enough *about* science, to support it. Attempts to gauge levels of public understanding date back to the early 1970s, when the US National Science Foundation conducted annual surveys to measure people's knowledge of scientific facts (for example, whether the earth goes round the sun or vice versa.) (Wynne 1995; Hornig Priest, Chapter 6.1). Walter Bodmer's 1985 report for the Royal Society placed PUS firmly on the UK agenda, and proclaimed 'It is clearly a part of each scientist's professional responsibility to promote the public understanding of science' (Royal Society, 1985, paragraph 6.3, p. 24). The Bodmer report gave birth to a clutch of initiatives designed to tackle the public's ignorance, including COPUS, the Committee on the Public Understanding of Science, supported by the Royal Society and the Government's Office of Science and Technology.[1]

Phase 2: From deficit to dialogue

For more than a decade, PUS was the dominant means of understanding and addressing the place of science in society. But scientists gradually discovered that the assumptions behind one-way science communication interrupted rather than encouraged a cordial relationship between science and its publics. PUS was built on questionable assumptions about 'science', the 'public' and the nature of 'understanding'. It relied on a 'deficit model' of the public as ignorant and it portrayed science as unchanging and universally comprehensible.[2] Unfortunately, the assumption that to know science was to love science was directly undermined by events.

Relations between science and society festered throughout the 1990s—a decade marked by uncomfortable showdowns involving science, politics and the public, starting with the BSE crisis, through GM crops and on to debates about the risks of mobile phones and the measles, mumps and rubella (MMR) vaccine. PUS was challenged in 2000, when an

---

1. The Office of Science and Technology became the Office of Science and Innovation and then the Government Office of Science, which now sits within the Department for Innovation, Universities and Skills.
2. For more on the 'deficit model' see Wynne (1991).

influential House of Lords report detected 'a new mood for dialogue' (House of Lords 2000). The PUS approach was condemned; even the government's Chief Scientific Adviser now acknowledged this was 'a rather backward-looking vision'.[3] In came the new language of 'science and society' and a fresh impetus towards dialogue and engagement.

From the perspective of a social scientist, it is tempting to argue that the conquest of the deficit model was achieved by social scientific evidence and insight. A more accurate story might be that social science was there to catch PUS when it fell, and provide a coherent alternative. Throughout the 1990s, social scientists in the UK and abroad had been applying work under the umbrella of 'sociology of scientific knowledge' or 'science and technology studies' to the relationship of science with various public groups. Academic work looked at the practice of science and concluded that it bore little relationship to the claims made in public by scientific institutions. Science, it seemed, was far more messy and uncertain than its public image suggested.[4]

At the same time, science's image of the public was misguided. Brian Wynne and others explored how scientists' 'deficit' model of the public was built and used in different contexts. By talking to members of the public in depth, starting from the publics' own perspectives, researchers found that people's relationship with science was far more active and sceptical than previously thought. People wanted to be able to ask questions of science and have their voices heard. As controversies over GM crops, nuclear power and other science-based issues become more politically significant, the louder became the call to heed public questions and incorporate public values into decision-making.

In cases where science, politics and public uncertainty are interwoven, social scientists provided a strong case for broader engagement with stakeholders, other experts and members of the public. This has been elegantly referred to as 'extended peer review' (Funtowicz and Ravetz 1992), where public science needs to be robustly tested in public, combining questions of scientific knowledge with those of values. And when the science is uncertain, as is so often the case with contentious issues (see Holliman and Scanlon, Chapter 6.3), this broader engagement becomes yet more important. There are many rationales for public engagement with science, and these are often blurred and confused in public discourse, but Daniel Fiorino thinks about them as being normative, instrumental or substantive. The *normative* reason is that representation of the public is in itself a good thing according to democratic ideals. The *substantive* reason is that lay assessments of risk are valid and useful in decision-making. And the *instrumental* one is that the appearance of greater public involvement makes decisions more legitimate (Fiorino 1990).

So we saw in the 1990s a series of experiments in dialogue on contentious and potentially contentious issues involving science. *GM Nation?* was perhaps the most visible—an innovative attempt by the government to stage a series of dialogues as part of a new approach to the management of what was at the time a highly contentious issue. Its process and outcome has since received a good deal of criticism, but its political importance lies in the attempt to begin a different sort of debate about science (UEA 2004).

3. House of Lords (2000, paragraph 3.9). This quote comes from Sir (now Lord) Robert May's evidence to the committee.
4. See Irwin and Wynne (1996) for a sample of this work.

We can learn plenty of lessons from *GM Nation?*. One is that dialogue needs to take place early, before it gets bogged down in vested interests and political argument. Another is that it is impossible to tell the public how they should talk about issues.

Most of the social science reflection on controversies between science and the public has looked at questions of risk—Is it safe? How safe? How safe is safe enough? Who decides what counts as safe? But subsequent work has suggested that risk is only part of the story. And public engagement still tends to reflect, as Brian Wynne puts it:

. . . the false assumption that public concerns are only about instrumental consequences, and not also crucially about what human purposes are driving science and innovation in the first place.

Wynne (2005, p. 67)

Science–public relations are crucially concerned with questions of innovation as well as regulation. So we need not just to talk about the products, uses and ends of science, but also its means. This takes the conversation upstream.

Phase 3: Upstream engagement

The aim of upstream engagement is to get some questions back onto the table in discussions about innovation:

Why this technology? Why not another? Who needs it? Who is controlling it? Who benefits from it? Can they be trusted? What will it mean for me and my family? Will it improve the environment? What will it mean for people in the developing world?[5]

A debate fuelled by these questions can be more proactive and positive than a narrow discussion centred on questions of risk. But for these questions to have any purchase, the discussion needs to take place before decisions have been made and before scientific areas have crystallized, which is not at all straightforward. Upstream engagement therefore demands real openness and genuine open-mindedness.

The language of upstream engagement features in several recent policy statements. The Royal Society (RS) and the Royal Academy of Engineering (RAE) recognized in their report on nanotechnology that: 'Most developments in nanotechnologies, as viewed in 2004, are clearly "upstream" in nature' (RS/RAE 2004, p. 64) and called for:

. . . a constructive and proactive debate about the future of nanotechnologies [to] be undertaken now—at a stage when it can inform key decisions about their development and before deeply entrenched or polarised positions appear.

RS/RAE (2004, p. xi)

The UK government agreed. Lord Sainsbury, then Science Minister, responded to the nanotechnology report by saying that:

We have learnt that it is necessary with major technologies to ensure that the debate takes place 'upstream', as new areas emerge in the scientific and technological development process.

DTI (2004)

---

5. These questions are taken from Wilsdon and Willis (2004), which provides a longer discussion of upstream engagement.

Most significantly, the government's recent 10-year strategy for science and innovation includes a commitment:

... to enable [public] debate to take place 'upstream in the scientific and technological development process, and not 'downstream' where technologies are waiting to be exploited but may be held back by public scepticism brought about through poor engagement and dialogue on issues of concern.

HM Treasury/DTI/DfES (2004, p. 105)

Upstream engagement, it seems, is here to stay.

Arguments against and barriers to engagement

The idea of early two-way engagement between science and its publics has had its detractors, however. Rumours of the death of the 'deficit model' have been greatly exaggerated and there are still those who maintain that the public are too ignorant to contribute anything useful to scientific decision-making. One of the most vocal is the Liberal Democrat peer Dick Taverne. In a letter attacking a *Nature* editorial supporting upstream engagement, Taverne rejects 'the fashionable demand by a group of sociologists for more democratic science'. He goes on: 'The fact is that science, like art, is not a democratic activity. You do not decide by referendum whether the earth goes round the sun.' (Taverne 2004, p. 271).

Taverne by no means represents the mainstream of critique, much of which is useful and thoughtful.[6] But his arguments have allowed a useful clarification of the case for engagement. Upstream engagement is not about members of the public standing over the shoulder of scientists in the laboratory, taking votes or holding referenda on what they should or should not be doing. It is about enriching and building more reflective capacity into the *practices* of science. So as well as bringing the public into new conversations with science, the aim is to bring out the citizen *within* the scientist—by enabling scientists to reflect on the social and ethical dimensions of their work.

In *The Public Value of Science* (Wilsdon *et al.* 2005) we discussed the vital role that scientists themselves play in social reflection on issues involving science. Fast-moving technologies and emerging areas of science, as we've seen with nanotechnology and more recently with synthetic biology, demand the involvement of cutting-edge scientists in public conversations about uncertainties and benefits. In the course of our own public engagement work, we have come across a number of scientists who see getting involved in such debates as part of their work. One, Alexis Vlandas, was a postgraduate at Oxford University when we met up with him. Alexis looks set to be a leading future nanoscientist. But he also grapples with some bigger issues outside his day job. He coordinates the Oxford branch of Pugwash, an international network for socially responsible scientists, and regularly organizes meetings on the social dimensions of nanotechnology or the dilemmas raised by corporate and military funding of research.

6. A more useful discussion of the benefits and risks of public engagement in practice can be found in Irwin (2006), Stirling (2005) and Wynne (2006).

But a lot of what Alexis does falls outside the prevailing model of what counts as good science. He told us that:

One of the senior scientists in my group said to me 'what kind of scientist do you want to be, a social scientist or a real scientist?' He's worried that I spend too much time on this stuff. . . . If you want to get a place at Harvard or any other top university, there's no incentive to engage in these debates . . . the funding system infantilizes you as a scientist. It pushes scientists to steer their research to fashionable areas. There's no space to say 'Wait a minute, should we be asking different questions?'

Wilsdon *et al.* (2005, p. 46)

Some critics have argued that public engagement leads to the 'demoralization of scientists' (Durodié 2003). But Alexis is invigorated by reflecting on the social dimensions of his work. He has to be, because the structures that surround him—for funding, research assessment and career development—often push in the opposite direction.

In 2006, a survey jointly commissioned by the Royal Society, Research Councils UK (RCUK) and the Wellcome Trust found that there are still barriers stopping scientists from getting as involved as they'd like in public engagement activities. Scientists on the whole feel that expectations from funders, colleagues and those who judge quality systematically ignore social aspects; so public engagement becomes a hobby rather than part of their everyday work. Almost half of the scientists spoken to would like to spend more time engaging with the public (Royal Society, RCUK and Wellcome Trust, 2006). But the system is getting in the way. While the rhetoric asks scientists and members of the public to talk, the system that decides what counts as good science often interrupts the conversation before it can begin.

Thankfully, in the last few years, we have been able to see public engagement in action, see its benefits and learn from the various forms it has taken.

Public engagement in action

Take the UK Alzheimer's Society.[7] Like many patient groups around the world, this group has become more and more closely involved with the science behind its members' disease (Rabeharisoa and Callon 2002). This has taken them upstream, to ask questions about what research is valuable and what research is necessary. They are interested in what scientists are researching across 'cause, cure and care'.[8] And they are interested in how science can learn from patients' and carers' experiences of the everyday reality of the disease. In 2000, the society created a Quality Research in Dementia (QRD) network, a groundbreaking example of upstream engagement. Through this scheme, patients and carer volunteers shape research priorities, review proposals from scientists, assess researchers and monitor research.

7. This case and the quotes within it are taken from Wilsdon *et al.* (2005).
8. 'Cause, cure and care are key'. *Living with Dementia* August/September 2006, the Alzheimer's Society (http://www.alzheimers.org.uk/downloads/living_with_dementia_august_2006.pdf).

Ted Freer is one of the carers involved. When Ted's wife was diagnosed with Alzheimer's disease, he started getting involved with the Alzheimer's Society, knowing that they invest over a million pounds in dementia research every year. Ted is one of the society's 150 QRD members. When we asked him what difference the QRD makes to science, he told us that: 'it provides a totally different viewpoint for researchers . . . if it wasn't for us, they'd only be able to discuss it with their peers'. The questions asked by the QRD network make a real difference to research. And a growing number of scientists are rising to the challenge.

James Warner is one such scientist. He's a psychiatrist specializing in dementia at Imperial College, London. Seeking funding for a project looking at alternative treatments for dementia, he approached the Alzheimer's Society. Used to the standard, traditional peer review process (see Wager 2009 for discussion), he was surprised to receive 67 comments from members of the QRD network. After the initial shock, he realized that the collected comments gave him important pointers that he could use to improve his research. He redrafted the proposal, sent it back, faced the QRD interview and secured funding. He has been converted:

As a one-time cynical scientist, I'm now signed up to the QRD idea. . . . It's not tokenistic. It's real, good quality help . . . a fantastic collaborative approach to research.

And there is no question that the research that is getting done is good, if not *better*, science. One of James's colleagues, Steve Gentleman, also holds a grant from the society. He has been similarly enthused:

I think there's no going back. I don't think you can take the public out of the science, and I don't think you should.

This upending of the peer-review process strikes many people—scientists and members of the public—as unusual. Ted told us that it takes some getting used to:

It is strange, yes. But watching it work in practice, I've been reassured. It's very effective.

And it seems to have a therapeutic benefit. Ted's engagement with the research programme gives him a new perspective and helps him carry the burden of caring for someone with Alzheimer's disease.

The conversations that take place within and around the Alzheimer's Society are driven by the immediate or future relevance of science to this particular public group. In other areas of science, it might not be clear who are the relevant publics and what their role should be. With issues such as nanotechnology, conversations about the means and ends of science are harder to arrange. At the think tank Demos, we have staged public engagement 'experiments' to explore public values as they apply to emerging nano-technologies. The aim has been to provide forums in which members of the public and scientists can discuss questions about the future of science before conversations narrow to questions of risks and benefits. Richard Jones, professor of physics at Sheffield University, is one of a growing band of scientists who are keen to get involved in such discussions. Reflecting on one engagement experiment, he argued that:

I think what's important is not the narrow issue of 'Do you do this piece of science and don't you do this piece of science?' Rather it's 'What kind of world do you want to live in?' The things that worried the people in my focus group were the things that worried me. I am uncertain about how lots of this stuff will turn out. I have a positive view of how I would like it to turn out but there are people who have opinions about how it ought to turn out that I really don't like at all. It's quite reassuring to think that I am not alone in worrying about the things I worry about.[9]

Nanodialogues: experiments in engagement

Discussions about the visions and values of science are important, but it is a common criticism of public engagement that it often fails to connect to the real decisions made by institutions. The aim of the *Nanodialogues* project was to connect public engagement to these institutions and the various contexts in which they operate. We ran four dialogues, each with an organizational partner. The first was with the UK's Environment Agency. The second was with both the Engineering and Physical Sciences Research Council and the Biotechnology and Biological Sciences Research Council. The third was with Practical Action. And the fourth was with Unilever.

In scientific terminology, the *Nanodialogues* were a 'proof of concept'—that public engagement can make a positive difference. By running our four experiments with partner organizations, all of whom were thinking through what nanotechnology means for them, we were able to flex and test the systems behind science. With that in mind, it is worth some exploration in more depth of two of these experiments—those involving the Environment Agency—a regulator of science—and Practical Action—a charity who are interested in how technology can benefit people in the developing world (Stilgoe 2007).

The first experiment didn't start well. An hour after arriving, the participants in our 'People's Inquiry' were stuck in a debate about Mork and Mindy ('nanoo, nanoo')[10] and the Teletubbies' vacuum-cleaning pet (NooNoo). One person admitted, 'I don't know. I don't do science'. Yet just 4 weeks later, after 15 hours of deliberation and input from a range of experts, this group of 13 Londoners had got to grips with the issue, realized the importance of engagement and produced a thoughtful set of recommendations.

Their task had seemed rather esoteric: to consider whether government should allow the use of nanoparticles as a method of cleaning up contaminated land. But for the Environment Agency this was a pressing question. A number of companies had applied to use nanoparticles in this way in the UK, requiring the release of nanoparticles into the environment. This was potentially controversial, as the influential 2004 report from The Royal Society and Royal Academy of Engineering had cautioned against any such releases (RS/RAE 2004). In this context, the aim of our People's Inquiry was to broaden societal discussion, so that new regulations for the use of nanoparticles could better reflect the values and interests of the wider public.

9. For more on this experiment see Kearnes *et al.* (2006).

10. *Mork and Mindy* was a US-produced science fiction sitcom from the late 1970s.

Nanoparticles have been tested in a number of countries to clean up pollution, particularly from chlorinated hydrocarbons (Zhang 2003). For a given mass, nanoparticles are more reactive than their bigger equivalents, and their size allows them to reach the parts of contaminated land that bigger particles cannot. Yet tests so far have analysed the extent to which the nanoparticles are doing their job, rather than any unintended environmental effects they may have.

This was the focus of our People's Inquiry, conducted in 2005 in partnership with the Environment Agency. The discussion took place in the light of: a technology on the horizon with suggested environmental benefits; expert advice that suggested we should prevent release until we knew more; a small group of experts around the world, most of whom openly acknowledged the inherent uncertainties; and virtually non-existent public knowledge about the technology, let alone its potential benefits and hazards.

Over 3 days of deliberation, the participants in the People's Inquiry generated countless questions, which they were able to pose to the expert witnesses they interrogated. Some of these were factual:

How do things actually stick to these nanoparticles? Is it that it's actually physically sticky or has it got little things like Velcro on it, or has it got sucky things that suck the contaminant out? Or is it a gluey thing?

Is the process of using nanoparticles for land remediation a quicker process than other methods?

Other questions echoed those of current scientific and regulatory concern.

How far can the nanoparticles travel?

Presumably nobody's actually looked at whether the things could be made to break down in cells?

But most were open questions with no easy answers. They highlighted the areas of concern that were likely to define the future public context of nanotechnologies:

Will there be any unanticipated effects?

Who has a say?

Would the fact that it's a quicker process mean that the safety issues may be overlooked?

What's the rush?

What about irresponsible companies?

Is information sharing too informal?

The conclusions of the People's Inquiry could be grouped into two areas. First, uncertainty was seen as a defining feature of this emerging regulatory debate. Broadly speaking, the participants agreed with the RS/RAE (2004) recommendation that nanoparticle release be prevented. In line with the precautionary principle (see Box 1), uncertainty was seen as sufficient to justify action.

A second theme was openness. Our participants, realizing that they would never be experts in the various areas required to grapple fully with the issue, demanded that any steps taken should be more accountable. They were supportive of plans by the UK government to develop a notification scheme for nanotechnology companies, but were sceptical of companies' willingness to declare data. Realizing that the most important challenges lay in the future, they demanded a more open approach to the application

BOX 1 THE PRECAUTIONARY PRINCIPLE

Developed in response to concerns about human impact on the global environment, which gained momentum in the 1960s, the precautionary principle is an important guiding principle that informs some decision-making about complex contemporary techno-scientific issues.

The precautionary principle, as defined in the World Charter of Nature adopted by the UN General Assembly in 1982, states that:

[. . .] lack of scientific certainty should not be used as a reason for postponing measures to prevent suspected or threatened environmental damage.

The precautionary principle has been included, with minor revisions, in various treaties and conventions governing environmental issues. In essence, it codifies a fundamental shift in decision-making about risk, requiring that, in the absence of full scientific evidence, a precautionary approach to risk should be adopted.

and governance of technology. They argued that regulation should be proactive, but also responsive to the changing social and economic context of technology, including the emergence of new concerns and uncertainties.

A few months after the inquiry was complete, four of the participants visited the UK's Department for Environment, Food and Rural Affairs for a meeting with the team of civil servants managing nanotechnology policy. The meeting involved a robust but constructive exchange of views, and positive outcomes can be taken from the experience. For example, one of the public participants concluded that it had convinced her of the value of such exercises. 'We can help policy-makers', she said. 'I feel like we have made some nanoscule contribution to society.'

Science, society and development

With Practical Action, our experiment was part of a growing debate about science and the developing world. With nanotechnology, the Royal Society and Royal Academy of Engineering report had spoken of the possibility of a 'nano-divide'. It pointed to

. . . repeated claims about the major long-term impacts of nanotechnologies upon global society: for example, that it will provide cheap sustainable energy, environmental remediation, radical advances in medical diagnosis and treatment, more powerful IT capabilities, and improved consumer products. . . . Concerns have been raised over the potential for nanotechnologies to intensify the gap between rich and poor countries because of their different capacities to develop and exploit nanotechnologies, leading to a so-called 'nano-divide'.

RS/RAE (2004, section 6.3)

Communities in the developing world rarely have much of a voice in debates about innovation. They are less likely to see the benefits of new technologies, and are more likely to have risks imposed upon them. Our modest contribution to the debate took the form of a stakeholder workshop in Harare, Zimbabwe. We chose the potential contribution of nanotechnologies to water purification as the focus of our 3-day experiment. Our dialogue aimed to bring the views and values of people for whom clean water is an everyday problem into debates about possible technical solutions.

Zimbabwe is a country with numerous problems. Its inflation and unemployment are, at the time of writing (2008), higher than anywhere else in the world, and its government has recently tried to move the problem off its doorstep by clearing out slums. In Epworth, a suburb of Harare, these problems only exacerbate an issue that for many is an everyday struggle—the search for clean water. Technology may be able to play a role in this, but not without an understanding of this context. On the first day of our workshop, people aired their views. Water was unaffordable, it was scarce, it was a long way away and it was normally collected by women and girls. Where wells exist, they are crammed next to latrines and difficult to seal off from contamination. As well as a recent cholera outbreak, there is chemical pollution from factories downstream.

The community in Epworth are sceptical of well-intentioned technological schemes. Too often in the past they have been let down by treatment techniques that have failed to do their job, or have broken and proved impossible to fix. Technology for these communities is as much about human capacity as it is about new widgets. As one participant put it: 'When the NGO goes away, who has the knowledge to run and maintain it?'.

For our participants, 'technology' in general and 'nanotechnology' in particular was understood as the system, of which the thing—the filter, the treatment plant or whatever —was a part. Sceptical of the West's assumed desire to impose technologies, they demanded some level of participation in the process. Features such as sustainability, maintenance, adaptability and extension into communities were seen as vital. One of the Zimbabwean scientists who took part told us the difference between technology and traditional aid: 'Technology can't be handed over to a community like a sack of mealy-meal'.

The experiment in Zimbabwe revealed the gulf between understandings of the benefits of nanotechnology in the 'rich North' and the 'poor South'. No-one should bemoan the growing interest in technologies with potential benefits for the developing world. But promises should be more measured. We need to ask 'What makes science and technology work for the poor?' (Leach and Scoones 2006). Conversations about technology need to include the voices of real human need.

Learning the lessons of upstream engagement

Alongside the growing enthusiasm for early dialogue about science and technology, we have seen an understandable questioning of its value and impact. It is clear that, for the people directly involved in dialogue, it can make a difference to how they think about science—whether they are scientists, policy-makers or members of the public. Evidence from the various experiments that have taken place around nanotechnology in the UK suggests that the impact on scientists in particular is marked (Doubleday et al. 2007). But if the aim is to affect just those people in the room, such initiatives seem awfully expensive. We would hope that the ripples spread further, to the decision-makers, institutions and systems where power lies. Once the dust has settled on public engagement, what can we say about the changes that it makes?

Advocates of public engagement have made it hard for themselves. They have sold engagement by reciting a litany of policy mistakes: GM, BSE, nuclear power and other science-based issues where the lessons are clear in hindsight. Upstream, where policy options have yet to be laid out, let alone chosen, engagement provides no easy answers. But it can ask some deep questions about how we do policy and who we involve.

Public engagement provides a lens through which policy-makers can see issues differently, focusing on contexts, uncertainties, alternatives and local concerns. This often leads to further debate and opens up new areas of policy. Once issues are illuminated by public values, the systems behind them start to look very different. Looking at the example of the Zimbabwe dialogue experiment we are forced to ask questions about why science in the developed world does not prioritize the needs highlighted by our conversation.

New tensions, new questions

Less than two decades ago the UK was known for its consensual technocratic approach to science (Jasanoff 2005). There can be little doubt that, through a combination of events, political and social change and new rationales, the UK has advanced a long way towards a more deliberative culture of science and society. Many other countries now see the UK as a leader in experimenting with new, deliberative techniques of governance. But for all of the current enthusiasm for public engagement, there are tensions and assumptions that remain unquestioned, leading to a lack of clarity.

With the move towards deliberation, across all policy areas, we have seen the emergence of consultants eager to deliver democracy. Nik Rose calls this group 'experts of community'. They come armed with 'devices and techniques to make communities real' (Rose 1999, pp. 189–90). But these 'technologies of elicitation', such as focus groups, surveys, citizens' juries and new online devices, can create a new form of technocracy by disguising the politics of both science and participation (Lezaun and Soneryd, 2007; Chilvers, 2008). The discussion of *how* frequently obscures the more fundamental discussion of *why*. One of the most important aims of upstream engagement is to encourage institutional reflection, to get decision-makers to question their own assumptions and consider a wider range of alternatives. If done disingenuously, engagement runs the risk of manipulating the public, which is worse than ignoring them (Cooke and Kothari 2001). So the challenge is to encourage a deeper debate about science's and government's relationship with the public. We need to keep asking why engagement should take place and why it should look a certain way. Off-the-shelf processes can exacerbate the distinction between the bits of issues that are considered 'scientific' and the bits that are considered 'social', leaving assumptions untouched.

For public engagement to make a difference, it must become part of the routine practice of good science. We have argued that this does not mean an endless stream of citizens' juries, but it requires us to think through the different forms that engagement will take at different points in the cycle of research, development and diffusion (Stilgoe 2007).

The aim should be to create an ongoing process of what one recent report calls 'collective experimentation' (Wynne *et al.* 2007, note 92).

A critique of public engagement allied to this is that organized deliberative processes can sideline equally important engagement with interest groups and NGOs (Stilgoe 2006). But according to one recent paper, the tendency is to see the silent majority as the true voice of the public (Lezaun and Soneryd, 2007). But if public engagement is to help us understand systems of science and technology, then interest groups need to be invited back in. We are starting to see, especially in areas of medical science, the emergence of public groups which are neither disinterested nor uninterested in science. As we saw with debates over the MMR vaccine, animal rights and nuclear power, 'engagement' can be uninvited but impossible to ignore. Patient groups in particular have demanded a say in scientific research.[11] In the future, such groups are likely to become more vocal and powerful. The challenge for institutions is to acknowledge the diverse interests that make up 'the public'; to learn from uninvited engagement, while making the most of organized engagement. We need to tie engagement to politics, rather than strip the politics away.

Putting the politics back into science

Despite a decade of real progress, public engagement is still splashing about in science's shallow end. But at its best, it can ask deep questions about the politics of science. At Demos, we have argued that public engagement cannot be an end in itself. But if it is a means to an end, what is that end?

We believe that the goal should be a renewed social contract for science based on a deeper appreciation of the politics of science. This means questioning flawed assumptions that still sit behind much science policy. One of these, the so-called linear model of innovation, 'outlives all falsification' (Wynne *et al.* 2007, note 34, p. 77). The idea that there is a simple line from science, through technology, to economic growth and progress has been constantly questioned by history, and yet we still see debates about science and innovation focusing on questions of scale—'how much?' and 'how fast?'. We need to also be able to talk about direction—the outcomes to which all of this investment and activity are being directed. We need to think about the directions of alternative innovation trajectories. In the global 'race' to compete in science and technology, the choice we are often presented with is faster or slower, but with no option to change course. We don't devote enough attention to considering the plurality and diversity of possible directions.

These possibilities are illustrated most vividly on the global stage. Few ministerial speeches about science and innovation are now complete without an obligatory reference to China and India. These two vast, heterogeneous nations—home to a third of the world's population—are perpetually conjoined in a form of political shorthand designed to convey the onward march of globalization.

11. Examples of this include the argument between Alzheimer's sufferers and the National Institute for Health and Clinical Excellence (NICE) about access to drugs and, further back, the involvement of AIDS patients in defining the early history of the disease (Epstein 1996); see also Stilgoe *et al.* (2006).

But we also need to be alert to the way these new 'science powers' are used to argue for a more relaxed stance on social, ethical or environmental issues here in the UK. Tony Blair's 'Science Matters' speech to the Royal Society in 2002 is a notable example:

The idea of making this speech has been in my mind for some time. The final prompt for it came, curiously enough, when I was in Bangalore in January. I met a group of academics, who were also in business in the biotech field. They said to me bluntly: Europe has gone soft on science; we are going to leapfrog you and you will miss out. They regarded the debate on GM here and elsewhere in Europe as utterly astonishing. They saw us as completely overrun by protestors and pressure groups who used emotion to drive out reason. And they didn't think we had the political will to stand up for proper science.

Blair (2002)

We need to resist such myths of the 'wild east' in the way we think about global science. Our first defence has to be that this is a counsel of despair, the logical endpoint of which is a set of lowest-common-denominator standards not just for science but also for labour rights, civil liberties and environmental standards. It is also misleading, not to mention patronizing, to pretend that people in India and China don't share many of these same concerns—albeit expressed in a variety of ways. Even in China, where there is less freedom to debate such issues in formal terms, the environmental and social consequences of rapid technological development are now becoming the focus of intense debate, and at times public protest.

The way our politics describes the relationships between science, globalization and competitiveness must start to reflect these subtleties. Instead of seeing Europe's progress towards the more democratic governance of science as a *barrier* to our success in the global knowledge economy, can it not become a different form of advantage? Might it not lead us down new—and potentially preferable—paths of innovation?

The evidence we have from the environmental sphere suggests that countries can gain competitive advantage from the adoption of higher standards (Porter and van der Linde 1995; Hawken *et al.* 2000). We need to explore whether similar patterns can emerge here. There may also be insights from scientific governance, ethics and public deliberation that we can exchange and export. We need to develop networks which allow policy-makers and scientists in Europe to forge common purpose and alliances on these issues with their counterparts in emerging economies.

The politics of science are subtle. There are questions about the science we need and the science we want; questions about uncertainty, evidence and burdens of proof; questions about ownership, access and control. We—scientists, social scientists, politicians, policy-makers and members of the public—need to learn how to open up and debate these questions in public.

■ **REFERENCES**

Blair, T. (10 April 2002). *Science Matters*. Available at **http://www.number-10.gov.uk/output/Page1715.asp**.

Chilvers, J. (2008). Deliberating competence: theoretical and practitioner perspectives on effective participatory appraisal practice. *Science, Technology and Human Values*, **33**, pp. 155–85.

Cooke, B. and Kothari, U. (eds) (2001). *Participation: the New Tyranny?* Zed Books, London.

DTI (Department of Trade and Industry) (29 July 2004). Nanotechnology offers potential to bring jobs, investment and prosperity—Lord Sainsbury. *DTI Press Release.*

Doubleday, R., Gavelin, K. and Wilson, R. (2007). *Nanotechnology Engagement Group Final Report.* Involve, London.

Durodié, B. (2003). Limitations of public dialogue in science and the rise of the 'new experts'. *Critical Review of International Social and Political Philosophy*, **6**(4), 82–92.

Epstein, S. (1996). *Impure Science.* University of California Press, Berkeley, CA.

Fiorino, D.J. (1990). Citizen participation and environmental risk: a survey of institutional mechanisms. *Science, Technology and Human Values*, **15**(2), 226–43.

Funtowicz, S. and Ravetz, J. (1992). Three types of risk assessment and the emergence of post-normal science. In: *Social Theories of Risk* (ed. S. Krimsky and D. Golding), pp. 251–73. Praeger, Westport, CT.

Hawken, P., Lovins, A.B. and Lovins, H.L. (2000). *Natural Capitalism: the Next Industrial Revolution.* Earthscan, London.

HM Treasury/DTI/DfES (2004). *Science and Innovation Investment Framework 2004–2014.* HM Treasury, London.

House of Lords, Select Committee on Science and Technology (2000). *Science and Society*, Third Report. HMSO, London.

Irwin, A. (2006). The politics of talk: coming to terms with the 'new' scientific governance. *Social Studies of Science*, **36**(2), 299–320.

Irwin, A. and Wynne, B. (eds) (1996). *Misunderstanding Science?* Cambridge University Press, Cambridge.

Jasanoff, S. (2005). *Designs on Nature: Science and Democracy in Europe and the United States.* Princeton University Press, Princeton, NJ.

Kearnes, M., Macnaghten, P. and Wilsdon, J. (2006). *Governing at the Nanoscale.* Demos, London.

Leach, M. and Scoones, I. (2006). *The Slow Race—Making Technology Work for the Poor.* Demos, London.

Lezaun, J. and Soneryd, L. (2007). Consulting citizens: technologies of elicitation and the mobility of publics. *Public Understanding of Science*, **16**, pp. 279–97.

Page, B. (2004). Public attitudes to science. *Renewal*, **12**(2), 20–31.

Porter, M.E. and van der Linde, C. (1995). Toward a new conception of the environment – competitiveness relationship. *Journal of Economic Perspectives*, **9**(4), 97–118.

Rabeharisoa, V. and Callon, M. (2002). The involvement of patients' associations in research. *International Social Science Journal*, **54**(171), 57–63.

Rose, N. (1999). *Powers of Freedom: Reframing Political Thought.* Cambridge University Press, Cambridge.

Royal Society (1985). *The Public Understanding of Science.* Royal Society, London.

Royal Society, RCUK and Wellcome Trust (2006). *Factors Affecting Science Communication: a Survey of Scientists and Engineers.* The Royal Society, London. Available online at: **http://www.royalsoc.ac.uk/downloaddoc.asp?id=3052**.

RS/RAE (Royal Society/Royal Academy of Engineering) (2004). *Nanoscience and Nanotechnologies: Opportunities and Uncertainties*, Royal Society Policy Document 19/04. Royal Society, London.

Stilgoe, J. (2006). Between people and power: nongovernmental organisations and public engagement. In: *Engaging Science: Thoughts, Deeds, Action* (ed. J. Turney), pp. 62–7. The Wellcome Trust, London.

Stilgoe, J. (2007). *Nanodialogues: Experiments in Public Engagement with Science*. Demos, London.

Stilgoe, J., Irwin, A. and Jones, K. (2006). *The Received Wisdom: Opening up Expert Advice*. Demos, London. Available online at: **http://www.demos.co.uk/publications/receivedwisdom**.

Stirling, A. (2005). Opening up or closing down: analysis, participation and power in the social appraisal of technology. In: *Science and Citizens* (ed. M. Leach, I. Scoones and B. Wynne), pp. 218–31. Zed Books, London.

Taverne, D. (2004). Let's be sensible about public participation. *Nature*, **432**, 271 [in response to: Going public, *Nature*, **431**, 883].

UEA (University of East Anglia Understanding Risk Programme) (2004). *A Deliberative Future? An Independent Evaluation of the GM Nation? Public Debate about the Possible Commercialisation of Transgenic Crops in Britain, 2003*. University of East Anglia, Norwich.

Wager, E. (2009). Peer review in science journals: past, present and future. In: *Practising Science Communication in the Information Age: Theorizing Professional Practices* (ed. R. Holliman, J. Thomas, S. Smidt, E. Scanlon and E. Whitelegg). Oxford University Press, Oxford.

Wilsdon, J. and Willis, R. (2004). *See Through Science: Why Public Engagement Needs to Move Upstream*. Demos, London.

Wilsdon, J., Wynne, B. and Stilgoe, J. (2005). *The Public Value of Science: or How to Ensure That Science Really Matters*. Demos, London.

Wynne, B. (1991). Knowledges in context. *Science, Technology and Human Values*, **16**(1), 111–21.

Wynne, B. (1995). The public understanding of science. In: *Handbook of Science and Technology Studies* (ed. S. Jasanoff, G.E. Markle, J. Ceterson and T. Pinch), pp. 361–88. Sage, Thousand Oaks, CA.

Wynne, B. (2005). Risk as globalizing 'democratic' discourse? Framing subjects and citizens. In: *Science and Citizens: Globalization and the Challenge of Engagement* (ed. M. Leach, I. Scoones and B. Wynne), pp. 66–82. Zed Books, London and New York.

Wynne, B. (2006). Public engagement as a means of restoring public trust in science: hitting the notes, but missing the music? *Community Genetics*, **9**(3), 211–20.

Wynne, B., Felt, U., Callon, M., Eduarda Gonçalves, M., Jasanoff, S., Jepsen, M., Joly, P-B., Konopasek, Z., May, S., Neubauer, C., Rip, A., Siune, K., Stirling, A. and Tallacchini, M. (2007). *Taking European Knowledge Society Seriously*. Report of the Expert Group on Science and Governance, to the Science, Economy and Society Directorate. European Commission, Brussels.

Zhang, W. (2003). Nanoscale iron particles for environmental remediation: an overview. *Journal of Nanoparticle Research*, **5**, 323–32.

■ FURTHER READING

- Stilgoe, J. (2007). *Nanodialogues: Experiments in Public Engagement with Science*. Demos, London. This pamphlet presents the findings of the *Nanodialogues* discussed in this chapter. The *Nanodialogues* were a series of experiments in upstream public engagement with different partners in different contexts.

- Wilsdon, J. and Willis, R. (2004). *See Through Science: Why Public Engagement Needs to Move Upstream*. Demos, London. This pamphlet explores the ways in which citizens can expose to public scrutiny the assumptions, values and visions that drive science. In so doing, it provides a rationale for moving public engagement upstream as novel science innovations begin to emerge, then continuing these deliberations downstream.

■ USEFUL WEB SITES

- **Demos:** http://www.demos.co.uk/. Demos is described on its web site as '. . . the think tank for "everyday democracy". [Their] aim is to put this idea into practice by working with organisations in ways that make them more effective and legitimate'. Copies of the various Demos pamphlets listed in this volume can be accessed through this web site under 'Creative Commons' licenses.

- **Involve:** http://www.involve.org.uk/home. Involve is an organization whose aim is 'To show how public participation can positively and productively change the lives of individuals and improve the working practices of institutions'. Involve ran the Nanotechnology Engagement Group (see Doubleday *et al.*, 2007).

- **Royal Society/Royal Academy of Engineering on nanotechnologies:** http://www.nanotec.org.uk. This website is linked to the RS/RAE (2004) report discussed in this chapter. It provides some background and rationale for this project, also a link to a copy of the final report.

1.3

(In)authentic sciences and (im)partial publics: (re)constructing the science outreach and public engagement agenda

Richard Holliman and Eric Jensen

Introduction

This chapter addresses the meanings of 'sciences' and their 'publics'. These definitions are particularly important given the recent calls to shift practices from deficit-led science outreach to dialogue-led public engagement (see, e.g., House of Lords 2000). In effect, such a shift demands that practitioners conceptualize publics in more sophisticated ways, acknowledging the knowledge, values, attitudes and beliefs that they bring to these more symmetrical and interactive exchanges, and relinquishing at least some control over what, when, where, how and with whom science-based issues are (not) discussed. In this chapter we argue that the way that practitioners of science outreach and public engagement (SCOPE)[1] conceive of sciences and their publics not only informs what is discussed and who participates, but also the subsequent choice of SCOPE activities, as well as when, where and how often they are conducted. In so doing, we draw on empirical

---

1. We use the acronym SCOPE—science outreach and public engagement—as shorthand to encapsulate the range of activities that can be characterized as falling into two broad approaches, i.e. ranging from deficit-informed science outreach (e.g. a public lecture) to dialogue-informed public engagement (e.g. a consensus conference). In so doing, we make no normative distinction between the benefits and drawbacks of these two different approaches. Rather, we argue that a single umbrella definition of public engagement masks (e.g. for practitioners seeking to change their practices) the important differences between deficit and dialogue-informed approaches.

evidence to illuminate postgraduate, early career and experienced scientists' self-imposed limits on their public engagement activities.

It is clear that practitioners of SCOPE make decisions about which sciences, which publics and through which mechanisms they engage each time they participate in an activity of this nature. However, it is less clear whether practitioners of SCOPE make these decisions implicitly or explicitly. That is, are practitioners making active choices when they define sciences, publics and the 'where', 'when' and 'how' related to the resulting SCOPE activity? Such factors are likely to be key determinants of the efficacy of particular science communication episodes.[2] Furthermore, are practitioners of SCOPE changing their pre-existing practices to meet, at least in part, the new dialogue-informed public engagement agenda? This chapter addresses these questions, drawing on data collected and analysed as part of an initial study conducted for the Informing Science Outreach and Public Engagement (ISOTOPE) project (see Holliman *et al.* 2007; Jensen and Holliman, Chapter 2.1 this volume).[3]

Background

There is currently a general rhetorical enthusiasm emanating from those with a responsibility for science governance in the UK, the European Union and internationally—in particular with respect to the relationship of sciences with their publics—that public engagement *per se* is a good thing (see Irwin 2006 and Chapter 1.1 this volume; see Jasanoff 2003 for discussion). For example, the UK government has recently endorsed ideas about 'upstream engagement' (e.g. POST 2006),[4] also providing financial resources to the Higher Education Funding Council for England (HEFCE)[5] to fund Beacons for Public Engagement over the next 4 years (2008–2012) (noted by Stilgoe 2007). Furthermore, in the last 5 years research councils have issued several reports related to aspects of public engagement (e.g. RCUK 2002, 2005, 2006; EPSRC 2003), The Wellcome Trust has published an edited collection based broadly on its public engagement programme (Turney 2006), and the Royal Society, RCUK and Wellcome Trust have jointly funded a study investigating scientists' and engineers' communication practices (Royal Society,

---

2. The importance of making active choices in science communication is supported by research on communication competence. Communication competence includes the effective adaptation of one's message and communication style to the particular social context at hand (Littlejohn and Jabusch 1982; Spitzberg and Cupach 1984). At a minimum, communication competence requires 'two things: (tacit) knowledge and (ability for) use' in a particular situation (Hymes 1972, p. 282).

3. ISOTOPE (http://isotope.open.ac.uk) is an interdisciplinary action research project, involving practitioners of SCOPE, social researchers and experts in the design, delivery and evaluation of web-based resources. The project is funded by the National Endowment for Science Technology and the Arts (NESTA) (http://www.nesta.org.uk/) Learning Award LP0286.

4. For a detailed discussion of the concept of upstream engagement see Wilsdon and Willis (2004).

5. With matched funding from Research Councils UK (RCUK), and additional funding from the Scottish and Welsh Funding Councils and the Wellcome Trust (http://www.hefce.ac.uk/pubs/hefce/2006/06_49/).

RCUK and Wellcome Trust 2006). Given this 'official' endorsement of the dialogue-informed agenda for public engagement, it could be tempting to accept that practices have changed in line with this rhetoric. However, as Irwin (2006) notes, deficit-informed activities are still being practised, although now under the ascendant and apparently conflated branding of public engagement. It follows that the policy rhetoric may be masking several important issues that are worthy of further exploration. For example, given the broad policy consensus regarding the new agenda it is worth noting there is much less agreement about what constitutes public engagement in practice, or how this might be judged to have been effective. In turn, it is also worth noting that, at least in part, this is reflected in the range of heterogeneous actors who have some responsibility for and/or interest in participating in public engagement, and their reasons for doing so.

Of course, these challenges have been actively debated for many years in the UK and elsewhere, not least by social researchers (see Irwin and Michael 2003 for an overview). From a UK perspective the Bodmer report (Royal Society 1985) has become the principal signifier for a deficit-informed public understanding of science (PUS) approach. This report, alongside the PUS approach that it promotes, has long been critiqued for its uncritical deference to scientists and scientific institutions, alongside the attendant (re)enactment of the 'empiricist repertoire' (see Burchell 2007 for discussion), or 'myth' (see Barthes 1977 for discussion) of scientific knowledge as a body of objective, universal, value-free facts that is epistemologically superior to other ways of knowing, and the *only* knowledge worthy of informing public debates (see Wynne 2003 for an extended critique). Furthermore, a reliance on scientists fulfilling their 'duty' as the *only* fully qualified mediators of scientific knowledge has been characterized as a 'top-down' approach; and an overly simplistic view of the public, conceptualized as an undifferentiated mass of scientific illiteracy (see Irwin 1995; Irwin and Wynne 1996; Wynne 1996; Gregory and Miller 1998; Irwin and Michael 2003 for extended critiques of the deficit model). Such an approach draws clear lines between scientific expertise and 'the public', the latter who are apparently expected to passively assimilate scientific knowledge as a body of uncontested facts, thereby increasing their levels of scientific literacy, awareness and appreciation of science, without any degree of control over the direction of techno-scientific development.

Irwin (e.g. 2001; see also Irwin and Wynne, 1996; Irwin and Michael, 2003) characterizes the deficit approach as 'first-order' thinking, arguing for the need to engender a more sophisticated relationship between sciences and their publics. In contrast, 'second-order' thinking can be characterized as a dialogic, 'two-way' or 'bottom-up' approach—sciences on one side, in a more symmetrical relationship with publics on the other—where openness and transparency are valued. This is a more accountable relationship between sciences and their publics than first-order thinking implies, reflecting the need to build trust, informed consent and consensus, made clear by high-profile episodes such as BSE/vCJD and the genetically modified crops controversy (see Irwin, Chapter 1.1). It follows that second-order thinking does not automatically privilege scientific knowledge and expertise over other ways of knowing and citizen expertise (see Wynne 1991, 1996, 2003; Collins and Evans 2002, 2003; Irwin and Michael 2003; Jasanoff 2003; Stilgoe *et al.* 2006 for an extended discussion of expertise), although the limited opportunities for publics to influence the framing of consultative exercises may result in this being the case.

'Third-order' thinking is a further development of first- and second-order thinking in terms of the conceptualization of sciences and their publics. Third-order thinking can be characterized as a contextual model, reflected in the contested framing of definitions and priorities for public engagement *with* sciences, as defined by sciences *and* their publics. This contestation yields heterogeneity in stakeholder positions and types of expertise. In third-order thinking active citizens have a more symmetrical role in deliberating the agenda for socio-technical change and a responsibility for engaging in these processes (Wilsdon *et al.* 2005). As Irwin and Michael (2003, p. 90) argue, in this approach:

... the public [are] social actors who can reflexively engage with science, making contingent judgements about the trustworthiness of experts and their knowledge in the light of their own cultural and social investments.

It follows that, from this perspective, rather than being conceived as problematic, disagreement and critical analysis focused on the social implications of science are seen as societal resources to be valued.

In this chapter we start from the assumption that first-, second- and third-order thinking can each be enacted by practitioners of SCOPE, and that it is possible for deficit-informed practitioners to shift their activities to dialogic and deliberative efforts. Starting from these premises, we investigate evidence of current and developing practices, and compare these findings with others who have conducted analogous studies (see, e.g., Burchell 2007). Moreover, in articulating first-, second- and third-order thinking we aim also to compare them with how postgraduate, early career and experienced scientists participating in the first phase of the ISOTOPE study conceptualized SCOPE and its aims and purposes, and, from there, which science and which publics they particularly valued. In short, does the practice match the policy rhetoric?

Data collection and analysis

In this chapter we provide a brief explanation of our data collection methods and analytical techniques. For a more detailed review of our research design, data collection methods and analytical techniques see Jensen and Holliman in Chapter 2.1.

In the initial phase of the ISOTOPE project we used focus group interviews in combination with questionnaires to collect data from a 'structured sample' (see Kitzinger and Barbour 1999; Holliman 2005 for discussion) of SCOPE practitioners. Focus group interviews were chosen because they have the potential to provide valid accounts of participants' views on a particular set of issues (ibid.). If designed and conducted effectively focus group participants generate rich data through peer interaction, using their own vocabulary and terminology (Kitzinger 1994), with less intervention from a moderator than would be the case for individual interviews with the same participants. However, there is a danger that individual perspectives may be lost during the peer interaction. To address this issue we introduced initial[6] and final questionnaires allowing

---

6. The initial questionnaire was completed online prior to the date of the focus group interview. These data were checked prior to the focus group interviews with a view to informing the moderator's questions.

participants to articulate their views of SCOPE individually prior to and following the peer interaction.

The overall sample for the study, which involved eight focus groups, is outlined in Table 1 (see Box 1 for further explanation). Initially, we decided to sample groups of scientists from a range of scientific disciplines who were all at a similar stage in their careers (see Table 1, Groups 1, 2 and 3). A key aim was to compare and contrast scientists' views on SCOPE across these groups. Data collected and analysed from Groups 1, 2 and 3 (italicized in Table 1) are discussed in this chapter.

BOX 1 SAMPLING SCOPE PRACTITIONERS

In addition to sampling postgraduate, early career and experienced scientists, we also sampled from groups of people who regularly work with SCOPE practitioners, for example science teachers (Table 1, Group 6), and two groups of Open University science staff tutors (Table 1, Groups 4 and 5). Open University science staff tutors are assigned to work in each of the university's 13 regional centres. Their main roles are related to teaching and managing part-time associate lecturers (tutors) working on science courses. However, many of these staff also have a role in organizing regionally based SCOPE events. These individuals were included in the sample for their: (1) experience in organizing SCOPE events; (2) experience in working with scientists interested in SCOPE; and (3) knowledge and understanding of organizing SCOPE events in a range of geographical locations.

Finally, we included a group of full-time professional science communicators and a 'mixed' group of participants, which included two professional science communicators, a science teacher, and a 'pro-am' with a particular interest in geology (see Leadbetter and Miller 2004 for an extended discussion of 'pro-ams').

Notwithstanding the aforementioned initial and final questionnaires, which were completed individually, each of the focus group interviews followed the same overall structure, beginning with a moderator-led briefing, followed by two 'focused' participant-led activities. The first involved discussion of a range of methods for SCOPE. To facilitate this activity participants were given 16 cards, each with a method for SCOPE listed on

Table 1 The sample for the study

Group ID	Description of participants
Group 1	*Postgraduate research students*
Group 2	*Experienced scientists*
Group 3	*Early career scientists*
Group 4	Open University science staff tutors
Group 5	Open University science staff tutors
Group 6	Science teachers
Group 7	Professional science communicators
Group 8	'Mixed' group

it, e.g. lecture demonstration, consensus conference, citizen jury, etc. Participants were asked to take each card in turn and discuss what they knew about these methods.

Having reviewed all the cards, participants were asked to conduct the second focused participant-led activity, where they chose one of the methods for SCOPE from the first activity, with a view to producing a 'storyboard' plan for how they would put this method into practice. To facilitate this discussion, participants were given a 'tip sheet' of questions to consider, addressing aspects of preparation, content and evaluation. The moderator gave no further advice or guidance. Having produced a storyboard each group was asked to 'present' this, leading to a general discussion, which was led by the moderator following a semi-structured format.

The data from the initial and final questionnaires were collated and analysed quantitatively or qualitatively depending on the question being asked and the types of responses that were collected. Where data were analysed quantitatively both researchers coded a subset and measures of inter-rater reliability were calculated. These are listed with the relevant results in the following section.

The focus group interviews, each of which lasted for over 3 hours, were recorded and full transcripts were produced. These transcripts were considered alongside field notes taken by the moderator and, where possible, an observer. In addition, the storyboard plans were collected for analysis. For the purpose of the analysis discussed in this chapter we started with the core concepts of 'sciences' and 'publics', analysing the data qualitatively following an inductive or 'grounded' approach (Strauss 1987; Strauss and Corbin 1990; Glaser and Strauss 2001). We used computer-aided qualitative data analysis software (ATLAS-ti)[7] to facilitate this approach, thus allowing the researchers to (re)code the same data, collapse and examine new concepts, thereby to investigate this rich data set more systematically (Hansen *et al.* 1998; Seale 2000).

Results and discussion

In presenting the results from Groups 1, 2 and 3, we draw on data from the initial questionnaires completed by postgraduate researchers, early career and experienced scientists ($n = 21$), and the first three focus group interviews ($n = 15$).[8]

Evidence of first-, second- and third-order thinking

Initially, we decided to explore participants' definitions of 'science outreach' and 'public engagement', respectively, with a view to investigating the apparent conflation of first-, second- and third-order thinking under the single 'umbrella' term of public engagement. To this end, participants were asked to recount their definitions of 'science outreach' and

7. For more information about ATLAS-ti see http://www.atlasti.com/.

8. The sample for the initial questionnaire ($n = 21$) differed from the focus group sample ($n = 15$) in that six of the respondents who completed the initial questionnaire were subsequently not able to attend the resulting focus group interview.

Table 2 Defining 'science outreach' and 'public engagement' (*n* = 21)

Type of definition	'Science outreach'	'Public engagement'
First order (deficit)	15	8
Second order (dialogue)	0	4
Third order (contextual)	0	2
No response or N/A	6	7

The second author (E.J.) was the primary coder for the preliminary questionnaire data. However, the first author (R.H.) recoded a randomly selected subsample of six extracts (29% of the total *n*, comprising 72 cases), yielding raw inter-coder agreement of 100%.

'public engagement' in the initial questionnaire.[9] We also sought evidence of first-, second- and third-order thinking in the focus group transcripts, in particular coding for extracts that were also explicitly linked to conceptions of 'sciences' in relation to 'publics'.

Where the questionnaire data referred specifically to respondents' definitions of 'science outreach' and 'public engagement' they were coded as either 'deficit', 'dialogue' or 'contextual' for each type of activity based on the definitions of first-, second- and third-order thinking as outlined in the introduction to this chapter (Table 2).

Table 2 shows that all 15 participant responses that could be coded under our scheme defined science outreach in a manner consistent with first-order thinking (i.e. the deficit model). The following initial questionnaire extract is an example of a first-order definition of science outreach:

Science outreach suggests a *unidirectional*, BBC-like, *broadcast* approach.

Experienced scientist (our emphasis)

Eight participants also defined 'public engagement' in first-order terms, suggesting that these participants had conflated first-order deficit thinking (science outreach) under the ascendant 'umbrella' branding of public engagement. The extract below was coded as a first-order definition of public engagement:

Public engagement would be *to* the general public with *little or no scientific knowledge*.

Early career scientist (our emphasis)

In contrast, four participants described public engagement in terms of second-order thinking, by attributing some value to dialogic approaches on scientific issues that involve citizen perspectives. The following definition of public engagement was coded as second-order thinking:

Public engagement suggests more of a *two*-way street with *reciprocal efforts and interest*.

Early career scientist (our emphasis)

9. Some responses indicated that participants did not know how to answer the question posed. For example, one respondent wrote '[I] don't know the difference between "outreach" and "engagement".' Other responses that were not included for further analysis include, 'Outreach encompasses all forms of communication'. And from another respondent: 'public engagement would not necessarily be face to face'.

Furthermore, we found evidence in two participant extracts—both postgraduate scientists—that contained aspects of both second- and third-order thinking (coded in Table 2 as contextual), acknowledging both the notion of a two-way dialogue between science and its publics, but also that such a process could be linked to specific outcomes related to issues of governance:

Public engagement is a more active relationship *between* the public and science community, there is *feedback in both directions* and *the scientists become more accountable for what they do*.

Postgraduate scientist (our emphasis)

These findings, which are consistent with the distribution in the full sample (see Holliman *et al.* 2007), suggest that many of the participants who completed the questionnaire had pre-existing knowledge of first-order (deficit) approaches to science outreach. A smaller subset of participants also defined public engagement in first-order terms, suggesting that the recent adoption of 'public engagement' as an 'umbrella' term may be confusing at least some practitioners of SCOPE. Having said this, we also found evidence that several participants had pre-existing knowledge of dialogic approaches to public engagement, with two participants also including elements of third-order thinking in their definitions.

Given this mixed pattern in terms of participants' knowledge of first-, second- and third-order thinking as identified in the initial questionnaire responses, we decided to search for equivalent comparative evidence in the focus group data. In so doing, we analysed the group's plans for planning, conducting and evaluating a SCOPE activity. What follows is an illustrative example from Group 2 (experienced scientists).

This group decided to develop a school visit linked to National Science Week,[10] possibly on a national scale, based on the popular BBC television series *Dr Who*,[11] but also drawing on ideas developed in the television series *Rough Science*.[12] Using these ideas, this group worked creatively to link the perceived out-of-school prior knowledge and experiences of their desired audience—children and young people who had yet to decide on a career path—with their proposed productive activity. Such an approach is consistent with calls for greater links to be made between compulsory science education and 'real-world' out-of-school sites for learning (see Braund and Reiss 2006 for discussion).

The activity itself, although based on scientific principles and concepts that could be described as 'ready-made science' (see Latour 1987 for discussion) was designed to introduce participants to some of the 'authentic' processes of scientific investigation (see Murphy *et al.* 2006; Braund and Reiss 2006 for discussion of authenticity). To this

---

10. For further information about National Science Week see http://www.the-ba.net/the-ba/Events/NSEW/index.html.

11. *Dr Who* is a recently resurrected BBC science fiction television series. For more information see http://www.bbc.co.uk/doctorwho/.

12. *Rough Science* was an Open University (OU)–BBC television co-production that aired on BBC2 over six series. For more information see http://www.open2.net/roughscience6/. The six series of the show were based on a similar format where a small group of scientists from different disciplines were set a series of challenges by a presenter. They were given 3 days to complete these challenges, using everyday materials and working outside of controlled laboratory conditions; e.g. in series 6 Mike Bullivant (an OU-based chemist) attempted to produce an antiperspirant from aluminium foil and bleach.

end, the children would be shown a video illustrating a *Rough Science*-style experiment and asked to work in teams, thereby introducing a competitive element, to try and repeat what they had viewed. To achieve this they would be provided with a 'kit' with the required equipment and materials, all of which would have been checked to ensure that they met health and safety requirements.

The science teacher would be given a high-profile role during the course of the activity to ensure that the children associated the teacher with a 'fun' activity, and not just with the requirements of the National Curriculum.

In the following extract from the focus group transcript participants were asked to reflect on decisions made when producing the SCOPE activity in the form of a storyboard, as well as the purpose for their activity.

1[13]   I think, at the end of the day, we decided that in any sort of science outreach one of the important things is engaging people while they are young, so that was the way to do that. If you can get them engaged and interested whilst they're still at school, then hopefully that will persist a few years after.

2   It's a very sort of public-spirited thing. Actually what we should be doing is trying to interest people like David Beckham in science; if he were to put a drop of his money that would make a heck of a difference.

3   That wasn't one of the options though. [Laughter]

Mod   So, what would you say was the purpose of the activity?

3   It was to enthuse young people into science. I think later on we got the idea that we were actually trying to build up the esteem of the science teacher. I think that didn't come right until the end.

2   No, but I think that's our strongest output, actually.

3   I think you're right; it just took a long time to get there, but we all agreed.

1   But even if that outcome is building up the esteem of the science teacher, then that then feeds forward to further enthuse, hopefully, the children in science.

3   Absolutely.

2   That's why it's time well spent. It adds tremendous value to what we do.

Mod   And what's the ultimate goal there, if the immediate goal is to enthuse kids?

4   Interest them, rather than just sort of educating them about one point, get them interested so they would be willing to go and find more areas interesting probably.

1   And to continue to be engaged with science as an adult then, because to a large extent many people, most people aren't. And I think that's not a good situation in the society in which we live, that has a large reliance on science and technology underlying it.

2   To remain open to what science can offer.

Group 2 (Experienced scientists)

This extract reinforces the findings from the initial questionnaire to the effect that, amongst these participants at least, there was some confusion over the terminology of

13. The numbers indicate which speaker was talking. 'Mod' stands for moderator.

SCOPE with evidence of conflation of first-order thinking under the banner of 'public engagement'. For example, in the first utterance listed in this extract, speaker 1 uses the terms 'science outreach' and 'engagement' to mean the same thing. What becomes clear as the participants continue to discuss their reasons for planning the activity in this way is that a primary deficit-informed objective is to represent science in such a way as to 'enthuse' and 'interest' participants; 'To remain open to what science can offer' as opposed to scientists *and* participants exchanging knowledge and ideas, and valuing each other's perspectives.

Given that 'official' efforts to shift practices from first-order thinking towards second- and third-order thinking are still relatively recent (see the discussion in the Introduction) it is possible to interpret this mixed picture as evidence of either success or failure depending on one's perspective. We do not intend to make a normative judgment of this nature here. Rather, we argue that the conflation of first-, second- and third-order thinking under the umbrella term of 'public engagement' may undermine active decision-making and therefore limit communicative efficacy for those who seek to become more reflective and engaged practitioners of SCOPE.

Taken together the plan for the activity and the extract listed above also illustrate the participants' active choices favouring particular formulations of 'ready-made' science— the re-enactment of the empiricist repertoire (as described by Burchell 2007)—albeit grounded within aspects of a real-world context and a productive, participatory activity. We argue that, if such a pattern were to be (re)constructed consistently and over time as is the case with Groups 1, 2 and 3, then publics would be limited to a partial, and therefore 'inauthentic', view of the complexities, indeterminacy and inherent uncertainties that relate to the practices of contemporary 'science in the making' (Latour 1987). In other words, choosing to consistently promote a partial view of science becomes self-limiting both for sciences and their publics.

Furthermore, in this example the group selected a preferred section of the public—in this case children and young people who had yet to make a choice over their future career —as the desired audience. This is a pattern consistent with the other two groups discussed in this chapter, and is linked with the goal of securing a future generation of science, technology, engineering and mathematics graduates to fulfil the requirements of a techno-science saturated (national) economy within a competitive globalized marketplace (see also Table 3).

Reasons for participation

Definitions of SCOPE are, of course, closely linked to practitioners' reasons for participating in this type of work. To investigate these issues in more detail, participants were asked as part of the initial questionnaire to describe why they undertake science outreach and/ or public engagement work. Participant responses were quantified under the categories listed in Table 3.[14]

---

14. Participants were invited to list as many reasons as they wanted. In the subsequent analyses multiple coding was therefore applied.

Table 3 Reasons for participating in SCOPE ($n = 21$)

Reasons for SCOPE participation	Frequency
Personal enjoyment	11
Professional responsibility	5
Recruiting new scientists/science students	4
Make science accessible	5
Create awareness	3
Combating stereotypes of science/scientists	3
Public accountability of science	2
No response/NA	1

Following the same procedure as for Table 2, raw inter-coder agreement was found to be 100%.

The most frequent answer ($n = 11$) was 'personal enjoyment'. The next most frequent response ($n = 5$) was that they participated out of a sense of 'professional responsibility'. The following extract exemplifies both of these codes:

I *enjoy* it and I consider it to be *part of my job*.

Experienced scientist (our emphasis)

The goal of 'recruiting new scientists or science students'—which featured heavily in the focus group interviews, as the discussion in the previous subsection showed—was mentioned by a further four participants in the initial questionnaires. This goal is implied in the following extract:

It is also good to break down the stereotypes many *children* seem to have of scientists. These stereotypes may play a part in the *lack of children entering careers in science*.

Postgraduate scientist (our emphasis)

The focus group data offered further elaboration of the goals of combating negative stereotypes and misconceptions about science as part of a perceived professional responsibility and desire to make sciences attractive to potential students and future scientists. The following extract first illustrates the 'personal enjoyment' dimension, describing SCOPE as 'fun' and 'interesting'. Second, the extract evinces a desire to represent a partial view of science, in this instance defined as being 'cool', as a way of 'securing the next generation', further investment in science, and designed to address this participant's view that the public have a negative perception of science and engineering:

Mod   Why do you get involved in science outreach and public engagement activities?

5   Because it's important.

1   It's fun.

4   Yeah, it's interesting.

5   Yes, it's fun. Also you're looking to try and . . . well, we're all enthused about science, and generally when you're enthusiastic about something you want to try and enthuse other people and say this is really cool. Certainly my perception of the public's perception of

science and engineering isn't that great. If you say I'm a scientist or an engineer they say oh, that's nice. It should be something really cool; we should be proud to be a scientist or an engineer, but we don't really get that. Hopefully it is changing.

2   I think it's important for securing the next generation. [Sounds of agreement from other participants] I think there is a danger, which [. . .] has alluded to more than once, about the fact that, as a nation, or even as a part of Europe, we need that investment in science to fuel the further development of society. This is a small contribution which we can make to that, but if we don't do it we are doomed.

Group 2 (Experienced scientists)

Such views were generally consistent with the other groups, where the desire to enthuse a future generation of scientists by raising awareness and challenging what were perceived to be negative stereotypes was often paramount. The following extract is illustrative of these arguments, also of the need to make science relevant to the desired audience of 'malleable' children (read potential recruits to science courses and careers):

4   In this setting, as well as teaching something specific in scientific terms, we're just generally raising awareness and trying to spark an interest in science generally to make it a little bit more accessible.

5   Yeah, make it less scary. And relevant as well.

4   Then again, we were talking about the bigger picture of that, and how it relates to their everyday life.

2   Maybe about helping to break down stereotypes as well and thinking that people can do science rather than, you know, 'you're not clever enough to do science' or 'it's just a load of geeks who are geniuses' and making it accessible for them as well as breaking down intimidating stereotypes.

3   The thing is, at that age, I think kids can . . . are still really malleable, you know, they can be shaped into . . . any of them can become scientists. You don't . . . I mean, like you said earlier about you do get the odd ones that are actually genuinely like super-brainy from the start kind of thing, but I think most of them, with the right teaching staff they can become what they want. It's just motivating and enthusing them to follow that path I think. Okay, so we've got raise awareness, more accessible, less scary, boring, how science relates to everyday life and breaking down stereotypes, as our list of what we want to achieve.

Group 1 (Postgraduate scientists)

We have documented evidence to suggest that participants in these focus groups were partly (if not primarily) motivated to participate in SCOPE by an active desire to recruit future generations of scientists, prioritizing school-age children and young people who had yet to decide on a career path at the expense of a whole range of other potential audiences. Whilst such arguments have their place as part of the overall landscape of SCOPE, we argue that such an all-encompassing instrumentalist approach has the very real potential to prioritize deficit-informed activities over other more engaged approaches, becoming self-limiting for practitioners of SCOPE when conceptualizing what they perceive to be 'authentic' representations of sciences for 'ideal' publics. Indeed, one of the groups specifically discussed the desire to prioritize working with 'talented and gifted children'. In the following extract, one participant described how such a partial approach had implications for them when faced with a different audience.

1 I tell you what I found really tough when I did my Research in Residence,[15] I actually was in a science class that wasn't a top tier; I think it might have even been the bottom tier and these . . . most of the kids were not going to go and do A-level and I found I got a bit stumped when I was trying to encourage them like, yeah your science, but oh, but you're really . . . I don't know. What do they . . . if you don't go and do A-level, what do you do with science?

2 You mean A-level at all or just A-level science?

1 A-level at all maybe for some of them I think, you know. And if you're faced with people who are going to drop out of school then . . . or take up some vocational thing which I think loads of people should, you know, but yeah, so I was bit like 'how do you teach . . . ?' I don't know, but I needed to excite them about science so they could pass and get a GCSE but, beyond that, it was difficult to say like, 'hey, you could do this and that with science in your career', and they'd be like 'what the f . . . ?' like 'what career?', so that was weird. So, okay, I was never in that position so perhaps there'd be . . . perhaps within your [ISOTOPE] project there'd be a way of speaking to people who aren't . . . haven't gone into an academic route, you know, and getting experience, because I suppose all the people that I hang out with and talk to about out-reach are academics, and so perhaps getting experience and advice off somebody else on how to enthuse kids who really don't see an academic future. It doesn't mean to say we should not, you know, help and enthuse and excite them.

2 Sometimes it's raising the awareness is a good thing, rather than just . . . rather than saying you're going to love science and going to be a scientist, just being able to say . . .

1 And like you said, knowing the technology and science and that works together, you don't have to be like 'I'm going to have a scientific career' but, yeah, I think it struck me that I went in there thinking 'yeah, I'm going to get more people to do science at A-level even, never mind university, never mind a PhD', and it was like, oh . . .

Group 1 (Postgraduate scientists)

This extract illustrates some of the challenges that result from the kinds of self-limiting choices illustrated above. If one consistently chooses to promote science in a particular way, prioritizing a partial section of the public, there is a danger that this will become seen as the 'accepted wisdom'. Where then do early career scientists gain the knowledge and experience to engage with other publics; citizens who may not become scientists, but who pay taxes, vote in democratic elections and may need to decide on whether to vaccinate a child, eat beef, recycle, etc.?

Conclusions

The findings from the initial questionnaires illustrate a mixed picture in terms of post-graduate, early career and experienced scientists' knowledge and experiences of SCOPE. Overall, the majority of these participants had a much clearer understanding of science outreach (first-order thinking) than public engagement (second- and third-order thinking), with some evidence that first-order thinking is being conflated with dialogue approaches

15. *Researchers in Residence* is a programme jointly funded by RCUK and the Wellcome Trust (see http://www.researchersinresidence.ac.uk/rir/).

under the umbrella definition of public engagement. These findings are also supported by the data collected during the focus group interviews where participants made active if self-limiting choices, valuing the promotion of a partial view of science to a partial section of the public, with the primary aims of improving the image of science/scientists and recruiting the next generation of scientists. Taken together, and if made consistent over time, these choices are likely to have the effect of (re)constructing a heavily mediated 'myth' (Barthes 1977) of science as a singular entity, scientific method as a single unified process, and scientific knowledge as value-free, whilst reinforcing the deficit-informed notion of a largely scientifically illiterate adult population. From these findings, it would be possible to argue that at least some of these participants' conceptions of sciences and publics have changed little from the much maligned deficit model.

However, we have also found evidence to suggest that a small number of participants were able to provide a definition of public engagement that was consistent with second-order thinking, with some indicative evidence of third-order thinking. If we also consider some of the dialogue-informed approaches suggested by the other groups in the study who are not discussed in this chapter (e.g. Table 1, Groups 4 and 6 both planned to run a Café Scientifique; see Grand 2009 for discussion) then it would appear that some practitioners of SCOPE are willing to shift their practices towards dialogic approaches that value public engagement *with* science. Having said this, we also note that we found no evidence of third-order thinking in the SCOPE activity plans developed by the groups.

In this chapter we have considered evidence of practitioners' understandings and experiences of science outreach and public engagement, also how these relate to shifting conceptions of sciences and their publics. What we have not considered in any detail is how publics' conceptions of sciences and SCOPE might also be shifting. What we can be confident about is that these conceptions do shift over time, e.g. in relation to conceptions of citizenship (Jenkins 2006). With the recent adoption of the new curriculum for science in England and Wales,[16] young people's perceptions may begin to challenge more traditional conceptualizations of science for those pupils who study science and its relationship with society in this more grounded, contextual way; '. . . what *we* count as "scientific" is also partly defined by *our* experiences as science learners' (Holliman and Thomas 2006, p. 193, emphasis in original). Furthermore, as more publics engage with sciences in dialogic and contextual ways, the expectations of practitioners may shift accordingly. As a result, publics for science may therefore demand more of SCOPE practitioners than deficit-led approaches can deliver. As the final extract from the focus group transcripts illustrates, engaging postgraduate, early career and experienced scientists with these issues may fulfil a demand to enable active choices and further extend the science outreach and public engagement agenda; in effect, to develop communication competence in these new areas (Hymes 1972; Littlejohn and Jabusch 1982; Spitzberg and Cupach 1984). Of course, at the same time we neither underestimate nor ignore those who are less supportive of dialogic and contextual approaches to SCOPE (e.g. Durodié 2003). Rather, we argue that much can be gained from exchanges between practitioners of SCOPE and social researchers. We hope that this chapter, and the resources being co-produced for the

16. For more information see http://www.21stcenturyscience.org/.

ISOTOPE web portal, will make a positive contribution to this shifting agenda, and the development of a reflective, self-critical community of SCOPE practitioners.

■ REFERENCES

Barthes, R. (1977). *Image–Music–Text*. Fontana, London.

Braund, M. and Reiss, M. (2006). Validity and worth in the science curriculum: learning school science outside the laboratory. *The Curriculum Journal*, **17**(3), 213–28.

Burchell, K. (2007). Empiricist selves and contingent 'others': the performative function of scientists working in conditions of controversy. *Public Understanding of Science*, **16**, 145–62.

Collins, H. and Evans, R. (2002). The third wave of science studies. *Social Studies of Science*, **32**, 235–96.

Collins, H. and Evans, R. (2003). King Canute meets the Beach Boys: responses to the Third Wave. *Social Studies of Science*, **33**, 435–52.

Durodié, B. (2003). Limitations of public dialogue in science and the rise of the 'new experts'. *Critical Review of International Social and Political Philosophy*, **6**(4), 82–92.

EPSRC (2003). *Partnerships for Public Awareness: Good Practice Guide*. People Science and Policy Ltd, London.

Glaser, B. and Strauss, A. (2001). The discovery of grounded theory and applying grounded theory. In: *The American Tradition in Qualitative Research*, Vol. 2 (ed. N.K. Denzin and Y.S. Lincoln), pp. 229–43. Sage, Cambridge.

Grand, A. (2009). Engaging through dialogue: international experiences of Café Scientifique. In: *Practising Science Communication in the Information Age: Theorizing Professional Practices* (ed. R. Holliman, J. Thomas, S. Smidt, E. Scanlon and E. Whitelegg). Oxford University Press, Oxford.

Gregory, J. and Miller, S. (1998). *Science in Public: Communication, Culture and Credibility*. Plenum Trade, London.

Hansen, A., Cottle, S., Negrine, R. and Newbold, C. (1998). *Mass Communication Research Methods*. Macmillan, Basingstoke.

Holliman, R. (2005). Reception analyses of science news: evaluating focus groups as a method. *Sociologia e Ricerca Sociale*, **76–77**, 254–64.

Holliman, R. and Thomas, J. (2006). Editorial. *The Curriculum Journal*, **17**(3), 193–6.

Holliman, R., Jensen, E. and Taylor, P. (2007). *ISOTOPE Interim Report*. The Open University, Milton Keynes.

House of Lords, Select Committee on Science and Technology (2000). *Science and Society*, Third Report. HMSO, London.

Hymes, D. (1972). On communication competence. In: *Sociolinguistics* (ed. J. Pride and J. Holmes), pp. 273–93. Penguin Books, New York.

Irwin, A. (1995). *Citizen Science*. Routledge, London.

Irwin, A. (2001). Constructing the scientific citizen: science and democracy in the biosciences. *Public Understanding of Science*, **10**, 1–18.

Irwin, A. (2006). The politics of talk: coming to terms with the 'new' scientific governance. *Social Studies of Science*, **36**(2), 299–320.

Irwin, A. and Michael, M. (2003). *Science, Social Theory and Public Knowledge*. Open University Press, Maidenhead.

Irwin, A. and Wynne, B. (eds) (1996). *Misunderstanding Science? The Public Reconstruction of Science and Technology*. Cambridge University Press, Cambridge.

Jasanoff, S. (2003). Breaking the waves in science studies: comment on H.M. Collins and Robert Evans 'The third wave of science studies'. *Social Studies of Science*, **33**(3), 389–400.

Jenkins, E. (2006) School science and citizenship: whose science and whose citizenship? *The Curriculum Journal*, **17**(3), 197–211.

Kitzinger, J. (1994). The methodology of focus groups: the importance of interaction between research participants. *Sociology of Health and Illness*, **16**(1), 103–21.

Kitzinger, J. and Barbour, R. (1999). Introduction: the challenge and promise of focus groups. In: *Developing Focus Group Research: Politics, Theory and Practice* (ed. R. Barbour and J. Kitzinger), pp. 1–20. Sage, London.

Latour, B. (1987). *Science in Action: How to Follow Scientists and Engineers Through Society*. Open University Press, Milton Keynes.

Leadbetter, C. and Miller, P. (2004). *The Pro-am Revolution: How Enthusiasts are Changing Our Economy and Society*. Demos, London.

Littlejohn, S. and Jabusch, D. (1982). Communication competence: model and application. *Journal of Applied Communication Research*, **10**, 29–37.

Murphy, P., Lunn, S. and Jones, H. (2006). The impact of authentic learning on students' engagement with physics. *The Curriculum Journal*, **17**(3), 229–46.

POST (Parliamentary Office of Science and Technology) (2006). *Debating Science*, Postnote Number 260. Available at: **http://www.parliament.uk/documents/upload/postpn260.pdf**.

RCUK (Research Councils UK) (2002). *Dialogue With the Public: Practical Guidelines*. People Science and Policy Ltd, London.

RCUK (Research Councils UK) (2005). *Evaluation: Practical Guidelines*. People Science and Policy Ltd, London.

RCUK (Research Councils UK) (2006). *RCUK Science and Society Strategy*. Available online at **http://www.rcuk.ac.uk/cmsweb/downloads/rcuk/scisoc/sisstrategy.pdf**.

Royal Society (1985). *The Public Understanding of Science*. Royal Society, London.

Royal Society, RCUK and Wellcome Trust (2006). *Factors Affecting Science Communication: A Survey of Scientists and Engineers*. The Royal Society, London. Available online at: **http://www.royalsoc.ac.uk/downloaddoc.asp?id=3052**.

Seale. C. (2000). Using computers to analyse qualitative data. In: *Doing Qualitative Research: a Practical Handbook* (ed. D. Silverman), pp. 154–74. Sage, London.

Spitzberg, B. and Cupach, W. (1984). *Interpersonal Communication Competence*. Sage, Beverly Hills, CA.

Stilgoe, J. (2007). *Nanodialogues: Experiments in Public Engagement with Science*. Demos, London.

Stilgoe, J., Irwin, A. and Jones, K. (2006). *The Received Wisdom: Opening up Expert Advice*. Demos, London. Available online at: **http://www.demos.co.uk/publications/receivedwisdom**.

Strauss, A. (1987). *Qualitative Analysis for Social Scientists*. Cambridge University Press, Cambridge.

Strauss, A. and Corbin, J. (1990). *Basics of Qualitative Research: Grounded Theory Procedures and Techniques*. Sage, Newbury Park, CA.

Turney, J. (ed.) (2006). *Engaging Science: Thoughts, Deeds, Action*. The Wellcome Trust, London.

Wilsdon, J. and Willis, R. (2004). *See-through Science: Why Public Engagement Needs to Move Upstream*. Demos, London.

Wilsdon, J., Wynne, B. and Stilgoe, J. (2005). *The Public Value of Science: or How to Ensure That Science Really Matters*. Demos, London.

Wynne, B. (1991). Knowledges in context. *Science, Technology and Human Values*, **16**(1), 111–21.

Wynne, B. (1996). May the sheep safely graze? A reflexive view of the expert-lay knowledge divide. In: *Risk, Environment and Modernity: Towards a New Ecology* (ed. S. Lash, B. Szerszynski and B. Wynne), pp. 44–83. Sage, London.

Wynne, B. (2003). Seasick on the third wave? Subverting the hegemony of propositionalism: response to Collins and Evans (2002). *Social Studies of Science*, **33**(3), 401–17.

■ FURTHER READING

- Irwin, A. and Michael, M. (2003). *Science, Social Theory and Public Knowledge*. Open University Press, Maidenhead. This book delivers a sophisticated critical account of the complexities of science–society relations, introducing the concept of ethno-epistemic assemblages. In so doing, it draws on a number of case studies, including: BSE/vCJD, genetically modified foods and xenotransplantation.

- Royal Society, RCUK and Wellcome Trust (2006). *Factors Affecting Science Communication: A Survey of Scientists and Engineers*. Royal Society, London. This study adduces data from a web survey of 1485 scientists based in higher education institutions and 41 qualitative interviews with stakeholders in science outreach and public engagement. The results reveal the methods scientists tend to employ, what percentage participate in science outreach activities and why, as well as both the barriers scientists face in engaging with the public and the ways in which the number of science outreach activities could be increased.

- Stilgoe, J., Irwin, A. and Jones, K. (2006). *The Received Wisdom: Opening up Expert Advice*. Demos, London. This pamphlet addresses the continued dominance of 'scientific expertise' in science policy-making. The authors argue that the current technocratic forms of science governance should be 'opened up', giving publics a greater role in determining what counts as 'acceptable risk' in the realm of techno-scientific development. As an analytical tool for assessing different orientations towards public involvement they identify two ideal types of science policy-making: 'technocratic' and 'open and diverse'. The authors argue persuasively for a move from the former towards the latter in UK science governance.

- Wilsdon, J., Wynne, B. and Stilgoe, J. (2005). *The Public Value of Science: or How to Ensure That Science Really Matters*. Demos, London. This pamphlet is an insightful sequel to Wilsdon and Willis's (2004) previous Demos publication *See-through Science*. It begins by reviewing the current state of public engagement with science. The authors emphasize how the scientific community's orientation has changed, at least at a surface level, yet deficit model thinking and scientism persist below the surface. They argue that what is needed to address the contemporary situation is a greater appreciation of the importance of the fundamental 'software' of public engagement; that is, the 'codes, values and norms that govern science' (p. 19). Examples are highlighted from consultation and public engagement exercises surrounding nanotechnology.

■ USEFUL WEB SITES

- **The ISOTOPE web portal: http://isotope.open.ac.uk/**. The UK-based ISOTOPE web portal provides practitioner-driven resources to facilitate the selection and implementation of a range of science outreach or public engagement activities. These resources are informed by empirical research and consultation with practitioners, as well as the academic literature on effective science communication.

- **Netscope: http://www.open.ac.uk/science/outreach/netscope.php**. The Netscope (Network for Science Outreach and Public Engagement) Initiative aims to raise the profile of science outreach and public engagement by providing a network where practitioners can exchange advice about theoretical, research and practical issues related to science outreach and public engagement. Anyone with an interest in science outreach and public engagement can register as a member of Netscope through the self-managed 'intelligent' data base.

- **Urban Research Program Toolbox: https://www3.secure.griffith.edu.au/03/toolbox/index.php**. This Australian-based website provides information about a range of 'tools' for engaging with publics on a topic, including citizen juries, search conferences and information repositories. There are case studies available on the site regarding different methodologies, as well as an annotated bibliography with further information about community and public engagement.

SECTION 2
Researching public engagement

Long before the official shift away from the deficit model, research at Lancaster, and elsewhere, contradicted the prevailing account of publics and their ways of reacting to science and technologies. . . . The commitment to 'public engagement with science' emerged, replacing the discredited but deeply entrenched 'public understanding of science' paradigm. . . . This reflected our own listening to ordinary citizens in qualitative fieldwork research situations: public meetings, structured focus group discussions, interviews, participant observation and so on.

Brian Wynne (2006). Afterword. In: Matthew Kearnes, Phil Macnaghten and James Wilsdon (ed.) *Governing at the Nanoscale*

2.1

Investigating science communication to inform science outreach and public engagement

Eric Jensen and Richard Holliman

Introduction

In recent years, research councils and organizations such as the Wellcome Trust and the Royal Society have joined high-level government officials in declaring a commitment to facilitating active public involvement in decision-making about techno-scientific developments and engagement with science (see Irwin 2006 and Chapter 1.1 this volume; Davies, Chapter 2.2 this volume; Holliman and Jensen, Chapter 1.3 this volume for a discussion of this new agenda for science–society relations). Notable in this regard is the House of Lords Select Committee on Science and Technology report on Science and Society, which concluded that:

> . . . direct dialogue with the public should move from being an optional add-on to science-based policy-making and to the activities of research organisations and learned institutions, and should become a normal and integral part of the process.
>
> House of Lords (2000, paragraph 5.48)

Such calls attempt to shift the practices of those scientists and other stakeholders who have interests in science–society relations from a deficit-informed to a more dialogue-informed agenda for public engagement with science. However, as Irwin (2006) has argued, deficit-informed activities are still being practised, 'although now under the ascendant and apparently conflated branding of public engagement' (also see Holliman and Jensen, Chapter 1.3 this volume p. 37).

Calls for public engagement with science have been met by 'as many as 1500 initiatives or programmes' (Mesure 2007, p. 8) in the UK conducted by a wide range of citizens (including scientists and professional science communicators) who have taken up the challenge of engaging with audiences of all kinds, from 'gifted and talented' schoolchildren to socially excluded minority groups of adult citizens. Today there is a heterogeneous community of practice operating in the space between what can be characterized as deficit-informed 'science outreach'—aimed primarily at increasing scientific literacy—and dialogue-informed 'public engagement' seeking to foster productive exchanges between scientists and other stakeholders (including members of the public). This variegated community, comprising scientists, science teachers, professional–amateur (pro-am) enthusiasts (see Leadbetter and Miller 2004 for explanation), and professional science communicators, is the focus of the ISOTOPE (Informing Science Outreach and Public Engagement) project.[1] This chapter will describe our approach to using action research methodology and how it has informed our practice of conducting research on science outreach and public engagement to elucidate the science communication practices contained therein. (For a discussion of some of our findings, see Holliman and Jensen, Chapter 1.3 this volume.)

Background

There have been numerous studies of science–society relations over the years, including investigations and essays focusing on the interface between scientific and other forms of (citizen) expertise (Irwin 1995; Wynne 1996, 2003; Collins and Evans 2002; Jasanoff 2003; Rip 2003). The ISOTOPE study draws upon insights from such previous studies, adopting an interdisciplinary collaborative approach to co-producing an open access web portal of empirically informed resources, written in part by, and for, practitioners of science outreach and public engagement. The research underpinning this action, as well as the web resource itself, critically engages with the contemporary context for science outreach and public engagement, providing practical solutions, in particular for early career scientists. The driving ethos behind the co-production of these resources is a commitment to transcending simplistic divisions between practitioners and social researchers. Rather, the project ethos is about facilitating informed choice in this important area; providing resources that are informed by scientists, science outreach and public engagement practitioners and social researchers, for use by scientists, science outreach and public engagement practitioners and social researchers. The design and implementation of this study follow rigorous standards of quality assurance and evaluation in line with the most recent methodological literature on qualitative research and analysis. The discussion that follows will include examples and 'lessons learned' from the ISOTOPE project, while delineating the methodological options available to researchers examining these forms of science communication.

---

1. ISOTOPE is a NESTA-funded Open University project, which began in October 2006. At the time of writing (February 2008) we have completed the initial study (see below for further discussion) and produced an interim report (see Holliman *et al.* 2007).

Designing science communication research and action

Designing a good science communication study requires careful planning and preparation, which takes time and resources. Quantitative researchers may even specify their expected findings at the outset of the research endeavour in the form of testable hypotheses based on previous research. For qualitative researchers, however, it can also be important to balance the construction of a comprehensive plan with the need to maintain a degree of flexibility in order to 'follow the data' where they lead.

This concern to inductively 'ground' qualitative research by beginning a study with empirical data has been most prominently promoted by Barney Glaser and Anselm Strauss (1967), who argued that researchers can miss important phenomena if they impose a priori assumptions and deductive theoretical models in advance of their observations. Rather Glaser and Strauss contended that models of social life should be developed from the data of everyday experience, and then integrated with existing academic literature and theory at a later stage in the process. This perspective has persuaded many qualitative researchers to adopt an inductive approach, though not necessarily adhering to all the specifics of grounded theory. Indeed, this inductive orientation is reflected in the qualitative methods of researching the particular forms of science communication that are reviewed in this chapter. However, we acknowledge that it is neither possible, nor desirable, to approach research with a *tabula rasa*. As we have already noted, our approach has also been informed by theoretical perspectives and methodological approaches that we have found to be persuasive.

Research goals

As with any social research topic, studies of science communication must first establish their primary goals in order to select appropriate methods of data collection, analysis, evaluation and dissemination. In traditional social science, the research question or central intellectual, empirical or theoretical problem to be addressed by the study provides this generative foundation for the research design. However, in action research the goal of increasing social scientific understanding on a particular topic runs in parallel with the additional objective of mobilizing some form of pro-social 'action' on the basis of this newly generated knowledge.

First elaborated by field psychologist Kurt Lewin (1946), action research methodology is based upon the dual aims of: (1) increasing knowledge or understanding about a particular field of practice; and (2) acting on the basis of that newly produced knowledge to effect positive change within this field. The relative emphasis within the continuum between 'action' and 'research' varies from project to project. The process of conducting action research is cyclical, creating a continuous dialogue between investigation and action. This cycle is evident in the design of the ISOTOPE project, which involves a continuous cycle of planning/design, action, empirical data collection and analysis, and evaluating/reviewing findings and resources for the web portal. Finally, action research involves both researchers and practitioners, in this instance including practitioners of science outreach and public engagement, web designers and social researchers.

The qualitative surveys and focus group interviews conducted for the initial ISOTOPE study encompassed two main goals:

- Research: to investigate participants' experiences and views about science outreach and public engagement.
- Action: to identify the functionality and types of resources that participants might value for inclusion in the ISOTOPE web portal, and to co-produce these resources.

The ISOTOPE 'research' results (see Holliman and Jensen, Chapter 1.3 this volume; Holliman *et al.* 2007) informed the subsequent structure, content and delivery media to be used in the prototype web portal. A second phase for both the research and action goals is planned. Once the web portal prototype is in place, evaluation research and participant observation will be used to assess the efficacy of the web portal and its tool-kits of informational resources in a second round of 'research'. This will be followed by a concomitant second round of 'action' to incorporate these research findings into a revised and updated web portal. Funding permitting, these iterative phases of action and research, research and action can continue indefinitely.

Research design

The fundamental principle guiding the design process must be the selection of methods of data collection, analysis and presentation that correspond with the research goals, as discussed above. For example, if one wants to learn about the perceptions of individual science communicators without restricting respondents' possible range of answers, then an open-ended qualitative survey may be appropriate.[2] If these science communicators tend to have internet access, and are spread out in geographically disparate locations, then a web-based survey instrument may be a suitable, if not unavoidable, choice. After navigating the 'Scylla and Charybdis' of the research ideal and practical constraints, the range of reasonable methodological passageways narrows significantly. Indeed, when practical constraints such as a lack of funding for recruiting focus group participants, postage costs for a questionnaire study or the researcher's limited range of methodo-logical expertise enter the equation there will sometimes only be one viable research design available. That said, at least some of these constraints can be ameliorated by con-ducting research as part of a team (ideally with research funding that at least covers direct costs). For example, in addition to including science outreach and public engagement practitioners upstream to inform the study design process, the ISOTOPE project team brought to bear an overlapping, interdisciplinary range of methodological expertise to tailor the research design to the emergent needs of science outreach and public engage-ment practitioners and social scientific questions of interest to science and technology studies scholars.

2. A closed approach is best used with populations that have been studied previously. That way previous research can be used to inform the selection and framing of survey questions. The results from closed-ended surveys are easier to manage as they are frequently in a numerical form ready for statistical analysis as soon as the survey is completed. The data from open-ended questionnaires, on the other hand, must still be coded before they can provide useful results.

Data collection

Science communication research relies upon empirical observation, that is, the collection of social scientific data. Ideally, data collection should be systematic and customized to the research question in order to provide valid and reliable information about the social world. Indeed, the concepts of validity and reliability can help to guide the selection of a particular subgroup or sample out of the larger population of interest, as well as the preferred means of collecting data from this sample (Box 1).

BOX 1 VALIDITY AND RELIABILITY IN SOCIAL RESEARCH

Validity refers to the degree to which the sample data authentically represent the concept or phenomenon under study. For example the ISOTOPE project sought to: (1) identify people engaged in science outreach and public engagement and (2) to record their authentic experiences and views on science outreach and public engagement.

Reliability typically indicates the degree to which the same results will be found if multiple iterations of the same method of data collection are employed. In other words, this concept could be viewed in terms of test, re-test repeatability. This criterion is most appropriate for psychometric tests aimed at certain characteristics within the same sample, whereas it is difficult to assess reliability for cross-sectional research on social scientific phenomena that are contingent and mutable from day to day or year to year.

For qualitative researchers validity is highly prized; whereas quantitative researchers tend to privilege reliability based on statistically representative samples. For example, a closed-ended quantitative survey of individuals' reported level of beef eating following the BSE/vCJD episode might yield statistical representativeness and therefore some level of reliability, but it might provide a less valid account than an inductive qualitative study of how and why individuals made these decisions.

The ISOTOPE project employed three different methods of qualitative data collection, each aimed at maximizing validity by accessing the authentic views and experiences of science communicators. Initially, participants completed questionnaires, providing basic demographic details and their individual perceptions in response to open-ended questions. Focus groups created a more naturalistic group setting in which participants co-constructed discursive representations of their experiences as science communicators through conversation (Kitzinger 1994; Holliman 2005). Participant observation will generate ethnographic data regarding the efficacy of particular science outreach and public engagement activities without the strictures of a formal interviewing structure. Together, these methods of data collection access multiple, overlapping dimensions of the social reality of science communication practitioners. However, this consistent emphasis on validity can come at the expense of the reliability associated with larger sample sizes, making research findings based on small samples contingent to the context within which they were generated. To an extent, however, the limited reliability of small-sample studies can be addressed by methodological triangulation—the use of multiple methods to address a single research question.

Methodological triangulation 'entails the use of different methods to collect information' in order to lessen the particular 'limitations . . . validity threats and distortions' that inevitably crop up in any single method (Tindall 1994, p. 147). Combining questionnaires with focus groups, and later participant observation, exemplifies methodological triangulation in which the weaknesses of each method are reduced due to their collaboration such that the whole is greater than the sum of its parts. For example, questionnaires offer each individual the space to express their particular point of view without interruption or disagreement. At the same time, focus groups provide naturalness and a dialogical dimension lacking from questionnaires. Thus, if used effectively these methods can overlap and complement each other, ultimately yielding greater validity for the study, and also a greater measure of reliability if the results of the mixed methods from the same sample yield analogous findings. Moreover, participant observation will provide direct access to the practices of people engaged in science outreach and public engagement, minimizing the reliance on potentially malleable self-report data and 'completing', at least in terms of the ISOTOPE project's research design, the methodological triangle.

Sampling

A complete census of a given population is rarely a feasible goal, given resource constraints and population variability; even the UK national census is only conducted once every 10 years. As such, social researchers (and opinion pollsters) must conduct their investigations with a smaller subgroup selected from within the larger population. This is known as sampling, a process that can take a number of forms, including random, quota[3] and qualitative or structured sampling. In quantitative data collection, such as large-scale social surveys, sampling would typically be random to maintain its fidelity to the rules and assumptions of inferential statistics (Blalock 1972). However, some quantitative studies merely seek to provide numerical descriptions of their samples without pretence to identifying patterns of statistical significance.

Results from random samples can be identified as statistically significant, for example if two sample means are dissimilar enough to reject the null hypothesis that there is no difference between them. A false positive in such a situation is known as a Type 1 or 'alpha' (α) error. The generally accepted convention is that if Type 1 error is less than a 1 in 20 probability, the result can be described as statistically significant. Thus $\alpha < 0.05$ means that there is a less than 5% chance that the apparent statistical difference is actually random and non-significant (based on all the assumptions of statistical theory).

There are numerous statistical tests available for this purpose, including t-, z- and F-tests. The first two of these—the t and z tests—assume that the sample represents a

---

3. Quota sampling typically involves some form of consistently applied rule, for example sampling all the even-numbered houses in a housing estate.

random selection from a normally distributed population. Such assumptions restrict the range of phenomena that can be addressed using quantitative, statistically representative sampling (also see Ritchie 2003).[4] Indeed, it is possible to argue that quantitative methods are most powerful for measuring central tendencies within a data set and macro-level social stability or change within a society. As such, they were deemed to be of less use to the requirements of the ISOTOPE project.

In contrast, much early qualitative research targeted borderline phenomena occurring within less well-defined, difficult to access or emerging populations unlikely to satisfy the requirements of random or quota quantitative sampling. The nascent community of practice around science outreach and public engagement is just such an ill-defined, hard to reach and still coalescing population. No one knows the boundaries of this population, nor its precise descriptive statistics (cf. Mesure 2007), a situation that is unlikely to change in the near future. Thus, obtaining a representative sample of this population in quantitative terms would be very challenging, requiring significant resources to do so effectively and with any confidence. Therefore, the ISOTOPE project employed qualitative or structured sampling techniques to examine the views of this heterogeneous population.[5]

The primary goal in qualitative sampling is to maximize validity within the bounds of practical constraints. One sampling technique used when practical constraints are high is *snowball sampling*. This involves identifying one member of a difficult-to-reach population and asking them to connect the researcher with other members of the group. This same procedure is repeated with the new contacts and so on, snowballing until the sample is complete. Another method of gathering participants is *convenience sampling*. As the label suggests, this refers to the selection of easily accessed participants, even if they are not ideal from a methodological or theoretical perspective. Both snowball and convenience sampling should only be used as a last resort. Each of these techniques may introduce substantial researcher-based bias, which can potentially compromise the validity of the study's results. That is, if participants are selected based on the researcher's convenience, the sample is generally less likely to be representative of the larger population under study.

The ISOTOPE project made limited use of convenience sampling by including Open University staff in the sample and holding one focus group at the main Open University campus in Milton Keynes. However, sampling for the ISOTOPE project was primarily conducted through open invitations e-mailed to universities, non-governmental organizations (NGOs), schools and amateur science groups asking to hear from people with

4. Ritchie (2003) argues that social researchers routinely employ these statistical tests in violation of such assumptions, suggesting that social scientists tend to use inferential statistics metonymically to communicate probabilistic metaphorical messages about their data just as one might say in normal conversation, 'I am 99% certain about this'.

5. We note, however, that it is possible to investigate clearly defined sections of this larger, diffuse community of practice using quantitative sampling techniques. For example, the jointly funded Royal Society, RCUK and Wellcome Trust (2006) survey investigated scientists' and engineers' perceptions of science communication, in particular focusing on aspects of the science outreach and public engagement agenda.

experience in science outreach or public engagement. This yielded a *self-selected* sample wherein individuals decided for themselves whether they met the criteria for inclusion in the study. From a quantitative perspective, self-selected samples are problematic as they are inherently non-random. However, for a qualitative study such as the ISOTOPE project, self-selection is preferable to researcher-imposed selections in most instances. This is because the goal of qualitative research is to gather authentic views and experiences from participants' perspectives; in other words, the aim is to maximize validity.

The range of available qualitative data collection methods

Having established that a qualitative study was most likely to be appropriate for exploring key questions about diffuse communities of practice such as science outreach and public engagement, the ISOTOPE project still needed to select the particular methods of data collection that would best access this population and address the dual research aims identified above. The main methods of qualitative data collection in the social science literature are open-ended questionnaires, interviews, focus groups and ethnographies, the last of which is often based on participant observation.

Using questionnaires

Also known as survey research, questionnaires are most frequently used as a means of obtaining breadth of knowledge about a research topic. They can be completed over the phone, in hard copy or online—and by respondents or (social/market) researchers. They provide a straightforward method for collecting demographic details, also relatively surface-level information from respondents, for example in the form of tick-box, multiple-choice or yes/no answers.

Questionnaires can also contain fixed response options for each survey item. For example, a survey question from one component of the UK *GM Nation?* government-sponsored public consultation (see Irwin, Chapter 1.1 this volume; Thomas 2009 for discussion) asked for respondents' level of agreement with the following statement: 'I believe GM crops could help to provide cheaper food for consumers in the UK'. The response options were 'Agree Strongly' to 'Agree' to 'Don't know/unsure' to 'Disagree' and 'Disagree Strongly'. This 'level of agreement' measure is known as a Likert scale, and it is a common structure for many questions within quantitative social surveys. For such fixed response questions, the phrasing of the statement or question is crucial, and should be carefully designed to minimize bias.

This pre-framed survey method is often aimed at producing quantitative data that can be subjected to statistical analyses—see above for discussion. This is an efficient method for producing generalizable findings about large, well-defined populations. Moreover, it tends to yield results that can be tested and re-tested to assess their reliability. Without a qualitative component though, the validity of the survey measures can be difficult to gauge. However, the questionnaire method can also be calibrated to assess participants' views in more depth by introducing open-ended questions, inviting respondents to provide more extended answers in the form of qualitative data. Instead of imposing fixed response options, open-ended questions relocate some of the framing power of the survey instrument by allowing participants to define the parameters of their responses—length, structure and content. This is particularly important in exploratory research, such as ISOTOPE, when there is a greater danger of prematurely foreclosing on specious conceptualizations of the population under study. In addition, it can infuse the results of accompanying quantitative measures with greater validity.

Using focus groups

Sociologist Erving Goffman (1961, p. 18) explicated the methodology of focus groups under the synonym 'focused gatherings', defining them in terms of their 'single cognitive focus of attention; a mutual and preferential openness to verbal communication . . . an eye-to-eye ecological huddle'. Kitzinger and Barbour (1999, pp. 4–5) extend this definition by arguing that:

Focus groups are group discussions exploring a specific set of issues. The group is 'focused' in that it involves some kind of collective activity – such as viewing a video, examining a single health promotion message, or simply debating a set of questions. Crucially, focus groups are distinguished from the broader category of group interviews by the explicit use of interaction to generate data. . . . Focus group researchers encourage participants to talk to one another: asking questions, exchanging anecdotes, and commenting on each others' experiences and points of view.

It follows that, if the structure and purpose of focus groups are carefully designed, they have the potential to facilitate analysis of the similarity and diversity of opinions on a particular issue from a variety of research participants (Kitzinger 1994). Thus, they were selected as the main research method for the initial ISOTOPE study because they allow a number of participants to discuss a particular issue—their experiences of science outreach and public engagement—in a supportive environment, using their own language and terminology (Kitzinger 1994; Holliman 2005). By setting the stage for primarily participant-generated data, focus groups can have clear benefits for increasing the validity of a study. (For a more detailed explanation for how focus groups were used in the ISOTOPE study, see Holliman and Jensen, Chapter 1.3 this volume.)

Using ethnographic methods: participant observation

Etymologically, 'ethnography' literally means writing about people. It is often based upon the data collection method of participant observation. The most vocal early advocate of this method was the anthropologist Bronislaw Malinowski (1922), who conducted his first ethnographic fieldwork in the Trobriand Islands near Australia around the time of World War I. Malinowski exhorted academics to immerse themselves in the environment of their subjects; to become an active participant within the community under study. The argument was that through extended first-hand participation and observation it might be possible to penetrate the surface level of a particular culture or subculture to address what Malinowski called 'the imponderabilia of actual life':

Here belong such things as the routine of a man's [and woman's] working day, . . . the subtle yet unmistakable manner in which personal vanities and ambitions are reflected in the behaviour of the individual and in the emotional reactions of those who surround him.

Malinowski (1922, pp. 18–19)

Moreover, 'thick description' of everyday life allows for a highly valid account of the behaviour of participants. Anthropologists conduct their research over extended time periods, often for years at a time. However, the underlying anthropological method of participant observation can be abstracted from this context and used in research conducted over any period of time—even just a few hours.

Rather than generating transcripts of spoken interactions or questionnaire data, participant observation tends to be recorded in the form of descriptive field notes, written by the researcher participating in the phenomenon under study. Within the spectrum of research methods, participant observation is therefore skewed towards the qualitative goals of 'depth' and fullness of description (as well as validity).

Participant observers seek to immerse themselves in the research context and to experience the phenomenon from the participants' point of view. While interviews and self-report data can still be important supplements to ethnographic field notes, the additional experiential dimension of direct observation lends a level authenticity and texture to the researcher's account that cannot easily be secured through other methods of data collection.[6] As with other depth methods, the trade-off is that the sample sizes tend to be small. Given the challenges of negotiating access, ethnographers frequently must rely on the less desirable methods of snowball and convenience sampling in order to recruit participants willing to be observed and recorded over long periods of time.

The ISOTOPE project will use an adapted form of participant observation to examine the efficacy of the prototype web resources, currently being commissioned based on the findings from the preliminary questionnaire and focus group study. One of the main web portal resources will be science outreach and public engagement (SCOPE) activity templates authored by practitioners to address the needs identified through the preliminary study. These templates are designed to act as 'off-the-shelf' recipes that SCOPE practitioners can download and use effectively. To evaluate these templates, ethnographic evaluation research will be conducted. First, a more or less typical SCOPE practitioner will be asked or commissioned to enact one of the activity templates created for the web portal. ISOTOPE researchers will then evaluate the efficacy of the activity using a combination of participant observation, questionnaires completed by other participants and interviews with practitioners about their experience using the template. The participant observations will provide direct evidence about the activity and its impact, unmediated by practitioner self-reports. This will also provide an interesting source of comparative data between the observers' field notes, audience questionnaires and practitioner perceptions of the SCOPE event.

---

6. Philosophically, the experiential dimension of ethnographic research reflects Heidegger's (2000/1962) insights regarding the uniquely human sense of 'being there' [*Dasein*], fully engaged and self-aware within the social world.

Quantitative versus qualitative data analysis

Under normal circumstances, each of the above methods of data collection is ultimately reduced to textual data in some manner.[7] Open-ended questionnaire responses, focus group transcripts and field notes are all forms of qualitative data, which should be analysed systematically in order to draw reliable and/or valid conclusions about the phenomenon under study.

Most qualitative data can be analysed using either a quantitative or qualitative framework, or both. Quantitative analyses will often involve counting words, concepts or other responses within a defined sample of textual data. Qualitative analyses could include forms of discourse analysis, metaphor analysis, ideological analysis, etc. Quantitative coding can offer a reliable, statistical description of the data in terms of the concentration of particular packets of meaning within the sample. This includes the calculation of *inter-coder reliability* (see Holliman and Jensen, Chapter 1.3 this volume). Based on a second coder reanalysing 10–20% of the data set, a *kappa* statistic or Pearson's *r* can be determined using a statistical software package such as SPSS (Statistical Package for Social Sciences). While unable to offer such statistical verifications of reliability, qualitative data analysis can provide 'thick description', in-depth understanding and highly valid accounts of participants' lived experiences. An additional alternative is the construction of a 'mixed' analysis utilizing both quantitative and qualitative analyses in concert. This approach was used in the ISOTOPE project to analyse the preliminary questionnaire study, first using quantitative analysis, and then developing these findings further through qualitative analysis. This mutually supportive analytic structure yielded many of the benefits of 'complementary assistance' discussed by methodologist David Morgan (1998, in press). Research motivated by 'complementary assistance' 'uses different strengths by connecting methods so that one contributes to the performance of another' (Morgan, in press, p. 75). Finally, whether quantitative, qualitative or mixed, most systematic approaches to data analysis rely upon a process of coding and thematization in order to make the data manageable and coherent.

Quantitative and qualitative coding of qualitative data

It is through rigorous and systematic analysis that research results can be derived from data in a reliable and valid manner. With qualitative textual data, most researchers begin their analyses by conducting line-by-line coding. Specifically, 'open coding' can be a useful way to begin when one is facing a corpus of unstructured text-based data. Strauss and Corbin (1990, p. 74) define open coding as 'the analytic process by which concepts are identified and developed in terms of their properties and dimensions'. This is accomplished by asking mental questions about the data, making comparisons, developing labels and groupings for similar phenomena, and beginning to form categories

---

7. If quantitative data have been obtained, such as responses to a closed-ended Likert scale, then data analysis will be relatively straightforward.

(e.g. Strauss and Corbin 1990). The next step is to apply 'axial coding' which essentially consists of reconstructing data 'in new ways by making connections between a category and its subcategories' (Strauss and Corbin 1990, p. 97).[8] In the ISOTOPE study both of these coding procedures were carried out on the entire sample using computer-aided qualitative data analysis software (CAQDAS). Such software performs a similar function for qualitative data analysis as a word processor does for writing.

Quality in qualitative research

There is a growing body of methodological literature advocating quality assurance techniques (cf. Kvale 1996) to facilitate 'distinguishing properly from improperly conducted qualitative research' (Thorne 1997, p. 117), and assessing the 'calibre of the study' (Johnson and Waterfield 2004, p. 129). Flick (2002, p. 280) argues that 'the failure of qualitative research is discussed much too seldom', thus giving the false impression that all issues related to the validity of qualitative research are already settled. Although further elaboration is certainly needed to develop firm, accountable criteria about what represents 'good practice' in qualitative analysis (Gaskell and Bauer 2000, p. 336), there is substantial agreement about some of the ways in which quality can be evaluated in a qualitative study. While some (e.g. Silverman 2000, p. 175) have attempted to apply the concepts of validity and reliability directly from the domain of quantitative methodology (albeit with a revised logic), many methodologists have begun using alternative criteria designed specifically for evaluating qualitative studies. Indeed, Thorne (1997, p. 118) argues that 'much of the "bad" qualitative research' in the social science literature is the result of 'the inappropriate application of quantitative quality measures'. As Gaskell and Bauer (2000, p. 342) note, 'sampling, reliability and validity have served quantitative research well, but are just not appropriate for the evaluation of qualitative inquiry'. The functionally equivalent evaluation criteria they recommend include: reflexivity, thick description, transparency and procedural clarity and deviant-case analysis (ibid.). The ISOTOPE project employed three of these techniques in order to maintain and self-assess quality in the preliminary studies: thick description, transparency and procedural clarity.

First, thick description involves the presentation of extended verbatim extracts from the data, which empower the reader to either agree with the researcher's conclusions or to come to different interpretations. 'The term "thick description" or "dense description" is used when context, meanings and interpretations that elucidate the research process are provided, rather than mere statement of facts independent of intentions or situations' (Johnson and Waterfield 2004, pp. 127–8). Done correctly, this procedure should bring readers 'into the social milieu of the social actors', providing 'insights into the local colour, the language and the life world' of the agents under study (Gaskell and Bauer 2000, p. 347). This form of quality assurance was thoroughly developed in the presentation of ISOTOPE research results elsewhere (Holliman *et al.* 2007; Holliman and Jensen, Chapter 1.3 this volume). Rather than proffering isolated quotes of one to three words

---

8. The third form is selective coding, which is used to deductively test a core category once it has been constructed.

or a single sentence, longer segments of text have been displayed wherever possible to exemplify more of the depth and richness of the original data set.

In addition to providing thick description within the research report, Johnson and Waterfield (2004, p. 127) identify the importance of maintaining a methodological 'audit trail'. This idea 'derives from a fiscal audit that looks for sources of error or deception by examining the way in which the accounts are kept'. This method of quality assessment was addressed in this study primarily through the use of CAQDAS. Gaskell and Bauer (2000, p. 346) argue that CAQDAS compels transparency and procedural clarity 'by technological fiat'. The use of CAQDAS software allowed each step of the analytical process to be digitally captured and maintained for later re-evaluation. This software-based capability can be invaluable when it comes to maximizing procedural clarity and transparency in the data analysis process, in particular when working with large data sets.

Incorporating research findings into an action plan

Within the action research framework, data collection and analysis must feed into some form of related action. This imperative can be explicitly designed into the data collection methodology, with specific questions or observations centred on potential options for action. However, action researchers do not always begin their study with specific actions in mind. In such cases the action plan is emergent from the data. That is, an initial phase of data collection is conducted to produce knowledge about the research topic. The results are then used to guide the selection, design and content of the researcher's intervention.

This emergent approach to action research was employed in the ISOTOPE project. Relatively broad and open-ended questions were put to science outreach and public engagement practitioners, the analysis of which yielded participant-driven specifications of the required actions. Several desired resources were identified by multiple participants within the study, providing the researchers with a clear direction for action (Holliman *et al.* 2007). These ideas were then considered in light of the existing academic literature in order to fashion an action plan that is both practitioner guided and critically informed.

The challenges and promise of investigating public engagement with science

There were few rigorous studies of the role of science communication in science–society relations until about three decades ago. Today, this is a flourishing field of social research (see Hansen, Chapter 3.1 this volume). As sciences change and expand in new and sometimes unexpected directions, social researchers should adapt their methods to develop studies that are both relevant to practice and revelatory of social scientific phenomena. The contemporary context offers analysts a rich buffet of important research questions to explore, not to mention sponsors of social research. At the same time, there are certain challenges specific to this field of study.

One key challenge is the problem of identifying and recruiting participants from amongst the population of non-institutional science communicators. Indeed in the ISOTOPE project, 'pro-ams' (Leadbetter and Miller 2004) proved to be one such difficult-to-reach population. Invitations were sent to dozens of amateur groups, including amateur ornithologists, astronomers and geologists. In the end though, there was only one positive response. Given the difficulties, such populations may require the use of sampling techniques originally developed for research on 'at risk' populations such as sex workers and drug users, who are difficult to reach for obvious reasons. One such technique is snowball sampling (reviewed above).

A second challenge in conducting studies of science–society relations concerns the potential problem of an often unspoken and uncomfortable lack of engagement between scientists and social researchers. Specifically, there is sometimes a lack of trust between these communities, exemplified most recently by the 'science wars' (Mellor 1999). This attitude can make recruitment of scientists to social research difficult at times, especially if the aims and objectives are not clearly apparent or discussed and agreed with research participants. Of course, discussion of topics not seen to benefit the 'image of science', e.g. by delving into a potentially controversial aspect of science such as nuclear weaponry, genetic engineering or vivisection, may also be met with a lack of enthusiasm from potential research participants.

If resistance is encountered consistently, one potential strategy is to employ a modified snowball sampling technique wherein a scientist is asked to serve as a gatekeeper to help recruit other scientists at her or his institution. Alternatively, the imprimatur of the head of department or some other high-status scientist indicated on the contact letter or advertisement can ameliorate such resistance. However, the shift towards the discourse of public engagement may be lessening historical resistances such as this.

The ISOTOPE project attempted to negate such issues in advance, both in how the research was initially conceptualized, by choosing an active research approach that incorporated a co-production dimension within the project (emphasized in recruitment letters), and second by the development of a core management team that included scientists from a range of disciplines and social researchers. Members of this team were based within the Science Faculty at the Open University, having worked together successfully on previous collaborative projects. As such, a level of trust and mutual respect was apparent within the team prior to the start of the project.

Final reflections

This chapter has emphasized the importance of making active research choices on the basis of one's goals, while identifying the potential pitfalls that may detract from these goals. By maintaining the research question(s) as one's methodological sextant, the research design will be able to withstand the inevitable adjustments and course corrections required as a study unfolds without veering off course. The ever shifting horizon of scientific practices in a globalized world requires vigilance and forward planning in order to develop high quality investigations of science–society relations in the 21st century.

■ REFERENCES

Blalock, H. (1972). *Social statistics*, 2nd edn. McGraw-Hill, New York.

Collins, H. and Evans, R. (2002). The third wave of science studies: studies of expertise and experience. *Social Studies of Sciences*, **32**(2), 235–96.

Flick, U. (2002). *An Introduction to Qualitative Research*, 2nd edn. Sage, London.

Gaskell, G. and Bauer, M. (2000). Towards public accountability: beyond sampling, reliability and validity. In: *Qualitative Researching with Text, Image and Sound* (ed. M. Bauer and G. Gaskell), pp. 336–50. Sage, London.

Glaser, B. and Strauss, A. (1967). *The Discovery of Grounded Theory*. Aldine, Chicago.

Goffman, E. (1961). Fun in games. In: *Encounters: Two Studies in the Sociology of Interaction* (ed. E. Goffman), pp. 15–81. Bob Merril, Indianapolis, IN.

Heidegger, M. (2000/1962). *Being and Time* (transl. J. Macquarrie and E. Robinson). Blackwell, London.

Holliman, R. (2005). Reception analyses of science news: evaluating focus groups as a method. *Sociologia e Ricerca Sociale*, **76–77**, 254–64.

Holliman, R., Jensen, E. and Taylor, P. (2007). *ISOTOPE Interim Report*. The Open University, Milton Keynes.

House of Lords, Select Committee on Science and Technology (2000). *Science and Society*, Third Report. HMSO, London.

Irwin, A. (1995). *Citizen Science*. Routledge, London.

Irwin, A. (2006). The politics of talk: coming to terms with the 'new' scientific governance. *Social Studies of Science*, **36**(2), 299–320.

Jasanoff, S. (2003). Breaking the waves in science studies: comment on H.M. Collins and Robert Evans, 'The Third Wave of Science Studies'. *Social Studies of Science*, **33**(3), 389–400.

Johnson, R. and Waterfield, J. (2004). Making words count: the value of qualitative research. *Physiotherapy Research International*, **9**(3), 121–31.

Kitzinger, J. (1994). The methodology of focus groups: the importance of interaction between research participants. *Sociology of Health and Illness*, **16**(1), 103–21.

Kitzinger, J. and Barbour, R. (1999). Introduction: the challenge and promise of focus groups. In: *Developing Focus Group Research: Politics, Theory and Practice* (ed. R. Barbour and J. Kitzinger), pp. 1–20. Sage, London.

Kvale, S. (1996). *Interviews: an Introduction to Qualitative Research Interviewing*. Sage, Thousand Oaks, CA.

Leadbetter, C. and Miller, P. (2004). *The Pro-am Revolution: How Enthusiasts are Changing our Economy and Society*. Demos, London.

Lewin, K. (1946). Action research and minority problems. *Journal of Social Issues*, **2**, 34–46.

Malinowski, B. (1922). *Argonauts of the Western Pacific: an Account of Native Enterprise and Adventure in the Archipelagoes of Melanesian New Guinea*. Dutton, New York.

Mellor, F. (1999). Scientists' rhetoric in the science wars. *Public Understanding of Science*, **8**(1), 51–6.

Mesure, S. (2007). *The CreScENDO Project: Final Report for NESTA*. National Endowment for Science, Technology and the Arts, London.

Morgan, D. (1998). Practical strategies for combining qualitative and quantitative methods: applications to health research. *Qualitative Health Research*, **8**(3), 362–76.

Morgan, D. (in press). *Combining Qualitative and Quantitative Methods: Practical Strategies*. Sage, Thousand Oaks, CA.

Rip, A. (2003). Constructing expertise: in a third wave of science studies? *Social Studies of Science*, **33**(3), 419–34.

Ritchie, D. (2003). Statistical probability as a metaphor for epistemological probability. *Metaphor and Symbol*, **18**(1), 1–11.

Royal Society, RCUK and Wellcome Trust (2006). *Factors Affecting Science Communication: a Survey of Scientists and Engineers*. The Royal Society, London. Available online at: **http://www.royalsoc.ac.uk/downloaddoc.asp?id=3052**.

Silverman, D. (2000). *Doing Qualitative Research: a Practical Handbook*. Sage, London.

Strauss, A. and Corbin, J. (1990). *Basics of Qualitative Research: Grounded Theory Procedures and Techniques*. Sage, Newbury Park, CA.

Thomas, J. (2009). Controversy and consensus. In: *Practising Science Communication in the Information Age: Theorizing Professional Practices* (ed. R. Holliman, J. Thomas, S. Smidt, E. Scanlon and E. Whitelegg). Oxford University Press, Oxford.

Thorne, S. (1997). The art (and science) of critiquing qualitative research. In: *Completing a Qualitative Project: Details and Dialogue* (ed. J.M. Morse), pp. 117–32. Sage, Thousand Oaks, CA.

Tindall, C. (1994). Issues of evaluation. In: *Qualitative Methods in Psychology: a Research Guide* (ed. P. Banister, E. Burman, I. Parker, M. Taylor and C. Tindall), pp. 142–59. Open University Press, Buckingham.

Wynne, B. (1996). May the sheep safely graze? A reflexive view of the expert–lay knowledge divide. In: *Risk, Environment and Modernity: Towards a New Ecology* (ed. S. Lash, B. Szerszynski and B. Wynne), pp. 44–83. Sage, London.

Wynne, B. (2003). Seasick on the third wave? Subverting the hegemony of propositionalism. *Social Studies of Science*, **33**, 401–17.

■ **FURTHER READING**

• Babbie, E. (2000). *The Practice of Social Research*, 9th edn. Thomson Wadsworth, Belmont, CA. This classic text provides a highly readable overview of social science research methods striking a perfect balance between breadth and depth in the descriptions of the various methodological options (as well as their benefits and drawbacks).

• Kitzinger, J. and Barbour, R. (ed.) (1999). *Developing Focus Group Research: Politics, Theory and Practice*. Sage, London. This edited collection provides an extended presentation of the methodology, theory and practices of focus group research, drawing on a series of case studies.

• Silverman, D. (2000). *Doing Qualitative Research: a Practical Handbook*. Sage, London. As the title suggests, this well-written book is particularly useful for providing specific, practical advice about conducting qualitative data collection, analysis and write-up.

• Somekh, B. (2006). *Action Research: Methodology for Change and Development*. Open University Press, Buckingham. In this book the author draws on over 15 years of experience of using action research to effect change, primarily in educational settings. In so doing, Somekh introduces eight methodological principles for action research that are applicable to anyone considering this approach.

◼ USEFUL WEB SITES

- **Online Questionnaire Builder and Host Site: http://wufoo.com/**. This site offers ready-to-use, flexible templates for constructing and hosting qualitative, quantitative or mixed questionnaires online, with no software or programming knowledge required. Results are automatically captured in a spreadsheet format downloadable directly into Microsoft Excel. There is an option to run a questionnaire using this site for free.

- *International Journal of Qualitative Methods*: **http://ejournals.library.ualberta.ca/index.php/ IJQM/index**. This free-to-view open access online journal provides easy access to high-quality articles about qualitative methodology, including both practical and philosophical issues.

- *Social Research Update*: **http://sru.soc.surrey.ac.uk/**. Using a newsletter format, this free-to-view open access online publication offers useful and interesting articles on various aspects of qualitative research practice and methodology, including most of the topics addressed in this chapter.

Learning to engage; engaging to learn: the purposes of informal science–public dialogue

Sarah Davies

Dialogue, public engagement, debate or even public engagement with science and technology (PEST): whatever you choose to call it, it seems to be all the rage. Dialogue approaches to science communication and policy are not just urged in formal documents such as the House of Lords Third Report on *Science and Society* (House of Lords 2000), but are—as a skim through the web sites of organizations such as science museums, learned societies and scientific charities shows—espoused by many public bodies (see also Holliman and Jensen, Chapter 1.3 this volume). This chapter starts to critically examine some of these kinds of activities. In particular, I want to focus on 'informal' dialogue— that carried out not by those bodies with formalized connections to policy processes [such as the government-sponsored initiatives *GM Nation?* (see Irwin, Chapter 1.1 this volume; Thomas 2009 for discussion) or 'sciencehorizons'] but by museums, science cafés or charities. Those organizations, in fact, as Smallman and Nieman (2006, p. 6) suggest, that were previously engaged in 'communicating and popularising science, whether for entertainment, education or other "deficit" reasons'.

In order to look at the practices of dialogue in these informal contexts I'll be focusing on one case study, the Dana Centre in London. The Dana Centre is part of London's Science Museum and was one of the first purpose-built spaces for science–public dialogue. Its remit has always been to be innovative, and as such it can be viewed as part of the 'cutting edge' of public engagement in the UK (McCallie *et al.* 2007). In this chapter I'll be using it as an example of one type of informal dialogue, and analysing particular aspects of it in order to make some broader suggestions about the current PEST movement.

A word on my own language use here. I differentiate (artificially, in many ways) between two types of public engagement: those which impact upon policy in some way, and those with no clear links to science policy (see Lehr *et al.* 2007). The first kind I refer to as *formal*—in that they have formalized connections to science or policy.

Government-organized processes such as the *GM Nation?* debate and a whole tradition of 'public participation' fall into this category, and there is a relatively large body of literature which classifies (Rowe and Frewer 2000, 2004, 2005) or analyses (for example Irwin 2001; Horlick-Jones *et al.* 2004; Davies *et al.* 2006; Goven 2006) these kinds of processes. Research on *informal* dialogue—that which does not even attempt to connect to policy or have large-scale outcomes—is, however, much more sparse. Informal public engagement, by this terminology, covers extremely broad territory: it includes not only events which look rather informal—a Café Scientifique, for example (see Grand 2009)—but those which take place in purpose-built spaces or which are run by large organizations such as museums or learned societies. The Dana Centre in my case study, for example, runs highly professional evening events in a specially designed building, but only occasionally are those events designed to feed into science policy.

Even more confusingly, formal and informal dialogue can often look very similar, and may take place in similar venues. Learned societies such as the Royal Institution or the Royal Society may run both informal public debates and events which are meant to feed into policy, for example, while the Dana Centre hosts formal and informal events in the form of small group discussions.[1] I would suggest, however, that a typical informal event generally takes the form of a panel debate: two or three invited, 'expert' speakers briefly present to an audience before the floor is opened up for questions and debate. While the Dana Centre, in particular, is constantly innovating with this format (for example, using theatre, smaller groups or 'workshops'; see McCallie *et al.* 2007), in many cases the panel debate remains a standard format (Reich *et al.* 2006) and one which is common in events such as Café Scientifique or those run by smaller museums or organizations.

This chapter will be exploring some aspects of this typical format for informal dialogue. Before I discuss my analysis in detail, however, I'll describe some of the background to this surge of interest in science–public dialogue. I'll then move on to focus on the somewhat vexed question of the purposes of informal dialogue. Given that it does not aim to influence science policy directly, and that it claims to have left the 'deficit model' behind, what should it be used for? Moving on to look at data from the Dana Centre, I'll start to apply my discussion so far to this case study. Using examples from Dana Centre events I'll argue that these kinds of dialogue activities are fragmented and unstable: framings of what the events are and what they are for are constantly shifting and their purposes seem multiple and at times contradictory. Finally, I'll suggest that it might be helpful to re-imagine informal dialogue as it occurs within science communication contexts. We need to acknowledge gaps between rhetoric and practice, but also to appreciate the value of flexibility and multiplicity in these kinds of events.

1. I have said that distinguishing between formal and informal events can be 'confusing': it should be clear that this typology is by no means perfect. These types of dialogue may overlap or run into one another. For example, the two types may use identical formats, with their only difference being whether there is any intended policy outcome from them. It should also be noted that informal dialogue may have general impacts on civil society which result—eventually—in impacts on policy (see Davies *et al.* in press for a further discussion). Here I am making use of this typology in order to emphasize the special nature of these more 'science communication' based activities.

The 'dialogic turn' in science communication

Context is everything in research, and particularly so when we're examining any kind of social trend. It is not possible to talk about dialogue in isolation—as if it sprang fully formed into being in the year 2000—rather we need to consider key moments in the history behind it. In the UK this is a history which is generally depicted (for example, see Miller 2001) as beginning in 1985 with the publication of a report, *The Public Understanding of Science*, by the Royal Society. Miller (2001) argues that this report, and its after-effects, very effectively mobilized the scientific community for 'public understanding of science' (PUS) work: suddenly there was money, opportunity and, to some degree, academic respect for scientists to leave their ivory towers and communicate to the public. More specifically, the emphasis was on educating the public with scientific knowledge. The Royal Society Report argues that:

. . . better public understanding of science can be a major element in promoting national prosperity, in raising the quality of public and private decision-making and in enriching the life of the individual.

Royal Society (1985, p. 9)

In other words, knowing about science is necessary to be a good citizen and cultured individual within a vibrant economy. In addition, the report and the flurry of PUS activity that followed it were shot through with the assumption that 'to know science is to love it' (see Turney 1998 for a critique)—that there is a direct link between understanding scientific information and trust and confidence in science (Sturgis and Allum 2004; Irwin, Chapter 1.1 this volume). Thus, much of the PUS programme throughout the 1980s and 1990s operated with the implicit expectation not only of altruistically improving citizens' lives, but of also increasing optimism, interest in and support for science.

This expectation was misplaced. PUS activities might well have proliferated, but the period was marked by scientific controversy and a deep distrust of science in at least some of its forms. Survey research also indicated that, despite the intensive 'education' of the public through PUS, public knowledge of science remained almost identical to 10 years previously (Miller 2001; cf. Durant *et al.* 1989). If PUS was meant to create public enthusiasm for science and a more knowledgeable citizenry, it certainly wasn't delivering on its promises. So what had gone wrong?

The work of a number of critical social researchers points to an answer to this. Brian Wynne, Alan Irwin and others analysed science and the PUS movement's approach to public communication and argued that it was misguided in several ways. For a start, it operated in terms that were too simplistic: what is the 'science' in the public understanding of science, for example (Wynne 1991)? They emphasized the importance of context—such as past experiences of particular institutions—for how people deal with scientific information, arguing that science is frequently just one possible resource amongst many (Ziman 1991; Layton *et al.* 1993; Irwin 1995). Their work critiqued science and PUS for using a 'deficit model' of the public, which focussed attention solely on 'public ignorance or scientific illiteracy' (Ziman 1991, p. 101). Gregory and Miller (1998) note that the deficit model:

. . . conceptualises the lay mind as an empty bucket into which the facts of science can and should be poured. . . . the deficit model locates knowledge and expertise solely with the scientists and keeps them at the top of the heap.

Gregory and Miller (1998, p. 89)

Thus the models and assumptions behind many PUS activities—that 'pouring in' scientific information into non-scientists' ignorant and empty minds would lead to a straightforward uptake of science—were overly simplistic. Critical researchers argued this throughout the 1990s, and it was shown in practice in the very public controversies around science—such as BSE and vCJD, or genetically modified crops—that emerged in the latter half of that decade. Simply telling people 'the facts' was not enough: science, and science policy, had to take into account public values, views and positions; and, importantly, public *knowledges*. These critical authors argued that people being ignorant of a particular aspect of science didn't mean that they were ignorant full stop: studies showed that non-scientists frequently had knowledge which was nuanced, contextualized and far more useful to particular situations than scientific facts could be (Wynne 1992; Layton *et al.* 1993; Michael 1996). Not least, laypeople evaluated information in terms of its source: is this institution trustworthy or not? What has our past experience of dealing with it been (Irwin 1995; Wynne 2001)?

If there is useful knowledge located within lay domains then presumably science can benefit from this. Non-scientists, it was argued, are not only competent to deal with scientific issues (Kerr *et al.* 1998; Durant 1999) but their involvement, as well as being democratically important (Marris *et al.* 2001) will be scientifically and economically productive (MacMillan 2004; Fischer 2005; Stirling 2006). We come, in fact, to the idea that a *dialogue* between science and its publics will be the most fruitful way for science and society to relate. Public debate would, in this dialogue, be part of an 'extended peer review' which both examines the framing and internal values of particular areas of science and adds to it other relevant forms of knowledge (Marris *et al.* 2001, p. 113).

'They just move on to aliens . . .': what is informal dialogue for?

The concept of dialogue was used in another report which acted as a key turning point in the UK history of PUS/PEST: the House of Lords Third Report on *Science and Society* (2000). This report, which took advice from critical researchers such as Brian Wynne, spoke of a 'new mood for dialogue' (ibid., p. 37) and crystallized existing trends towards more interactive and contextual approaches to science–society relations (Wynne 2005). It formalized a new language of dialogue, public engagement and consultation (Irwin 2006), arguing, for example, that:

. . . direct dialogue with the public should move from being an optional add-on to science-based policy-making and to the activities of research organisations and learned institutions, and should become a normal and integral part of the process.

House of Lords (2000, paragraph 5.48)

Significantly, although much of the writing of the critical authors and the *Science and Society* report itself were primarily addressed to science policy processes, the language and to some extent the practices of dialogue were also rapidly taken up and used by informal bodies and organizations. The Dana Centre's website (http://www.danacentre.org.uk/), for example, talks of it being a space for 'debate' and 'dialogue' where you can 'challenge leading experts face to face'. What remains unclear, however, is both the extent to which this kind of dialogue actually occurs in informal contexts, and exactly how this multi-way dialogue can be worked out and intermingled with the other purposes of science communication activities. We can perhaps explore this further by examining one scientist's comments on his experience of taking part in a Dana Centre event. The quote below comes at the end of the speaker's (very positive) discussion of the event:

So I think the event's very good it's great for dialogue I think it's great for feedback, but what I'm not certain is what they're going to do with it. You know wh-what happens now you know okay you've got all this opinion but – they just move on to aliens . . .

'Aliens', it should be noted, refers to the topic of the next event, which was advertised at the end of the one that the speaker had just participated in. This scientist—while having thoroughly enjoyed his part in the event—is confused. What is the point of all the great 'feedback' and 'opinion' that the event has just generated if nothing is going to be 'done with it'? The purpose of the event is lost, he implies, if all that is going to happen is that participants move on to the next topic. Comments like these highlight the still confused nature of informal dialogue. Some have suggested that it is simply the deficit model 'in disguise' (see Lehr *et al.* 2007; Davies *et al.*, in press). Others argue that the primary intention is to facilitate equitable dialogue, but that enabling this is a difficult and therefore slow process (McCallie *et al.* 2007). And as in Irwin's (2006) analysis of 'dialogue' as a whole, the language of the deficit model—learning *from* experts and communication *to* the public—remains caught up and intermingled with the new language of dialogue and debate.[2] There is no clear consensus on what informal dialogue should aim to do, or on to what extent the old frameworks of 'learning *from* experts' remain valuable.

At this point it might be fruitful to move on to looking at what we might call 'internal' evidence about the purpose and practice of these events: how they are framed and understood as they are carried out. Within the next section I draw on recordings of Dana Centre events to suggest that there are a whole variety of competing frameworks and purposes that may be applied to a science dialogue event.

2. The Royal Institution (Ri), for example, writes on its 2005 web pages: 'During these 200 years [since its opening], the Ri has continued to communicate scientific issues to the general public through its high calibre events that break down the barriers between science and society. It acts as a unique forum for informing people about how science affects their daily lives, and prides itself on its reputation for engaging the public in scientific debate.' While the language of one-way education remains present ('informing people about how science affects their daily lives'), that of engagement and participation is also explicit ('engaging the public in scientific debate').

Mixed messages: dialogue as fragmented

The events which I'll be using as examples in my discussion took place through 2005–6, and took the form of the panel debates I described above. Rather than focusing on the format, however (although this might also suggest particular ways of interpreting the event[3]) I'd like to examine the language of events. In particular, I will look at framings in talk of what the event *is* and what it is *for*. These framings will—I suggest—go on to shape the dialogue process, giving participants a sense of exactly what they are involved in and what they will be expected to do. They are particularly important because informal dialogue is a relatively new kind of process. Most visitors to Dana Centre events, for example, have not been there before (Brehaut and Simonsson 2006). In such situations participants will be unsure of exactly how they should act: they are involved in a new social process and may use prior experiences or draw upon known genres to guide their behaviour (Davies *et al*. 2006). Crucially, participants will also take framing cues from event facilitators.

Firstly, then, how does the talk of Dana Centre events construct the nature of the event? The key point here is that there turns out to be *diversity* in how events are framed. We can perhaps see this most clearly by examining one segment of talk in detail—while this will not show all the different framings that are present in the data, it does illustrate key points. The extract below, which is part of a relatively long speech by an event facilitator, occurs near the start of the event. (I have numbered the lines for clarity, and transcribed movement or actions in italics[4].)

 1:   Good evening. Welcome. We've got a um looks like pretty much a full house tonight, and

 2:   we're looking forward to a very thought provoking evening. We still have some seats up in

 3:   the front (*pointing*) if anyone wants to move forward um how many of you have been to

 4:   the Dana Centre before for one of these one of these talks? (*Hands are raised in audience.*)

 5:   Okay. So you know that basically this is not the kind of event where our eminent experts sit

 6:   here and expound and you receive. This is the kind of evening where what makes it fun

 7:   what makes it interesting is that we have an engagement so we'd like to see an

 8:   engagement between audience members and we'd like you to feel free to express your

 9:   views with the speakers and the speakers may even be expressing views with each other.

10:   So it's very open, and um I'm sure we're going to have a really good time.

This extract does a large amount of the work to define the process participants find themselves in. The event is, for example, to be 'thought-provoking', 'fun', 'interesting'

---

3. For example, 'panellists' are invited to speak for anything up to 15 minutes at the start of events. Audience members are then invited to briefly comment or 'ask questions'. At the Dana Centre, panellists have head microphones which are permanently switched on, while audience members have to raise their hands and wait for a microphone to be brought to them if they want to comment. We might query the extent to which an equitable dialogue, in which all participants have fair access to the floor, is possible under these conditions.

4. 'Fac' indicates that the event facilitator is speaking. All other names have been anonymized.

and a 'really good time'. We are also told that there is almost a 'full house', and it appears that latecomers are directed to the few remaining seats 'in the front': such language connotes the theatre, with its overtones of performance from some and passivity from others. Similarly, the speaker asks if anyone has attended one of 'these talks' (line 4) before. The implication is that they—the facilitator's audience—will not be talking. One goes to a talk to listen to someone else, after all. This interpretation is borne out by the description of the invited scientists as 'speakers' (line 9; contrasted with others present as 'audience members', line 8). Their role, as summed up in this description, is to speak, and it is left to the 'audience' to listen.

However, this view of the event is mixed together with language that argues the very opposite: the evening is to be about 'engagement' (lines 7 and 8), 'expressing views' (lines 8 and 9), and being 'open' (line 10). We are told emphatically that it is 'not the kind of event where our eminent experts sit here and expound and you receive' (lines 5–6). There are therefore at least two ways of framing the evening (the 'event'—a suggestive word in itself) present within this short extract: as a 'talk', where the emphasis is on the provision of information to the audience; and as an interactive debate, where both audience members and invited speakers 'express views' with one another.

This analysis is borne out by other data from Dana Centre recordings collected during my research. In fact, there appear to be three key frameworks used to talk about events:

- Information plus a question and answer session (a *public lecture* format). Talk relating to this framework is close to a traditional PUS (deficit) framework: the event is about giving information, and audience participation is framed as their having 'questions'.
- *Interactive debate*. Hence talk about 'participation', being 'provocative', having 'lively debate', and 'dialogue'. The audience is—as in the segment above—encouraged to 'comment' or 'express views'.
- Adversarial debate or 'competition'. This framework constructs the event as a *traditional debate*: the speakers compete with one another in seeking to 'persuade' their audience to their point of view.

Closely linked with talk about what an event *is* is that about what the event is *for*: its purposes and aims. (You may have noticed this in the facilitator's talk above—she is talking about the nature of the event, but also about what she wants it to be.) Again, we find a range of different purposes presented for the event, several of which we can infer from the frameworks above. Thus a framework of a PUS-style lecture will have the purpose of providing information, while an adversarial framing might produce the expectation of disagreement, debate and potentially one speaker 'winning'. And, indeed, these are ideas that we find in the data—for example in the short extract below:

Luke: . . . I think this is something on which the – the three speakers the three-we are probably in pretty close agreement! (*laughs*)

Fac: Right we'll have to find something else to get them to disagree on then.

Here the speaker, Luke, shows surprise at speaker agreement (indicated by his laughter), which is reinforced by the facilitator's statement that 'we'll have to find something else to get them to disagree on': agreement is thus not the desired outcome, and disagreement is implied to be a good thing. Similarly, participation from the audience is frequently an aim (linking to the 'interactive debate' framework described above), and a substantial amount of work is done in encouraging this:

Fac: . . . But–   [but before I let you carry on I will come back out-

Jon: [But–[5]

Fac: –to the audience and I've seen a couple more hands but is anyone who hasn't spoken yet like to say something (*2 second pause*) Come on don't be shy cos we will run out of time! And then you'll be kicking yourselves.

Jon, one of the speakers, has just been talking in response to a question and—as we can judge by his attempted interruption—does not feel he has completed his turn. The facilitator, however, ignores his interjection and re-opens the floor to the audience, prioritizing coming 'back out' to them over letting Jon 'carry on'. In particular, she seeks to let someone who 'hasn't spoken yet' talk before they 'run out of time'. There is similar talk constructing audience participation as a desired outcome throughout all the events I studied: the emphasis is on 'sharing', 'expressing your views', being 'interactive' or having a 'debate'. We also find that enjoyment is constructed as a desired outcome. Events are meant to be 'fun' and 'fun', like participation, is seen as a good thing: we have seen this already, in the longer extract from a facilitator at the start of this section, where she talks of participants having a 'really good time'.

Thus the nature and purposes of events are mingled together within talk, and can be unpicked to produce a range of framings for the informal dialogue process: as learning *from* experts, as competition between speakers, as entertainment, or as an interactive debate involving all present. It is worth noting, however, that these desired outcomes, as constructed during the event itself, are entirely related to process ('sharing' or 'winning') and to impacts on the individual ('learning' and 'having fun'). None of the talk within the event refers to, for example, more formal outcomes. The focus remains on the activities of the evening itself or on the impacts on audience members—which might in itself lead us towards the pessimistic 'deficit model in disguise' understanding of the function of informal dialogue.

This reading of the data is certainly possible. As we have seen, ideas of education and learning, as applied to the audience, remain important within the discourse of dialogue events. A deficit model emphasis on non-scientists learning *from* scientists does seem to be present. What I would like to focus on as perhaps more significant, however, is the diversity of the framings we have seen applied and the fact that these seem to be readily interchangeable. We saw this most clearly in the first extract of this section: almost within the same breath, the facilitator could speak of the event as a 'talk', with 'speakers' and an 'audience', and as an evening which was *not* about the audience sitting and listening! The very meaning of the event seems to shift from moment to moment as language moves from that of PUS and the deficit model to that of dialogue and debate. The events are thus fragmented, rather than coherent wholes: participants experience a torrent of different

---

5. Here overlapping talk is represented using square brackets—thus Jon briefly interrupts the facilitator here.

messages through the evening about what the process they are involved in is and what it is for. Even those who are running the events—such as facilitators—seem unclear as to what they are involved in.

Identifying informal dialogue as fragmented in this way is not necessarily a critique. All of our talk is situated, complex and at times contradictory (for a science-specific example of this, see Gilbert and Mulkay's 1984 analysis of scientists' talk). But it is significant because it indicates on a micro-level (within a single dialogue event) what other authors have observed on the macro-level (the UK science dialogue movement as a whole). Irwin (2006), for example, has argued that the talk of both 'old' and 'new' modes of governance—the deficit and the dialogic—remains present within public discourse on science–society issues. All forms of public engagement, and the talk in and around them, are torn by complex histories and will bear the traces of previous models and current ideals. Tension and multiple discourses are therefore to be expected: he talks of a 'discursive struggle' (ibid., p. 315) being present within scientific governance. Similarly, Burchell (forthcoming), in an analysis of UK government-sponsored public dialogue on science over the decade 1997–2007, suggests that these processes may be characterized as hybrid in several ways. Their practice, he suggests, can be a mix of 'academic'-led (or more formal and deliberative) and 'practitioner'-led (more informal) techniques (cf. Smallman and Nieman 2006). Thus tensions, hybridity and contestation appear to exist on multiple levels within UK dialogue, from the talk from moment to moment to the framings of the entire movement. It seems that what dialogue is for remains open to debate; or, rather, is constantly shifting.

Finding a future for informal dialogue

What have we seen so far in this brief survey of informal dialogue? I have described the background to the 'turn to dialogue' in PUS: critiques of the deficit model approach, and the acknowledgement of relevant public knowledges by critical authors. I suggested that the adoption of dialogue approaches by informal institutions that followed this remains somewhat problematic, in that the purposes of these approaches frequently remain unclear. Indeed, my analysis of dialogue events has indicated the use of shifting and multiple framings of the event process. It also appears that the dialogue movement as a whole (rather than merely the discourse of events) is riven with these tensions (Irwin 2006; Burchell, forthcoming).

Informal dialogue therefore seems to be mired within a mass of divergent meanings, purposes and framings. It is unclear which of the framings present within events (lecture, debate, entertainment) should be understood as key. As I end, however, I want to suggest that the answers to some of these questions might become clearer if we start to think of informal dialogue as an endeavour completely divorced from formal processes. Rather than mimicking the language and practices of formal dialogue, informal events need—it seems to me—to reconsider from scratch what they are and what they are trying to do.

Lehr *et al.* (2007) and Davies *et al.* (in press) start to suggest ways in which this might be done: in particular, they argue that a shift in perspective from an interest in the large-scale (the level of organizations—a focus inherited from formal dialogue), to interest in the small scale (that of individuals or small groups) is key. With this shift in perspective, important outcomes can be located within individuals, rather than needing to involve changes in organizations or bodies. (This would, for example, enable us to respond to the scientist quoted earlier discussing his very positive experience of participating in a Dana Centre event: the organizers of dialogue do not need to 'do' anything with the opinions voiced at the event, as these opinions will have impacts and effects in the lives and experiences of those who participate—a valuable outcome in and of itself.) Thus, we are able to ignore the fact that in informal dialogue we have no 'large-scale' impacts on science or science policy, and rather aim to have outcomes and impacts on individuals— for example, learning (see Lehr *et al.* 2007) or, more generally, a 'social learning' (see Laird 1993; Limoges 1993; Davies *et al.* in press) which involves augmentation of knowledge but also 'knowledge of knowledge'. Learning might therefore not merely be of 'facts'— whether scientific or lay—but also involve a deepened understanding of the positions and motivations of others or a heightened ability to deal with the complexity of the topic under discussion. In addition, these authors emphasize the need for any such learning to be *mutual*: for, in fact, a genuine, multi-way dialogue to occur. If this is not the case—if the scientific experts involved are not as engaged in learning as those they are interacting with—then we do, inevitably, return to the concept of a deficient lay audience and the deficit model.[6]

If we follow the thinking of these authors, then informal dialogue starts to become a place where individuals from within science and its publics can engage in conversations that will lead to learning on all sides. (This is, of course, not far from the original vision described by critical authors (see Marris *et al.* 2001) but on a much smaller scale: individuals rather than institutions.) Exactly what this will look like remains up for grabs. Informal dialogue needs to provide scope for sustained engagement and allow all present to voice their views, knowledges and experiences. Perhaps this will mean much smaller events—six people from differing backgrounds meeting over a meal, for example. Or perhaps organizers will have to let go of some control of dialogue processes by issuing open invitations to all those with a stake in an issue and allowing them to present their views in whatever format they choose. Dialogue might not just mean discourse, but also include theatre, art, music or storytelling. It will almost certainly involve engaging with the technical experts involved in a different way: treating them not as the only experts present, but as those with access to a particular kind of knowledge. Indeed, understanding informal dialogue in this way may lead to a new emphasis in the scientific community on dialogue; one which seeks to enable their learning as much as that of non-scientists. And this kind of dialogue will certainly require new resources of imagination and creativity from those who organize such events, as they seek to identify topics and formats where mutual learning can productively occur. In short, there is plenty of scope for imaginative practitioners of these kinds of events to stabilize the move from deficit to dialogue.

6. Of course, there are times and places where deficit model approaches can be valuable and useful in and of themselves (see Dickson 2005; Sturgis and Allum 2004 for further discussion). My point here is only that these times and places are emphatically *not* those which go under the rubric 'dialogue event'.

A final point. I have argued, following Lehr *et al.* (2007) and Davies *et al.* (in press), that we could reconceptualize informal dialogue as having mutual, individual learning as its purpose. But what of some of the other framings found in data from the Dana Centre? Entertainment—in the form of 'having fun' or a 'good time'—seemed to be important, as did information provision to lay audiences, and 'sharing' or participation. What status do these have as valid purposes behind dialogue events? It is important to note that, in my data, they are important framings given to the dialogue process by participants. Clearly these things are *valued* by those who participate in Dana Centre events—and they will presumably be part of motivations to attend and to return. While we may wish to promote other purposes and meanings, ultimately what participants understand by dialogue and wish to get out of it is always going to be key for the practices of such events. Entertainment, sociality and learning therefore must remain key aspects of any conceptualization or analysis. Multiple framings and purposes for dialogue processes are inevitable; given this, it seems right that we acknowledge and celebrate the many meanings of dialogue.

Acknowledgements

Thanks are due in particular to the Dana Centre and its staff for its practical support of the research presented here; also to Kevin Burchell and Ellen McCallie. The empirical research was supported by the Arts and Humanities Research Council.

■ REFERENCES

Brehaut, J. and Simonsson, E. (2006). *Dana Centre Audience Profile Report 05–06*. Science Museum Visitor Research Group, London.

Burchell, K. (forthcoming). UK governmental public dialogue on science and technology, 1998–2007: governmentality and boundary work. *Social Studies of Science*.

Davies, C., Wetherall, M. and Barnett, E. (2006). *Citizens at the Centre: Deliberative Participation in Healthcare Decisions*. Policy Press, Bristol.

Davies, S., McCallie, E., Simonsson, E., Lehr J.L. and Duensing, S. (in press). Discussing dialogue: perspectives on the value of science dialogue events that do not inform policy. *Public Understanding of Science*.

Dickson, D. (2005). The case for a 'deficit model' of science communication [Editorial]. *SciDev.Net*. Available online at: **http://www.scidev.net/Editorials/ index.cfm?fuseaction=readEditorials&itemid=162&language=1**.

Durant, J. (1999). Participatory technology assessment and the democratic model of the public understanding of science. *Science and Public Policy*, **26**, 313–19.

Durant, J., Evans, G. and Thomas, G. (1989). The public understanding of science. *Nature*, **340**, 11–14.

Fischer, F. (2005). Are scientists irrational? Risk assessment in practical reason. In: *Science and Citizens: Globalisation and the Challenge of Engagement* (ed. M. Leach, I. Scoones and B. Wynne), pp. 54–65. Zed Books, London.

Gilbert, N. and Mulkay, M. (1984). *Opening Pandora's Box: a Sociological Analysis of Scientists' Discourse*. Cambridge University Press, Cambridge.

Goven, J. (2006). Processes of inclusion, cultures of calculation, structures of power: scientific citizenship and the Royal Commission on genetic modification. *Science, Technology and Human Values*, **31**(5), 565–98.

Grand, A. (2009). Engaging through dialogue: international experiences of Café Scientifique. In: *Practising Science Communication in the Information Age: Theorizing Professional Practices* (ed. R. Holliman, J. Thomas, S. Smidt, E. Scanlon and E. Whitelegg). Oxford University Press, Oxford.

Gregory, J. and Miller, S. (1998). *Science in Public: Communication, Culture and Credibility*. Plenum Press, New York.

Horlick-Jones, T., Walls, J., Rowe, G., Pidgeon, N., Poortinga, W. and O'Riordan, T. (2004). *A Deliberative Future? An Independent Evaluation of the GM Nation? Public Debate About the Possible Commercialisation of Transgenic Crops in Britain, 2003*. University of East Anglia, Programme on Understanding Risk, Working Paper 04–02. University of East Anglia, Norwich.

House of Lords, Select Committee on Science and Technology (2000). *Science and Society*, Third Report. HMSO, London.

Irwin, A. (1995). *Citizen Science*. Routledge, London.

Irwin, A. (2001). Constructing the scientific citizen: science and democracy in the biosciences. *Public Understanding of Science*, **10**, 1–18.

Irwin, A. (2006). The politics of talk: coming to terms with the 'new' scientific governance. *Social Studies of Science*, **36**(2), 299–320.

Kerr, A., Cunningham-Burley, S. and Amos, A. (1998). The new genetics and health: mobilizing lay expertise. *Public Understanding of Science*, **7**(1), 41–60.

Kerr, A., Cunningham-Burley, S. and Tutton, R. (2007). Shifting subject positions: experts and lay people in public dialogue. *Social Studies of Science*, **37**(3), 385–411.

Laird, F.N. (1993). Participatory analysis, democracy, and technological decision making. *Science, Technology and Human Values*, **18**(3), 341–61.

Layton, D., Jenkins, E., Macgill, S. and Davey, A. (1993). *Inarticulate Science? Perspectives on the Public Understanding of Science and Some Implications for Science Education*. Studies in Education Ltd, Nafferton.

Lehr, J.L., McCallie, E., Davies, S.R., Caron, B.R., Gammon, B. and Duensing, S. (2007). The role and value of dialogue events as sites of informal science learning, *International Journal of Science Education*, **29**(12), 1–21.

Limoges, C. (1993). Expert knowledge and decision-making in controversy contexts. *Public Understanding of Science*, **2**(4), 417–26.

McCallie, E., Simonsson, E., Gammon, B., Nilsson, K., Lehr, J.L. and Davies, S. (2007). Learning to generate dialogue: theory, practice, and evaluation. *Museums and Social Issues*, **2**(2), 165–84.

MacMillan, T. (2004). *Engaging in Innovation: Towards an Integrated Science Policy*. Institute of Public Policy Research, London.

Marris, C., Wynne, B., Simmons, P. and Weldon, S. (2001). *Public Perceptions of Agricultural Biotechnologies in Europe. Final Report of the PABE Research Project*. Lancaster University, Lancaster.

Michael, M. (1996). Knowing ignorance and ignoring knowledge: discourses of ignorance in the public understanding of science. In: *Misunderstanding Science? The Public Reconstruction of Science and Technology* (ed. A. Irwin and B. Wynne), pp. 107–25. Cambridge University Press: Cambridge.

Miller, S. (2001). Public understanding of science at the crossroads. *Public Understanding of Science*, **10**(1), 115–20.

Reich, C., Chin, E. and Kunz, E. (2006). Museums as forum: engaging science centre visitors in dialogues with scientists and one another. *Informal Learning Review*, **79**(July–August), 1–8.

Rowe, G. and Frewer, L.J. (2000). Public participation methods: a framework for evaluation. *Science, Technology and Human Values*, **25**(1), 3–29.

Rowe, G. and Frewer, L.J. (2004). Evaluating public-participation exercises: a research agenda. *Science, Technology and Human Values*, **29**(4), 512–56.

Rowe, G. and Frewer, L.J. (2005). A typology of public engagement mechanisms. *Science, Technology and Human Values*, **30**(2), 251–90.

Royal Society (1985). *The Public Understanding of Science*. The Royal Society, London.

Smallman, M. and Nieman, A. (2006). *Discussing Nanotechnologies*. Think-Lab, London.

Stirling, A. (2006). *Gover'Science Seminar 2005—Outcome: from Science and Society to Science in Society: Towards a Framework for 'Co-operative Research'*. Office for Official Publications of the European Communities, Luxembourg.

Sturgis, P. and Allum, N. (2004). Science in society: re-evaluating the deficit model of public attitudes. *Public Understanding of Science*, **13**(1), 55–74.

Thomas, J. (2009). Controversy and consensus. In: *Practising Science Communication in the Information Age: Theorizing Professional Practices* (ed. R. Holliman, J. Thomas, S. Smidt, E. Scanlon and E. Whitelegg). Oxford University Press, Oxford.

Turney, J. (1998). *To Know Science is to Love it? Observations from Public Understanding of Science Research*. Committee on the Public Understanding of Science, London.

Wynne, B. (1991). Knowledges in context. *Science, Technology and Human Values*, **16**(1), 111–21.

Wynne, B. (1992). Misunderstood misunderstanding: social identities and public uptake of science. *Public Understanding of Science*, **1**(3), 281–304.

Wynne, B. (2001). Creating public alienation: expert cultures of risk and ethics on GMOs. *Science as Culture*, **10**(4), 445–81.

Wynne, B. (2005). Risk as globalising 'democratic' discourse? Framing subjects and citizen. In: *Science and Citizens: Globalisation and the Challenge of Engagement* (ed. M. Leach, I. Scoones and B. Wynne), pp. 66–82. Zed Books, London.

Ziman, J. (1991). Public understanding of science. *Science, Technology and Human Values*, **16**(1), 99–105.

■ FURTHER READING

- Irwin, A. (2006). The politics of talk: coming to terms with the 'new' scientific governance. *Social Studies of Science*, **36**(2), 299–320. Irwin's 2006 paper analyses the concept of dialogue on a broad level, noting that it remains intermingled with more traditional language of scientific governance. In particular, he uses the *GM Nation?* debate as a case study.

- Kerr, A., Cunningham-Burley, S. and Tutton, R. (2007). Shifting subject positions: experts and lay people in public dialogue. *Social Studies of Science*, **37**(3), 385–411. This lengthy paper describes several dialogue processes in detail, before analysing the roles and positions participants take up. They suggest not only that most of these positions are 'hybrid' and shifting, but that for much of the time scientific expertise remains unchallenged as the key knowledge source within debate.

- Lehr, J.L., McCallie, E., Davies, S.R., Caron, B.R., Gammon, B. and Duensing, S. (2007). The role and value of dialogue events as sites of informal science learning. *International Journal of Science Education*, **29**(12), 1–21. Lehr and her co-authors focus on dialogue in the context of 'informal science institutions'—primarily museums. They give an overview of these processes, arguing that viewing them as sites of learning is key to understanding and studying them.

- McCallie, E., Simonsson, E., Gammon, B., Nilsson, K., Lehr, J.L. and Davies, S.R. (2007). Learning to generate dialogue: theory, practice, and evaluation. *Museums and Social Issues*, **2**(2), 165–84. This paper provides an overview of dialogue, written largely from a practitioner's perspective. Based on practices at the Dana Centre, it describes how events are evaluated and how this is fed back into further event development.

■ USEFUL WEB SITES

- **The Dana Centre: http://www.danacentre.org.uk** The Dana Centre web site gives a sense of the kind of activities it runs, as well as having occasional live web casts of events. It also has a section on 'Event DIY' which introduces how to run a dialogue event—including their definition of 'dialogue' and pointers on evaluation.

- **sciencehorizons: http://www.sciencehorizons.org.uk/.** sciencehorizons was a recent, large-scale public engagement exercise on future technologies, sponsored by the UK government. Its web site introduces the process, hosts the resources used in it, and gives an overview of its findings.

2.3

Engaging with interactive science exhibits: a study of children's activity and the value of experience for communicating science

Robin Meisner and Jonathan Osborne

Introduction

Scientific ideas are rarely simple, straightforward or commonsense (Wolpert, 1992). One of the common cultural arenas for communicating ideas about science is science centres and museums and, whilst the educational mission of various informal science institutions, including science museums and science centres, might vary depending on the specific remit of each establishment, one of the primary goals shared by all such institutions is to inspire individuals to engage with science (Semper 1990). For without engagement, there can be no communication. Engaging individuals with science means stimulating their curiosity, generating a sense of wonder and helping them to develop some sense of meaning or understanding of the explanations that science offers of the material world, and sometimes, the means by which those explanations were derived. Developing such an understanding is dependent on a communicative process where the ideas to be presented have been carefully chosen and, likewise, the means by which such understanding might be developed. Individuals working in science centres have to consider both what is to be communicated and how it is to be communicated. Research shows that visitors recognize the educational experience that museums and science centres offer for, when asked, both those who come alone and those who come in groups cite education, or more specifically 'discovering something new', as one of the primary reasons for their visit (Falk *et al.* 1998). Teachers give similar reasons for bringing their pupils, listing making connections to the science curriculum and increasing student motivation as important reasons for bringing their classes to a museum (Kisiel 2005). In response to such expectations, an ever-growing trend in the way in which many institutions attempt to engage their predominately school-aged audiences is through the deployment of 'interactive' exhibits—exhibits which are designed to be explored and enjoyed, and which respond to visitors' actions on them (Rennie 2007).

While museums have sought to capitalize on this aspect of visitor enjoyment to encourage engagement, which is seen as a necessary pre-requisite if any learning is to occur (Hughes 2001; Witcomb 2006), critiques have emerged (Shortland 1987). Most often questioned is whether the visitors, who appear to be 'just playing' at interactive exhibits, are actually learning anything from their engagement. However, exhibits are chosen specifically because they afford the opportunity to engage and experiment with a range of important phenomena which both stimulate the process of asking questions and permit a degree of experimentation. The visitor is encouraged to develop explanatory hypotheses and test these against the behaviour of the material world. In so doing, an understanding of science may be constructed and communicated. There are, moreover, a range of theoretical arguments and empirical studies which would suggest that learning is indeed occurring at interactive exhibits, not only on the cognitive level, but with such experiences having significant affective and social outcomes as well (cf. Wellington 1990; Falk and Dierking 2000). However, the nature of the experience and its effect on the individual is still not well understood. Measuring learning is a difficult task, even in formal environments let alone in informal contexts. Hence, a better question might be to ask— what is the nature of the activity and how might the experience engender learning? In particular, what do learning and/or engagement look like at the exhibits when examined in detail? For if we had a stronger sense of what learning behaviours are it would help the process of exhibit design and development and, as a corollary, the communication of science. Therefore, this chapter seeks to address such questions. First, drawing on psychological and philosophical perspectives on learning relevant to the context and the existing literature on interactive exhibits, it will develop theoretical arguments on the nature of experiences at interactive exhibits and how they might support learning. Second, it will illustrate how such arguments have been used to develop, and implement, a theoretically driven coding schema to look at video data of children's activity at specific interactive exhibits to see if evidence of such behaviours can be observed and documented. And finally, it will offer a reflection on the value of both the methods used and the value of children's experiences at such exhibits within the current information age.

How might experience at interactive exhibits engender science learning?

Philosopher of science Rom Harré (1986) has argued that science, and any understanding of science, is grounded in life-world experiences. Scientists construct theories or representations of how the world might be. To achieve this end, Harré's argument is that scientific theories begin on the macroscopic and observable level—theories of phenomena that can be seen, felt, heard, smelt and tasted within 'the realm of actual and possible objects of experience' (Harré 1986, p. 70). As phenomena become more complex, abstract or too large or too small to be observed readily, individuals rely on the knowledge collected and internalized from their everyday experiences to construct metaphors, representations or models of such phenomena. Thus, a child experiencing the flow of water in a stream (or perhaps a stream table in a museum gallery) is developing representations or macroscopic theories about the behaviour of water such as flow and current that can later be employed towards understanding more complex phenomena such as electricity. In this sense, metaphors act as ladders connecting ideas about the microscopic world with ideas about the macroscopic world (Harré 1986).

From this perspective, experience of the material world and the ways in which it is represented are central to the development of scientific understanding as these experiences form the vocabulary from which new understandings can be constructed. As Dewey (1938) suggests, therefore, experience is a necessary basis for education and, in the context of this argument, specifically science education. According to Dewey, the nature of experience is such that an individual does something to the world, and the world then does something to the individual. Experience is both reciprocal and transformative, shaping the quality of future experiences by changing the collection of prior knowledge upon which one might draw during all future endeavours. While Dewey suggests that not all experience is educative (Dewey 1938), he argues that in order to aid children to utilize the value of their experiences for learning, one should present them with problematic experiences—or, problems to solve. Problematic experiences are those wherein individuals are unable to match their prior understandings to the situation at hand (Roschelle 1995). Observations then help clarify the nature of the problem and lead towards the generation of solutions. The individuals then mentally test their ideas before proceeding to verify them through experiment. This process, for Dewey, is inquiry—a tool for interpreting and making use of experiences, or, in short, for constructing new knowledge. Communicating scientific ideas is only possible if individuals have a repertoire of experiences which can be shared and from which a common understanding can be constructed. Whether it is listening to sound echo (which demonstrates the finite speed of sound), exploring the variation of temperature on your body shown by a heat camera (which demonstrates that heat 'rays' are similar to light) or watching carbon dioxide sublimate (which shows phase changes in a dramatic form), interactive science centres offer a plethora of experiences on which meaning can be constructed and communication can occur.

Piaget (1951) adds another perspective—one that provides insight into the processes whereby a child develops knowledge both in and through experience. Piaget's particular argument is that play occupies a central role in the development of knowledge. Reilly (1974, p. 74) summarizes Piaget's argument as one in which:

... an object is known ... only to the extent that it is acted upon or forms part of an action sequence. Cognitive functions not only subserve action, but are themselves forms of action ... thinking is not 'for' action, it 'is' action.

That is, play is a child's attempt to assimilate the material world to his or her own understandings—'to "mentally digest" novel situations and experiences' (Reilly 1974, p. 80)—through the active repetition and experimentation that it affords. Thus, for

Piaget, play's value lies in its ability to provide children with the tools and representations they need to construct knowledge from experience. Such an argument is similar to Dewey's notion of inquiry whereby in, and through, the process of inquiry children are seen to make sense of problematic experiences. In other words, for children to be able to assimilate new knowledge into their repertoires, they need to be allowed to act on the world in a manner whereby they can manipulate and transform their representations—testing them against their observations of the world.

This brief review of this body of literature provides an argument for the value of experience in learning and constructing understandings of the material world. Moreover, it demonstrates that 'play' is an essential and effective mode of providing the experiences that can lead to learning and assist communication. As yet, however, none of it has explored the social aspects of engaging with experiences. One key concept emerging from sociocultural theory, and specifically Vygotsky's (1978) work, is that social factors change the human world from being one defined by nature into one created by culture (Wertsch and Tulviste 1996). That is, as much as people shape culture, culture shapes people and, more specifically, culture shapes a person's development. By constructing psychological tools, including language, individuals influence the behaviour and minds of others (Gredler 2001). For example, words shape activities by providing them with a structure. Language leads to associations; associations lead to abstractions; and abstractions lead to higher-order thinking. As such, within the setting of a science centre or museum, language and the interaction of people with each other act as the necessary mediators for development to occur providing the tool by which salient features are shared and new meanings constructed.

In summary, this brief review suggests that if science is to be communicated: (1) experience is essential to children's development of scientific knowledge, and more generally their learning; (2) play and exploration are necessary means whereby a child begins to make sense of such experiences; and (3) social interaction is critical to learning. What is known, however, about the nature of experiences that children have at interactive science exhibits and the types of behaviour they afford? What kinds of learning behaviours and interactions do they engender?

The nature of the activity at interactive exhibits

Interactive exhibits within informal learning institutions are designed objects (Allen 2004), constructed with the specific purpose of allowing and encouraging children to explore, play and experiment with natural phenomena within a science centre or museum. An interactive gallery is typically a collection of exhibits that are designed to encourage engagement and experimentation with scientific phenomena, particularly those within the physical sciences that visitors might not have the opportunity to experience and manipulate (Feher 1990). Rennie and McClafferty (1996, p. 56) characterize such environments as exhibiting 'ideas and concepts, rather than objects,' in a decontextualized manner—separating the science presented by the exhibit from its application in the real world.

Children are seen to 'play' with such exhibits—they explore exhibit components, test ideas and try things out again and again. Feher (1990, p. 36) calls such activity 'playful exploration' where, as Semper (1990, p. 52) describes, 'the ideas and objects in a museum are discovered by the active process of people moving about'. Such play may be guided either through the judicious use of labels or by the thematic presentation of ideas. Many researchers have recognized the value of play and exploration as a mode of learning in an interactive gallery (cf. Lucas *et al.* 1986; Diamond 1996; Rennie and McClafferty 2002). Yet, the existing research does little beyond describing episodes of activity or labelling children's actions and interactions as play and/or exploration. That is, with the rare exception, play and exploration are noted but not given sufficient analytic attention (cf. Spock 2004). Moreover, as Adams *et al.* (2004) suggest, there are few attempts to assess the value of play and exploration.

Additionally, the opportunity for social interaction that interactive galleries offer has long been considered a strength of such environments (Semper 1990; Falk and Dierking 2000). And, with the growing focus on social interaction as one of the primary modes of learning within informal environments (Schauble *et al.* 1997; Rennie *et al.* 2003), the nature of the discourse and interaction at exhibits itself has become a popular topic of study. For, as Blud (1990, p. 43) has argued, 'interaction between visitors may be as important as interaction between the visitor and the exhibit' for learning. In many of the more recent studies highlighting social interaction, a strong emphasis has been placed on the role of conversations and, in particular, learning conversations (Crowley and Callanan 1998; Ash 2003). This is perhaps not surprising given the use of a socio-cultural framework within such studies that values talk as the primary mode of learning (Leinhardt and Crowley 2002). However, upon observing individuals in an interactive gallery, it is evident that there is relatively little talk occurring. A few researchers (Carlisle 1985; vom Lehn *et al.* 2001; Rahm 2004) have begun to explore the non-verbal yet highly social behaviours that do occur. Given the findings of this work, which would suggest that it is the non-verbal rather than verbal behaviours which predominate, a deeper understanding of the value of exhibits for learning might come through a more detailed study of what young children do, both verbally and non-verbally, in essence, the focus of this work.

Identifying learning behaviours at the exhibit face

In order to begin to unpack and identify the specific behaviours within what is generally called 'play', this chapter now turns to examining theoretically the types of behaviours that one might expect or desire to witness of young children at the exhibit face. Such theory will be used to develop a set of criteria for investigating children's engagement with interactive exhibits. The focus is on young children, aged 4–12, because they are the predominant visitors to and users of such exhibits. Two models of engagement—that of Hutt (1981) and Ohlsson (1996)—are of interest. First, Hutt has developed a taxonomy of children's play behaviours in which she differentiates exploration, or epistemic behaviours, that is 'what can this object do?', from play, or ludic behaviours—'what can I do with this object?'. Her work implies that when children encounter a novel object they will engage firstly in epistemic and then in ludic play. For this chapter, both types of play are of interest—epistemic play because it is both what a child does when faced with a novel object and leads to knowledge acquisition, and ludic play for its connection to fostering creativity (Lieberman 1977). Second, Ohlsson distinguishes between 'learning to do', which can be easily observed, from 'learning to understand'. The latter Ohlsson has defined in terms of a set of seven discourse-based 'epistemic tasks' that can be used to recognize both the processes and products of higher-order learning. Such a list is valuable within the particular context of this chapter because it provides a basis for identifying epistemic activity at the exhibit face.

Implicit within the work of both Hutt and Ohlsson is the notion that, in developing skills and/or higher-order knowledge, children are refining and building their concepts of what something might do, how it might work, what it might mean and even, perhaps, their ability to predict. A significant body of research has been conducted around such concepts—or mental models—within science education (Gilbert and Boulter 2000). Thus, drawing on the existing literature briefly discussed above (see Meisner 2007 for a more extensive review), a set of features has been defined to examine the types of behaviours one might expect or desire to witness at the exhibit face if such exhibits are to promote learning. This schema has enabled us to investigate the experiences that interactive exhibits offer for young children, and examine the extent to which the specific exhibit might afford the following four types of behaviours: (1) epistemic behaviours, (2) ludic behaviours, (3) skill acquisition and (4) the development of mental models.

A fifth feature of children's behaviour arises from notions of everyday creativity or 'possibility thinking'—that is the extent to which exhibits offer novel and valuable insights, problem finding, problem solving and play (Craft 2000). The very core of the implications of creativity for education is nicely summed up by Boden (2001, p.102) in her statement that, 'creativity is not the same thing as knowledge, but is firmly grounded in it'. She draws on the work of Sutton-Smith (1988, p. 22) who argues that, 'every subject matter requires its own "what if" speculation, its own plane of imagination'. In other words, it is a child's imaginative searching that leads him or her to seek new knowledge or understanding. Thus, finding creativity at the exhibit face is of interest because creative behaviours—including asking questions and problem-finding—particularly in science, contribute to the drive for understanding and, moreover, the attempt to interpret experience. The sixth, and final, feature emerges from the role of social interaction in learning. Given its significance for learning, the question of interest is to what extent do specific exhibits enable social interaction?

In summary, the argument so far would suggest that when investigating activity at the exhibit face, it is of interest from a theoretical perspective to look at the extent to which a specific exhibit:

1. engenders epistemic behaviour
2. stimulates ludic play
3. facilitates skill acquisition
4. encourages the development of mental models
5. fuels opportunities for creativity
6. enables social interaction.

In compiling these six features of behaviour, three caveats become necessary. Firstly, attention to both affective behaviours and pre-existing knowledge has not been given any priority. While highly valuable (Packer 2006), the research presented in this chapter is about what children do, not what they might think, feel or the connections they might be making. Such a focus is in no way meant to devalue either affective behaviours or prior knowledge; rather it places analytic attention on action and interaction as indicators of engagement. Secondly, while recognizing that not all interactives need to do all things, the review of the literature would suggest that in order to increase both the range and value of experiences at interactive exhibits, the overall experience should attempt to incorporate as many of these features as possible. And finally, this set of features of behaviour is intended to be a theoretical guide for unpacking children's activity at the exhibit face and has not developed as a tool for evaluation either for the children's actions and interactions or the exhibits themselves, as there is no suggestion that it is a comprehensive list. However, it is an attempt to elucidate a set of principles which might guide the design of such exhibits.

Looking at children's activity through a theoretically and empirically driven lens

Using video recordings of children's actions and interactions recorded at interactive exhibits as principal data, the empirical research to be presented in this chapter builds on a tradition of employing relatively unobtrusive methods for capturing naturally occurring activity in order to provide a detailed picture of visitors' behaviour in informal learning environments (Stevens and Hall 1997; vom Lehn *et al.* 2001). Video data offer significant advantages over other forms of collecting observational data. In particular, video data are repeatable—they allow researchers to review the activity and to share the data with other researchers (Knoblauch *et al.* 2006).

Therefore, video data were recorded at 10 mechanical interactive exhibits in three different museums—a science centre, a science museum and a children's museum. Using a camera set upon a tripod and a radio microphone fixed to the exhibits, a total of over 1000 minutes of data were collected during school holiday weeks, weekend days and weekday afternoons to ensure a diverse mix of visitors. Because of ethical concerns about the use of video in public places, a sign was placed near the camera to gain the informed consent of visitors, i.e. to inform them of the recording and its purpose with the under-standing that if they chose to use the exhibit they were giving their consent to be filmed (Gutwill 2003). All procedures were conducted in a manner similar to those described in Heath and Hindmarsh (2002) and were agreed with museum staff before filming began. Supplemental data—including field observations and focus group interviews—were collected to augment the analysis of the primary video data.

The activity recorded has been analysed using a theoretically and empirically driven coding schema, which draws upon both the set of features outlined in the previous section and repeated viewing of the videos. The coding schema consists of five broad coding categories—'epistemic' (EPI), 'model generating' (GM), 'social' (SOC), 'creative' (CRE) and 'ludic' (LUD)—with a total of 33 specific behaviours falling into the categories. Using the detailed coding schema, the video data were systematically coded (Callanan *et al.* 1995). Behaviours were coded using time-based sampling. Specifically, the activity of two children between the age of 4 and 12 was coded every 15 seconds during 30 minutes (totalling 120 coding periods) of video at each exhibit. The systematic observations were then quantified (Chi 1997) and analysed through statistical methods (for details see Meisner 2007).

Results of applying the coding schema to one exhibit

The results of the coding procedure outlined above and aspects of the analysis for one exhibit —the *Leaning Tower* (Figure 1) at London's Science Museum—are shown below (Figure 2). *Leaning Tower* is a simple exhibit at which visitors are asked to build a tower using large solid foam blocks. Each block is shaped slightly differently with sides of differing lengths and slopes; and there are two sets of differently coloured blocks for visitors to use. Selected findings from the data at *Leaning Tower* are presented in this chapter. These have been chosen not because they are representative of all of the exhibits studied—for none of the results of the overall study are generalizable—but rather because they are provide an interesting picture of children's activity at one exhibit.

Figure 1 The *Leaning Tower* (photograph courtesy of Robin Meisner and Jonathan Osborne, Kings College London).

As can be seen in Figure 2, the least frequently observed category of behaviours is 'ludic' and the most frequent is 'epistemic'. The foam blocks of *Leaning Tower* tend to be used as directed in the exhibit label to actually build a tower, rather than providing children with a springboard for imaginary play. The prevalence of 'epistemic' behaviours is not surprising and might even be expected. As was highlighted by Hutt's research (e.g. Hutt 1981) reviewed earlier, upon encountering a novel object or experience children engage with the question of 'what does this object do?'. In general, the most common question children ask when approaching an interactive exhibit such as *Leaning Tower* is some variant of 'what do I have to do here?' (Carlisle 1985; Rix and McSorley 1999), which reflects notions of 'what is this?' and 'how does it work?'. While such questions are often silent, observations of children's exploratory activities support the notion that they are indeed trying to figure out how to manipulate the exhibit.

As seen in this example, the theoretically driven coding categories have been used as a way to generate a fairly comprehensive analysis of the children's behaviours. They are, however, just one set of lenses through which to examine the data and should not be seen as a rigid structure for analysis. Rather, they are a starting point for the grouping of

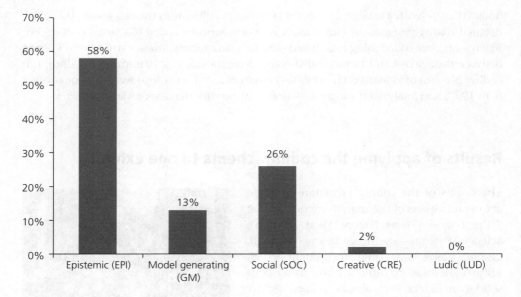

Figure 2 Percentages of theoretically driven category use at the *Leaning Tower*. Percentages do not total 100% because of rounding effects.

the behaviours—the detailed individual codes—based on the literature reviewed earlier. In other words, the specific individual behaviours are free to be treated as separate units —or building blocks—that can be arranged and rearranged to explore different aspects of the data. For example, one might wish to further investigate the 'social' behaviours witnessed. Indeed, within the theoretically driven coding categories there is a separate category for 'social' behaviours. However, not all behaviours that are social have been placed into that category—individual behaviours, such as describing (EPI-desc), have been placed into categories other than 'social' (SOC) but are indeed also social in nature. Thus, one might rearrange the behaviours to investigate the breadth of social learning behaviours in more detail.

In the following section, all behaviours that are social in nature were grouped together into one category—'acting with' (WITH) where a child acts with another. The remaining behaviours were then divided into three categories—'acting on' (ON) where a child acts on an object, 'affective' (AFF) where a child is expressing aloud their feelings, and 'observing' (OBS) where a child is inspecting an object or observing another person.

Applying the redefined categories

The results of the coding and aspects of the analysis for *Leaning Tower* using these categories are shown below (see Figure 3).

As can be seen in Figure 3, the least frequently observed category of behaviour is affective (AFF), whilst 'acting with' (WITH), 'acting on' (ON) and 'observing' (OBS) are approximately similar. The low occurrence of behaviours classified as 'affective' is not

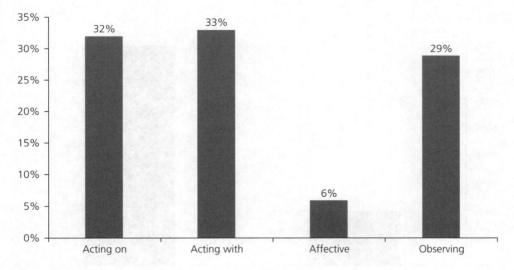

Figure 3 Percentages of redefined category use at the *Leaning Tower*.

surprising. The behaviours identifiable by observation as 'affective' are limited and only those accompanied by a verbal exclamation were coded. Thus, the picture shown here might not fully represent the affective experience at these exhibits, and points to the difficulty of capturing such data. What is of interest is the prevalence of observational behaviours. Here, the findings suggest that observation occurs nearly as frequently as either 'acting on' or 'acting with' behaviours. In looking at the behaviours of family groups at museums, Diamond (1986) found that children observe others as often as they manipulate the exhibits. Although Diamond was looking specifically at family groups and employed a less detailed coding schema than that used in this research, her findings reflect those found in this study. Moreover, though not reported in this chapter (see Meisner 2007 for details), the individual behaviour 'observing' (EPI-obs) was the most frequently witnessed of all the behaviours coded. This finding suggests that a significant element is communicated by the opportunity to watch the behaviours of others as it is by interacting with the exhibit itself. This points to the importance of deliberately facilitating such activity in any exhibit.

Yet another way in which the specific behaviours can be organized is to group them into one of two categories—'verbal' and 'non-verbal' behaviours. As mentioned previously, a current trend in research on learning in museums focuses on using conversations as a tool for assessing learning (cf. Leinhardt and Crowley 2002). Hence, it is of interest to separate the behaviours displayed at each exhibit into 'verbal' and 'non-verbal' categories to see the relative weight of conversation compared to non-verbal behaviours.

Applying the verbal/non-verbal categories

The results of the coding and aspects of the analysis using these categories are shown below (see Figure 4).

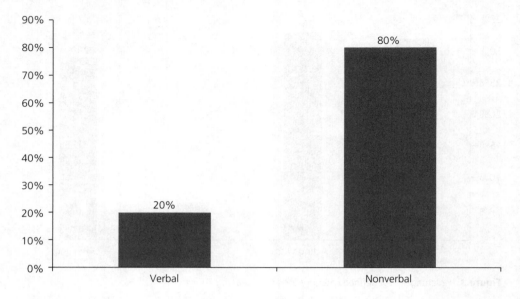

Figure 4 Percentages of verbal/non-verbal category use at the *Leaning Tower*.

As can be seen in Figure 4, 'non-verbal' behaviours are observed more than four times as often as 'verbal' behaviours. This result is not unexpected. *Leaning Tower* is designed to be touched and manipulated. Previous research has shown that children engage in conversation less frequently at interactive exhibits than at other types of exhibits, such as those displaying live animals (cf. Allen 2002) and these results support such observations. However, it calls into question the dominance of the interest in verbal behaviours in the research community suggesting that it may be the non-verbal behaviours encouraged at an exhibit which may be as, or more, important in facilitating communication. Moreover, it is of interest that while *Leaning Tower* was shown previously to afford many 'social' (SOC) or 'acting with' (WITH) behaviours, children do not necessarily display many 'verbal' behaviours. This finding points to the reality that not all social behaviours are verbal behaviours, and that in looking at social interaction, it is important to look beyond conversation.

Overall, the findings presented from the three different ways of categorizing the behaviours provide empirical data on children's activity at one interactive exhibit. They do indeed suggest that non-verbal actions and interactions predominate. But perhaps of greater interest for this chapter is that the findings illustrate how the application of a theoretically driven lens enables the production of detailed analyses of learning and/or engagement at exhibits. In short, how the visitors construct meaning. The analysis shows also that as a communicative event, the outcome is dependent on both the intent of the designer and, more importantly, the actions of the visitor. Communication is not, as such, a simple one-way transmission.

Secondly, based on our analysis of the literature which led us to suggest a set of six types of behaviours which one might expect or desire an interactive exhibit to engender in young children, we have argued that what is frequently referred to as 'play' is, in reality, a set of complex actions and interactions at, with and around an exhibit. Thus the argument advanced by Postman (1987) that such exhibits are merely sources of ephemeral amusement does not hold—these data suggest that children are engaged in learning behaviours. Whether they learn or how much they learn will always remain an open question. As Dewey argued, not all experiences are educative—the question for exhibit designers is how the child's interaction can be developed in a productive manner.

Reflecting on the coding of behaviours

Part of the value of this chapter has been in its attempt to use an innovative methodological approach to study behaviour at interactive exhibits. How appropriate are the methods adopted in this study? The first consideration is that the codes look broadly across a considerable amount of data. They are useful for reducing the data to manageable 'inventories' of behaviours (Mercer *et al.* 2004). Such inventories allow for a flexibility of analysis as, once quantified, the frequencies of the individual behaviours are essentially building blocks that can be arranged and rearranged into various categories to explore different approaches to the research question. Adjusting the organization of the codes brings certain features of the data to the fore. The codes and the coding process allow one to give names to the phenomena observed. And, moreover, they do so in such a way that another researcher can look at the codes and their definitions, and apply them to the data in a manner that closely resembles that of the original researcher.

In addition, the use of theoretically driven codes allows one to gain an understanding of both what is occurring and what is not. Knowing what is not happening enables an exploration of why it might not be happening—what it is about a certain exhibit that might encourage specific behaviours over others. Yet, on the other hand, we recognize that drawing on theoretically derived codes might be restrictive. The codes do not provide access to individuals' prior experience, their intended use of the exhibits or details of their affective experience and they may not be comprehensive. Given the observational nature of this method, this comment is perhaps not surprising, i.e. most of the research on such factors relies on interviews, questionnaires and the like, and even then is forced to make substantial inferences about the ways individuals' knowledge or feelings have changed. Moreover, the codes fall short in being able to describe in detail how or why the actual activity occurs. That is, the behaviours alone do not adequately depict the unfolding context of action and interaction that define in detail how individuals do what they do—how, for instance, specific behaviours, like observation, are central to the production of activity. To this end, and to complement the research presented in this chapter, an analysis that draws on an interactional microanalytic perspective is offered in Meisner (2007) which specifically explores the role of observation and imitation in the children's activity at the exhibit face. This takes an ethnographic approach which analyses specific selected actions in detail to show how such actions enable communicative performances—that is those actions undertaken by the visitor which enable and facilitate communication in social contexts.

Considering experience at the exhibit face in the information age

Finally, it is worth stating what may almost seem self-evident—children's activity at interactive exhibits is complex. What some might term 'play' is actually a wide range of valuable learning behaviours. It is highly non-verbal. It is highly social. And, it involves much exploration but fewer of the more creative, open-ended or higher-order behaviours that educators or museum curators/designers might wish to witness.

On some levels, one might say that the types of behaviours observed at the exhibits —the exploratory and social behaviours—are not unexpected. When faced with some degree of novelty within a social environment, it is not surprising that so much time and activity are devoted to figuring out what to do and working with others in some manner of doing it. Yet, is this enough? Or, perhaps, a better question is—should exhibits attempt to do more? And, moreover, in the current context, where much attention is placed on digital technologies, are such experiences worth creating and studying? The wide-ranging theories drawn upon to develop the coding schema were brought together in an attempt to make sense of what one might desire or expect at the exhibit face based on notions of what experience—of acting on the world—can offer science communication and, specifically, science education for young children. But how important are such behaviours. Some answer to that question comes from the recent findings of Shayer *et al.* (2007). This group of researchers compared children's performance on logico-mathematical tasks in the present decade with their attainment 30 years ago and found that children then were significantly more successful on tasks that required them to deal with the concepts of volume, weight and other phenomena that rely on everyday experiences. Their interpretation of the findings was that one of the major differences between the children of today and children then is that the latter had more opportunities to play with objects spending their time after school outside exploring the world, building and creating physical objects, taking mechanical objects apart and reassembling them. In contrast, today, children lives are dominated by virtual experiences. Even in school, the focus is on mathematics and literacy rather than the skills and knowledge that come from operating on the world. As a consequence, they suggest that children lack the repertoire of experiences necessary to construct scientific knowledge— that is the range of interactions with a diversity of concrete objects out of which models and representations can be constructed—an activity which lies at the core of science. Thus, the chance to 'mess about in science' (Hawkins 1965), and in particular with science exhibits, is possibly more important than ever. For without such experiences, access to the models and ideas that science offers may well become harder if not altogether inaccessible. Interactive science exhibits can, and do, offer such opportunities and thus may be key to communicating many scientific ideas. Nevertheless, having examined the details of children's activity at some such exhibits, there is undoubtedly considerable work yet to be done to improve the range of 'messing about' and 'playing' that we might desire exhibits to engender.

■ REFERENCES

Adams, A., Luke, J. and Moussouri, T. (2004). Interactivity: moving beyond terminology. *Curator*, **47**(2), 155–70.

Allen, S. (2002). Looking for learning in visitor talk: a methodological exploration. In: *Learning Conversations in Museums* (ed. G. Leinhardt and K. Crowley), pp. 259–303. Lawrence Erlbaum, Mahwah, NJ.

Allen, S. (2004). Designs for learning: studying science museum exhibits that do more than entertain. *Science Education*, **88**(S1), S17–S33.

Ash, D. (2003). Dialogic inquiry in life science conversations of family groups in a museum. *Journal of Research in Science Teaching*, **40**(2), 138–62.

Blud, L.M. (1990). Social interaction and learning among family groups visiting a museum. *Museum Management and Curatorship*, **9**(1), 43–51.

Boden, M.A. (2001). Creativity and knowledge. In: *Creativity in Education* (ed. A. Craft, B. Jeffrey and M. Leibling), pp. 95–102. Continuum, London.

Callanan, M.A., Shrager, J. and Moore, J.L. (1995). Parent-child collaborative explanations: Methods of identification and analysis. *Journal of the Learning Sciences*, **4**(1), 105–29.

Carlisle, R.W. (1985). What do school children do at a science center? *Curator*, **28**(1), 27–33.

Chi, M.T.H. (1997). Quantifying qualitative analyses of verbal data: a practical guide. *Journal of the Learning Sciences*, **6**(3), 271–315.

Craft, A. (2000). *Creativity Across the Primary Curriculum: Framing and Developing Practice*. Routledge, London.

Crowley, K. and Callanan, M.A. (1998). Describing and supporting collaborative scientific thinking in parent-child interactions. *Journal of Museum Education*, **23**(1), 12–20.

Dewey, J. (1938). *Experience and Education*. Touchstone, New York.

Diamond, J. (1986). The behavior of family groups in science museums. *Curator*, **29**(2), 139–54.

Diamond, J. (1996). *Playing and Learning*. ASTC, Washington, DC.

Falk, J.H. and Dierking, L.D. (2000). *Learning from Museums: Visitor Experiences and the Making of Meaning*. Altamira Press, Walnut Creek, CA.

Falk, J.H., Moussouri, T. and Coulson, D. (1998). The effect of visitors' agendas on museum learning. *Curator*, **41**(2), 107–20.

Feher, E. (1990). Interactive museum exhibits as tools for learning: explorations with light. *International Journal of Science Education*, **12**(1), 35–50.

Gilbert, J. and Boulter, C. (eds) (2000). *Developing Models in Science Education*. Kluwer, Dordrecht.

Gredler, M.E. (2001). Lev S. Vygotsky's cultural-historical theory of psychological development. In: *Learning and Instruction: Theory into Practice* (ed. M.E. Gredler), pp. 275–311. Prentice-Hall, New Jersey.

Gutwill, J.P. (2003). Gaining visitor consent for research II: improving the posted-sign method. *Curator*, **46**(2), 228–35.

Harré, R. (1986). *Varieties of Realism*. Blackwell, Oxford.

Hawkins, D. (1965). Messing about in science. *Science and Children*, **2**(5), 5–9.

Heath, C. and Hindmarsh, J. (2002). Analysing interaction: video, ethnography and situated conduct. In: *Qualitative Research in Action* (ed. T.E. May), pp. 99–121. Sage, London.

Hughes, P. (2001). Making science 'family friendly': the fetish of the interactive exhibit. *Museum Management and Curatorship*, **19**(2), 175–85.

Hutt, C. (1981). Toward a taxonomy and conceptual model of play. In: *Advances in Intrinsic Motivation and Aesthetics* (ed. H. Day), pp. 251–98. Plenum Press, New York.

Kisiel, J. (2005). Understanding elementary teacher motivations for science fieldtrips. *Science Education*, **89**(6), 936–55.

Knoblauch, H., Schnettler, B., Raab, J. and Soeffner, H. (eds) (2006). *Video Analysis: Methodology and Methods*. Peter Lang, Frankfurt am Main.

vom Lehn, D., Heath, C. and Hindmarsh, J. (2001). Exhibiting interaction: conduct and collaboration in museums and galleries. *Symbolic Interaction*, **24**(4), 189–216.

Leinhardt, G. and Crowley, K. (eds) (2002). *Learning Conversations in Museums*. Lawrence Erlbaum, Mahwah, NJ.

Lieberman, J.N. (1977). *Playfulness: its Relationship to Imagination and Creativity*. Academic Press, New York.

Lucas, A.M., McManus, P. and Thomas, G. (1986). Investigating learning from informal sources: listening to conversations and observing play in science museums. *International Journal of Science Education*, **8**(4), 341–52.

Meisner, R.S. (2007). Encounters with exhibits: a study of children's activity at interactive exhibits in three museums. Unpublished PhD Thesis. Department of Education and Professional Studies, King's College London.

Mercer, N., Littleton, K. and Wegerif, R. (2004). Methods for studying the processes of interaction and collaborative activity in computer-based educational activities. *Technology, Pedagogy and Education*, **13**(2), 195–212.

Ohlsson, S. (1996). Learning to do and learning to understand: a lesson and a challenge for cognitive modelling. In: *Learning in Humans and Machines: Towards an Interdisciplinary Learning Science* (ed. P. Reimann and H. Spada), pp. 37–62. Elsevier Science, Oxford.

Packer, J. (2006). Learning for fun: the unique contribution of educational leisure experiences. *Curator*, **49**(3), 329–44.

Piaget, J. (1951). *Play, Dreams and Imitation in Childhood*. The Free Press, New York.

Postman, N. (1987). *Amusing Ourselves to Death: Public Discourse in the Age of Show Business*. Methuen, London.

Rahm, J. (2004). Multiple modes of meaning-making in a science center. *Science Education*, **88**, 223–47.

Reilly, M. (1974). Defining a cobweb. In: *Play as Exploratory Learning: Studies of Curiosity Behavior* (ed. M. Reilly), pp. 57–116. Sage, Beverly Hills, CA.

Rennie, L.J. (2007). Learning science outside of school. In: *Handbook of Research on Science Education* (ed. S.K. Abell and N.G. Lederman), pp. 125–67. Lawrence Erlbaum, Mahwah, NJ.

Rennie, L.J. and McClafferty, T.P. (1996). Science centres and science learning. *Studies in Science Education*, **27**, 53–98.

Rennie, L.J. and McClafferty, T.P. (2002). Objects and learning: understanding young children's interaction with science exhibits. In: *Perspectives on Object-centered Learning in Museums* (ed. S.G. Paris), pp. 191–213. Lawrence Erlbaum, Mahwah, NJ.

Rennie, L.J., Feher, E., Dierking, L.D. and Falk, J.H. (2003). Toward an agenda for advancing research on science learning in out-of-school settings. *Journal of Research in Science Teaching*, **40**(2), 112–20.

Rix, C. and McSorley, J. (1999). An investigation into the role that school-based interactive science centres may play in the education of primary-aged children. *International Journal of Science Education*, **21**(6), 577–93.

Roschelle, J. (1995). Learning in interactive environments: prior knowledge and new experience. In: *Public Institutions for Personal Learning: Establishing a Research Agenda* (ed. J.H. Falk and L. Dierking), pp. 37–51. American Association of Museums, Washington, DC.

Schauble, L., Leinhardt, G. and Martin, L. (1997). A framework for organizing a cumulative research agenda in informal learning contexts. *Journal of Museum Education*, **22**(2/3), 3–8.

Semper, R. (1990). Science museums as environments for learning. *Physics Today*, **43**(11), 50–6.

Shayer, M., Ginsburg, D. and Coe, R. (2007). Thirty years on—a large anti-Flynn effect? The Piagetian test volume and heaviness norms 1975–2003. *British Journal of Educational Psychology*, **77**(1), 25–41.

Shortland, M. (1987). No business like show business. *Nature*, **328**(6127), 213–14.

Spock, D. (2004). Is it interactive yet? *Curator*, **47**(4), 369–74.

Stevens, R. and Hall, R. (1997). Seeing tornado: how video traces mediate visitor understandings of (natural?) phenomena in a science museum. *Science Education*, **81**(6), 735–47.

Sutton-Smith, B. (1988). In search of the imagination. In: *Imagination and Education* (ed. K. Egan and D. Nadaner), pp. 3–29. Open University Press, Milton Keynes.

Vygotsky, L.S. (1978). *Mind in Society: the Development of Higher Psychological Processes*. Harvard University Press, Cambridge, MA.

Wellington, J. (1990). Formal and informal learning in science: the role of the interactive science centres. *Physics Education*, **25**, 247–52.

Wertsch, J.V. and Tulviste, P. (1996). L.S. Vygotsky and contemporary developmental psychology. In: *An Introduction to Vygotsky* (ed. H. Daniels), pp. 53–74. Routledge, London.

Witcomb, A. (2006). Interactivity: thinking beyond. In: *A Companion to Museum Studies* (ed. S. Macdonald), pp. 353–61. Blackwell Publishing, Malden, MA.

Wolpert, L. (1992). *The Unnatural Nature of Science*. Faber and Faber, London.

■ FURTHER READING

- Falk, J.H. and Dierking, L.D. (2000). *Learning from Museums: Visitor Experiences and the Making of Meaning*. Altamira Press, Walnut Creek, CA. This book brings together research within the field of visitor studies to examine why people go to museums and what they do once there. In addition, the authors develop a theory of learning in museums which synthesizes various theories and research to suggest that such learning occurs over time within the personal, sociocultural and physical contexts of the museum.

- Lucas, A.M., McManus, P. and Thomas, G. (1986). Investigating learning from informal sources: Listening to conversations and observing play in science museums. *International Journal of Science Education*, **8**(4), 341–52. In this seminal article, the authors argue for conducting research in informal contexts in order to focus on how individuals develop scientific concepts. By presenting two case studies of research at the exhibit face, they illustrate that different methods reveal various aspects of learning, and highlight the importance of both context and social interaction in learning and studying how learning occurs.

- Paris, S.G. (ed.) (2002). *Perspectives on Object-centered Learning in Museums*. Lawrence Erlbaum, Mahwah, NJ. This book presents a collection of research on object-based learning—children's interaction with objects, ranging from traditional museum objects in cases to interactive science exhibits. It introduces a variety of theoretical perspectives and methodological approaches to studying such interaction.

- Rennie, L.J. (2007). Learning science outside of school. In: *Handbook of Research on Science Education* (ed. S.K. Abell and N.G. Lederman), pp. 125–67. Lawrence Erlbaum, Mahwah, NJ. This chapter provides a current discussion of the various definitions and characteristics of out-of-school learning, followed by a detailed review of the research literature within three areas—museums, community and government organizations, and media. It offers implications from research for both practice and policy.

■ USEFUL WEB SITES

- **The Center for Informal Learning and Schools: http://www.exploratorium.edu/cils/.** The Center for Informal Learning and Schools (CILS) is a collaboration between King's College London, the University of California, Santa Cruz and the Exploratorium in San Francisco, which conducts research on informal learning and seeks to bridge research and practice. Details of various research studies and an annotated bibliography are provided on the web site.

- **Informal science: http://www.informalscience.org/.** The informal science site through the University of Pittsburgh is designed as a place to share knowledge about informal science learning. It offers a data base of research and evaluation projects within informal learning institutions.

- **Learning in Informal and Formal Environments (LIFE) Center: http://life-slc.org/.** The Learning in Informal and Formal Environments (LIFE) Center is a collaboration between learning scientists at the University of Washington, Stanford University and SRI International, which seeks to understand and advance learning in informal and formal environments. Details of various research studies can be found on the web site.

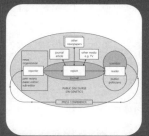

SECTION 3

Studying science in popular media

There is no such thing as *the* communication of science. Neither science nor the media environment is a unified phenomenon. Scientists disagree; the media present different accounts; receivers of scientific communication interpret each set of them in different ways, which may result in distinct, even disjointed, understandings.

Roger Silverstone (1991) 'Communicating science to the public'
(emphasis in original)

3.1

Science, communication and media

Anders Hansen

The first major review of 'science and the media' was published over 40 years ago (Krieghbaum 1967), and has been followed by a steady stream of books and articles on the subject (e.g. Goodell 1977; Goodfield 1981; Gardner and Young 1981; LaFollette 1982; Silverstone 1985; Friedman *et al.* 1986; Nelkin 1995). Increasing concern in the last 20 years about public understanding of science has also led to a significant increase in social science research on the communication of science and the role of the mass media in this context. Major book-length overviews and discussions of media and science, medicine and environment in the last 20 years include the revised edition of Nelkin's (1995) pioneering book *Selling Science* as well as edited collections (e.g. Wilkins and Patterson 1991; Hansen 1993b; Friedman *et al.* 1999) and synthesizing overviews (e.g. Lewenstein 1995b; Gregory and Miller 1998; Bucchi 1998; Weigold 2001; Allan 2002; McComas 2006).

The aim of this chapter is to review major trends in media and communication research on the role of the media in the increasingly overlapping domains of science, risk, health and environment communication. The considerable and expansive growth in this field over the last 40 years means that a brief overview such as this must of necessity be selective, and it is worth stressing from the outset that the 'perspective' guiding my eclecticism has been that of media and communications research with a principally sociological inclination. Starting with a brief overview of the development of concern about public understanding of science and the role that media and communication is seen to have in this context, and with a plea for 'differentiation' (particularly with regard to 'media'), the chapter proceeds to examine the traditional three primary foci of communication research by looking: first, at the production of science in the media, including studies of the relationship between sources and media professionals; second, at the contribution from studies of media content or representations (with three subsections devoted to the most pronounced areas of content research: media and nuclear technology; media, science and environmental issues; and medicine, biotechnology and genetics); and third, the contributions from studies of media audiences. The concluding section stresses the need for a flexible and dynamic perspective on the media's role in science communication and points briefly to the challenges—for science communication research—of an increasing diversification/fragmentation of media forms and modes of communication.

Communication, media and public understanding of science

Concern about the public's awareness and understanding of science and its role and contribution to society has a long and distinguished pedigree in British history. It is associated with such august institutions as the Royal Society (established 1660), the Royal Institution (established 1799) and the British Association for the Advancement of Science (established in 1831) as well as with famous laments about the polarization of society into 'Two Cultures' (Snow, 1993, originally voiced in the late 1950s): the humanities/literary intellectuals and science/scientists.

The immediate post-war optimism which characterized much of public and political debate about science, technology and progress had, by the late 1960s and 1970s, turned to increasing concerns about misuses of science, the unanticipated and devastating side-effects and by-products of science, the perceived inability of science and technology to solve major social problems and international conflict, and the pollution and destruction of the natural environment. Low levels of 'scientific literacy' in the general public were seen, by the early 1980s, as at least partially responsible for a growing public scepticism and mistrust, sometimes downright opposition, with regard to science, scientists and—not least—science-based high-tech industries such as the nuclear power industry.

In Britain, the starting point for what later came to be known as the 'public understanding of science movement' was the publication in 1985 of the Royal Society report *The Public Understanding of Science* (Royal Society 1985). The report articulated longstanding concerns about science education, funding for science and the decline of scientific literacy in the population, but of particular interest in the context of this chapter is its section on science communication and the important but, in its view, failing role of the mass media.

While by no means the only, or indeed the most prominent, focus, the emphasis on the importance of 'communicating' science, of communication and understanding between the scientific community, industry, government and the public, and of the key role for the mass media in this context, has remained prominent in a string of reports and guidelines published from the late 1980s to the present.

Important shifts have also occurred, however, in the conceptualization of the 'public understanding of science' (see Bauer *et al.* 2007). Where the focus of the 1980s was on educating the public to a higher level of 'understanding' of science, the emphasis in the late 1990s shifted towards increased public engagement with science. The latter concept —public engagement with science—acknowledges an increasing recognition that public attitudes to new scientific and technological developments were not simply a matter of improving the communication or popularization of scientific information—something that had long been recognized by social scientists (e.g. Dornan 1990; Hilgartner 1990)—but often articulated deep public anxiety and opposition as well as controversy, uncertainty and disagreement within the scientific community.

This shift in thinking is particularly significant in the context of how media roles and communication processes in relation to the communication and public understanding of science have evolved. Essentially, and inevitably oversimplifying, the 'deficit model' (for an early critique of this, see Ziman 1992), which was core to early concerns about the lack of 'public understanding', had much in common with the early 'transmission' model of communication: a model which saw 'the media' as simply a vehicle or conduit for the transmission of a message from a sender to an audience. While media and communication research had, from about the mid-1900s, moved towards much more complex—and importantly much more interactive and multi-directional—models of communication (see McQuail and Windahl 1993, for an excellent overview), many of the perceived problems with science communication and the 'misunderstanding' which continues to feature as a central theme in government and other institution/agency reports on public understanding of science still derive, at least in part, from a pervasive but outdated and overly simplistic model of communication as a simple linear process. Any critical analysis of official reports and their recommendations about 'science communication', 'public understanding' and 'media roles' would therefore do well to start with fundamental questioning of the explicit, or more often implicit, communication models and thinking underpinning such recommendations.

Differentiation: publics, media, sciences

A key problem in much writing about 'communicating science in the media' has been a relatively undifferentiated view of the 'mass media'. Whilst recognizing at an early stage that it makes little sense to talk about 'the public' for science communication as if the 'public' were a homogeneous entity, it has taken rather longer to recognize that, in an increasingly diverse media landscape, it makes equally little sense to talk about the mass media as a single homogeneous entity.

Differentiating between different media (television, radio, magazines, 'quality' press and 'popular' press, internet, etc.) and between different genres (news, editorial, documentary, opinion-piece/feature, drama series, comedy, film, etc.) is crucial to understanding the very different extent and nature of science coverage and science images across these media/genres. Media and genres are governed by format constraints and conventions that impinge on the extent and nature of science coverage (which in turn links closely to how we, as publics/audiences, make sense of or 'understand' science and its role in society).

The need for greater differentiation extends further still, beyond publics and media, to questions about what 'science in the (news) media' means (Weigold 2001). Thus, a narrow definition of 'science (news) coverage' as including only those articles, programmes or sections which are specifically about science or scientific research is tantamount to ignoring some of the most powerful images of science and its social role (Weiss *et al.* 1988).

In fact, research shows that in news coverage alone almost half of all portrayals of science or scientists appear in items which are not specifically about science (Hansen and Dickinson 1992).

Scientists and science are drawn into a wide range of news coverage, from crime and social problems to politics and consumer affairs. The argument applies even more emphatically when moving beyond news and associated factual media genres to drama, entertainment, advertising and film, where the most powerful images of scientists and science are rarely communicated in programmes/films advertised as being about science as such, but where science and 'sciency' imagery are central components of narrative and characterization (see Whitelegg *et al.* 2008). Prominent examples include popular 'medical' drama series (e.g. *ER*), forensics-based crime series (e.g. *CSI*) and even television comedy series (e.g. the BBC's *Supernova*). It follows that the ways in which communication researchers sample media and define what qualifies as 'science coverage' impinges directly on what they find (see Carr *et al.*, Chapter 6.2 this volume). Differences in sampling decisions and related definitions in turn often go a long way toward explaining the diverse and sometimes seemingly contradictory findings from research on 'science and the media'.

In summary, advancing our understanding about the role of media and communication processes in relation to 'public understanding of science' then requires a greater differentiation of 'media', of 'publics' and of what we mean by 'science in the media'.

Producing/managing science in the media

Much of the research on media and science has focused on the communication practices of scientists and scientific institutions as sources of scientific knowledge, expertise and information, as well as on the communications and media professionals, from public relations professionals to science correspondents/editors, 'covering' science. In tandem with the changing notion of 'public understanding' described above, research on sources and journalists in the science communication process has witnessed a gradual change, spurred on by general paradigm shifts in communication research, from the dominant concern in earlier research with 'accuracy', 'objectivity' and 'bias/distortion' in media reporting (e.g. Tichenor *et al.* 1970; Tankard and Ryan 1974; Borman 1978; Moore and Singletary 1985; Singer 1990) towards a more constructivist perspective on how news and other media coverage of science is driven by complex processes of claims-making, organizational arrangements, economic imperatives and cultural values (Schudson 1989, 2000; Hansen 1991). Originally focused very much on individual scientists as sources and on the practices of science correspondents and other media professionals, research on the production of science news and science communication has in the last one or two decades also increasingly turned to exploring the ways in which pressure groups, social movements and activists, as well as large companies, scientific institutions, government departments and political parties seek to actively manage and influence communication about science and science-based social issues (Rogers 1986; Hansen 1993a; Davis 2000; Hargreaves and Ferguson 2000).

Studies of news, news production, news organizations, sources and journalistic practices have helped explain much about the processes and structures determining which issues get reported and which do not. News, as numerous key news-studies from the last 50 years or so have amply demonstrated, is made, created and selectively reported (Tumber 1999). The making of news is a complex process of interaction between, on the one hand, institutions and individuals in society who act as sources or subjects of news and, on the other hand, the news media whose own organization and professional practices influence which institutions, events and individuals get reported.

Studies of science journalism have pointed to the importance of a range of media-related factors, including: the size of media organization (Friedman 1986); the professional beliefs, 'status' ('specialist' or 'general' reporter) and qualifications of journalists covering science (Dunwoody 1979; Lowe and Morrison 1984); the professional beliefs, policies and practices of editors (Endreny 1985); the growth of information officers or media liaison officers acting as middlemen between scientists/science institutions and journalists/media institutions (Rogers 1986; Dunwoody and Ryan 1983); scientists' stereotypes of journalists and vice versa (Dunwoody and Stocking 1985); and the images which media professionals have of their audiences (Tannenbaum 1963; Hansen 1994).

Research on science journalism has often revolved around the notion that science journalism/news is 'different' from other types of news in at least two respects: (1) that the agenda for public discussion of science is predominantly set by the sources, i.e. the scientists or the scientific community, and (2) that science journalism is by and large an uncritical and deferential celebration of science and scientists because the relationship between science journalists and their scientist sources is one of 'symbiotic' interdependence.

Goodell (1987) thus argues that the close, symbiotic, relationship between science reporters and their scientist sources lead to deferential reporting, which '. . . fails to acknowledge the politicised role of scientists in modern society'. However, she also recognizes that this type of source–communicator relationship is not unique to science reporting, but 'may have counterparts in other specialised areas of reporting such as business or the arts' (Goodell 1987). Indeed, we only need to look at the voluminous literature on news production in other 'specialist' areas such as crime reporting (e.g. Ericson *et al.* 1989) or education (Hansen 2007) to see that science journalism is not unique in this respect. While the 'symbiotic relationship' argument thus seems to have been unduly exaggerated in relation to science journalism, studies have pointed to the importance of 'trust' in the relationship between scientist sources and science journalists (Geller *et al.* 2005).

In terms of 'difference' from other types of journalism, studies have shown, *inter alia*, that: science correspondents tend to remain much longer with their specialism as science correspondents than other types of journalist; they are much more likely than other journalists to have a science degree, although this is rarely seen by the journalists themselves as a particular advantage in their day-to-day task of reporting—a recurrent refrain from science journalists themselves is that their training as journalists is what matters, not their science training (Hansen 1994; Hargreaves and Ferguson 2000); and they often have more contact with fellow science correspondents in competitor media than they do with colleagues in their own medium or organization—referred to in the literature as a 'competitor–colleague' relationship, or as the 'inner club' (Dunwoody 1980).

Much of the work of science correspondents consists of attending to routine news forums, whose agendas are set not by the news organizations but elsewhere, and of reacting to incoming news. Indeed, studies have pointed to the significant agenda-setting role of specialist science journals (de Semir *et al.* 1998), of press releases, press and publicity officers (Rogers 1986), of scientific meetings/conferences and of scientists themselves (Hansen and Dickinson 1992) in influencing the media and public agenda for science coverage.

Studies of environmental and science journalism have confirmed the findings from the sociology of journalism more generally, that one of the key journalistic strategies for coping with information which is not only conflictual, but which may, equally significantly, be of a highly complex scientific/technical nature, is to rely heavily on established, authoritative, recognized sources (Dunwoody and Peters 1992; Hansen 1994; Peters 1995). There is also interesting historical evidence of the increasing 'scientization', i.e. the increasing reliance on scientists and other experts, in a broad range of media coverage over the last 40 years or so (Albaek *et al.* 2003).

A key concern in much of the literature on media coverage of controversial scientific and environmental issues has focused on the extent to which interest groups of various kinds—but especially pressure groups who attempted to mobilize public opposition to particular technologies, scientific developments/procedures and environmental issues—have succeeded in 'managing' or influencing the media agenda. Interestingly, evidence from analyses of whose voices are heard and quoted in media coverage of controversial science or science-related issues has tended to show an overwhelming emphasis on government sources and independent scientists, with far less prominence given to either industry/business sources or to pressure groups (Greenberg *et al.* 1989; Priest *et al.* 1991; Trumbo 1996). It is also clear, however, that the influence of interest/pressure groups on media coverage varies significantly from issue to issue: Hargreaves *et al.*'s (2004) recent analysis, for example, shows pressure groups to be much more prominent in media coverage of climate change than in media coverage of the measles–mumps–rubella (MMR) vaccine controversy, or of cloning and genetic medical research. The literature on pressure groups and media strategies (e.g. Hannigan 2006) has provided useful lessons on the key tasks involved in successful campaigning (defined by Solesbury 1976 as 'commanding attention', 'claiming legitimacy' and 'invoking action'), while also showing that success in terms of commanding media attention does not always translate into success in claiming legitimacy, let alone success in invoking action (Hansen 2000). Animal rights groups, for example, are often successful in gaining short-term media coverage, but rarely in claiming legitimacy. Furthermore, in their much celebrated Brent Spar campaign of 1995, Greenpeace were highly successful in gaining and sustaining media coverage, but were 'endorsed' only by some media while other media distanced themselves from and were highly critical of the claims made by Greenpeace (Hansen 2000).

Media representations of science

When Hargreaves and Ferguson (2000) lamented that there was little published research on the extent of science coverage they were right in some ways, but rather spectacularly and expediently oversimplifying in other respects. It is indeed the case that general panoramic overviews of the extent and type of 'science' coverage in the mass media have been relatively few and far between, but this should not obscure the fact that there is now a very large body of communications research on specific types or aspects of media coverage of science, health and medicine, environmental issues and risk.

There is a relatively long tradition of studies of the extent and nature of science news coverage in a range of mass media (see, e.g., the early review by Cronholm and Sandell 1981). In Britain, an early study by Jones *et al.* (1978) found that approximately three–quarters of all science coverage concerned 'medicine and human biology'. A study by Hansen and Dickinson (1992) again confirmed the dominance of 'health/medicine' as the single most prominent category of science coverage in newspapers and on radio and television alike. They also demonstrated the very different prominence of different types of science and technology (and some variation depending on type of medium: newspapers, radio or television): for example, in addition to 'health and medicine', 'environment' and 'life sciences' featured prominently, while 'earth sciences', 'space' (in contrast to its prominence in the 1960s and 1970s), 'computing' and 'science policy' received very little coverage in print and broadcast media. These general trends have also been confirmed by studies taking a longer retrospective look at science coverage (Bauer *et al.* 1995, 2006). Bauer thus argues that the dominance of medical science in media coverage toward the end of the 20th century reflects the increasing 'medicalisation of modern society' (Bauer 1998).

The most comprehensive recent analysis of science coverage in the British media is Hargreaves *et al.*'s (2004) report of their Economic and Social Research Council (ESRC)-funded research, although they deal only very superficially with the measurement of science coverage generally, choosing instead to focus on three selected high-profile issues: MMR, climate change and cloning/genetic medical research. Like many studies before, their analysis shows the extraordinary dominance of medical science: '. . . on television, . . . nearly half of all science stories (44 per cent) are about medical developments or issues.' (*ibid.*, p. 10).

While sweeping general surveys of science coverage in the media are comparatively few in number, the last 50 years or so have witnessed a tremendous growth in studies of media coverage of specific types of science or, as is more often the case, specific high-profile issues or public controversies which involve a major science or technology component. Rather than attempting to provide an exhaustive catalogue, the objective here is simply to indicate some of the most high-profile foci of research on media and science issues in the last few decades. In very general terms these can be divided into those with a predominantly nuclear technological focus, those with a broadly environmental focus, and those with a medical–biotechnology–genetics focus; although, as always, many issues, controversies and types of coverage often span several categories, whilst a small proportion, e.g. planetary and space sciences, fall outside these categories.

Common to media and communication research in all three areas is the predomin-ance of studies of media content, with relatively much fewer studies of either the pro-duction of media coverage (i.e. studies of sources, source publicity practices, journalistic practices or media ownership and control) or the social and political implications of media coverage, whether in terms of influences on public and political opinions or in terms of how audiences use, appropriate or negotiate public information and representation of science.

The vast majority of social science studies of media coverage of science have used some form of systematic content analysis (a method that lends itself very well to measuring, in a reliable and quantitative way, the extent and general nature of science coverage), but many studies have gone considerably beyond simple 'inventories' of science coverage to show, for example, the 'frames', metaphors, narratives, themes and actors/voices that make up such coverage. Prominent theoretical models guiding these kinds of inquiries have ranged widely from broader sociological theories such as social constructionist theory and objectivism/realism (Spector and Kitsuse 1977/1987; Schneider 1985) to a range of more specifically media-oriented theories such as agenda-setting and framing theory (Reese *et al.* 2001; McCombs 2004).

There are considerable imbalances in terms of the types of media that have been subjected to analysis. The majority of studies of 'media and science' in fact focus solely on the print media—and here almost exclusively on newspapers (although there have been some notable studies of magazine reporting of science, e.g. LaFollette 1989; Lewenstein 1989; Condit *et al.* 2002a; McInerney *et al.* 2004; Clarke and Everest 2006). By contrast, the number of studies of science on television is smaller (although some of the most interesting in-depth analyses have focused on television programmes, see below), and studies of radio science are virtually non-existent.

The 1990s saw an emerging attention to internet or internet-related forms of communication (e.g. e-mail), but research on the role of the internet in relation to science communication and public images of science is still in its infancy. Lewenstein's (1995a) analysis of the use of multiple media and forms of communication in the cold fusion debate of 1989 onwards offers a sophisticated alternative to traditional linear 'transmission' models of science communication, while Rogers and Marres' (2000) work on interlinking web sites in the climate change debate demonstrates how key players—pressure groups, companies, government—seek to manage their own image as well as to frame the positions of their opponents through their web sites and associated hyperlinking practices. More recent work focusing on the internet as a tool and source in science communication includes several studies in thematic issues of the journal *Science Communication* (2001, **22**(3) and 2003, **24**(3)) as well as elsewhere, e.g. Byrne *et al.* (2002) and Weigold and Treise (2004).

By contrast, there is a longer tradition of research literature on representations of science and scientists in film, including several book-length studies (e.g. Tudor 1989; Lambourne *et al.* 1990; Schelde 1993) and significant journal articles (e.g. Jones 2001; Weingart *et al.* 2003; Steinke 2005).

While most studies of media representations of science have used content analysis as their principal method—albeit ranging from the very rudimentary to the highly complex and sophisticated types of content analyses seen in much of the research on biotechnology/genetics coverage and on climate change coverage for example—only a much smaller number of studies have been devoted to a detailed exposition of how television constructs science/scientists and scientific expertise. Roger Silverstone's book *Framing Science* (1985), drawing principally on semiotic and narrative methods of analysis (see also Mellor, Chapter 5.2 this volume), offered one of the first detailed analyses of the construction of science in television documentary.

Other exemplary studies of a comparable kind include Nimmo and Combs' (1985) narrative study of television coverage of the Three Mile Island nuclear incident, Collins'

(1987) study of the production of scientific certainty/uncertainty in television documentary, Murrell's (1987) ideology-analysis of representations of science in the BBC's *Tomorrow's World*, and Corner *et al.*'s (1989) analysis of television representations of nuclear power. Drawing closely on tools from semiotics, structuralism and narrative analysis, Priest's (1990) study of PBS's *Nova* offers a comprehensive unpacking of the construction of science and scientists in television documentary, as well as serving as a useful analytical template for anyone wishing to engage in the analysis of 'television science'. These studies show how the narrative and genre conventions of television documentary serve to reinforce the notion of science as a product rather than a process of inquiry; and where the process of science is shown this is highly truncated, portraying science as an orderly process working inexorably towards clear-cut solutions/answers and towards the production of 'certainty'. Scientists are constructed as omniscient, authoritative and credible.

Media and nuclear technology

Much research on 'science and the media' in the 1970s and 1980s focused on (news) media coverage of nuclear power and growing public opposition. This coverage, and the subsequent analyses, was concerned with the environmental disaster scenarios that would potentially result from a major nuclear accident (emphasized by the Three Mile Island incident in 1979 and more particularly by the Chernobyl disaster of 1986), and with risk (mis-)information, seen as essential in the context of public unease about and opposition to nuclear issues (Nimmo and Combs 1985; Friedman *et al.* 1987, 1992; Friedman 1989; Corner *et al.* 1990; Wober 1992; Nohrstedt 1993; Triandafyllidou 1995). The (mostly implicit) model underlying much of the American research on media and nuclear technology was an 'accuracy' model (and the concomitant 'deficit' model of public understanding of science, i.e. if only the public could be made to understand 'accurate' descriptions of the comparatively low risks associated with nuclear technology, then public fear and opposition would evaporate) particularly pronounced in studies such as those by Rothman and Lichter (Rothman and Lichter 1982, 1987; Rothman 1990). Predominantly focused on analysis of media content and on general trends in public opinion, few of these studies offered much explanation of how audiences interpret and act on mediated risk information.

One of the most interesting 'models' to emerge around this time was Alan Mazur's (1981, 1990) proposition that public opinion/opposition was directly related to the quantity of coverage and much less so to the balance of 'positive' and 'negative' coverage. Longitudinal historical studies of media reporting and public opinion on nuclear issues provided further much-needed context for understanding the dynamic interplay between media coverage, public opinion and popular culture (see particularly Weart 1988 and Gamson and Modigliani 1989). While the 1980s and early 1990s saw a large number of studies on media coverage and public opinion on nuclear technology, communication research in this area has been more muted in the most recent decade, but this may well be about to change as public/political debate about civil nuclear power is moving to the fore again in the climate change debate. The emerging reframing of the nuclear debate thus offers a particularly interesting opportunity for longitudinal research to examine the complex dynamics of what Downs (1972) called the 'issue-attention cycle' as well as the ways in which issues (here, civil nuclear power and climate change) interact in the media arena and the public sphere generally.

Media, science and environmental issues

While the rapidly growing body of communication research on media and environmental issues (overviews include Schoenfeld *et al.* 1979; Hansen 1991, 1993b; Anderson 1997; Shanahan and McComas 1999; Allan *et al.* 2000; Cox 2006; Hannigan 2006) may not at first sight appear to be 'science communication', science and scientists are a central component of most media coverage of environmental problems and issues. This is clear from a wealth of communication research which has focused on the question of who the key 'claims-makers', 'voices' or 'primary definers' are in public 'mediated' debate about environmental issues and controversies. Research has examined whose definitions receive (media-) prominence in the public sphere and whose definitions become 'successful' accounts of the causes and solutions to environmental problems. Research on media coverage of environmental issues has also sought to understand why some environmental issues are successfully constructed as issues for public concern, while other—seemingly equally serious or important—issues quickly vanish from the media agenda and from public view.

Not only has such research demonstrated the prominent role of 'independent' scientists (as opposed to government or industry scientists) as authoritative voices, but they have also shown how much less prominent voices, such as environmental pressure groups, have increasingly sought to enhance their definitional power in the public sphere by an 'alliance with science' (Hansen 1993a). Nor are pressure groups alone in this respect: government, organizations, industry and other claims-makers are similarly and increasingly deploying 'evidence-based' arguments in public debate, in what sociologists have referred to as the increasing 'scientization' of society (e.g. Nowotny *et al.* 2001).

While media and communication research on environmental issues has comprised a wide and diverse range of environmental problems, the most prolific area of research in the recent period has been that of 'climate change' (e.g. Bell 1994; Boykoff and Boykoff 2004; Antilla 2005; Carvalho 2007; Trumbo 1996; Zehr 2000). This is not entirely surprising, as the 'climate debate' itself has increasingly assumed the position of a 'master-discourse', i.e. an 'umbrella' concept subsuming a whole variety of hitherto relatively disparate, separate and distinct discussions and issues.

Medicine, biotechnology and genetics

Science in the media, as noted previously, is to a very large extent dominated by media coverage of health and medicine, and, in the last 20 years or so, by biotechnology and genetics. The coverage of genetics is itself characterized by a significant polarization into red (biomedical science) and green (agricultural biogenetics), a polarization that results —as communication and public opinion research has demonstrated—in very different valorizations: generally positive in the case of biomedical science and generally negative/ critical or sceptical in the case of biogenetic sciences dealing with genetic modification of plants, food and animals.

Communication research on health and medicine has focused on media reporting of AIDS and cancer or on so-called 'public health scares' in which the media were thought to have played a major role by fuelling public fears through selective or poorly balanced reporting of medical evidence. While the media have traditionally been an easy target for criticism—criticized by politicians, scientists and media researchers alike for imbalanced, inaccurate and sensationalizing coverage—perhaps the key characteristic that many of the so-called public health scares have had in common is not so much poor or sensational media reporting, but rather an acute absence of scientific consensus about anything from the size/prevalence of the problem to questions about epidemiology and effective public measures for tackling the problem.

The most prominent examples from the last quarter of a century or so include numerous public scares about food safety [*Salmonella* in eggs and other food (Fowler 1991); bovine spongiform encephalopathy (BSE), or mad cow disease and associated concerns (Miller and Reilly 1995; Kitzinger and Reilly 1997; Brookes 1999; Rowe *et al.* 2000)]; the safety of the contraceptive pill (Weatherall 1996; Hammond 1997); the safety and side-effects of vaccinations (Leask and Chapman 2002); about severe acute respiratory syndrome (SARS) (Washer 2004; Luther and Zhou 2005; Tian and Stewart 2005; Wallis and Nerlich 2005) and other 'germ panics' (Tomes 2000); and about avian flu or bird flu (although little media research has focused on this as yet, see Nisbet 2006).

A prominent trend in much of the communication research on health and medicine has been the focus on analyses of metaphor, discourse and framing in coverage of infectious diseases, notably AIDS. In a sophisticated analysis of metaphor-use and framing in media coverage of SARS, Wallis and Nerlich (2005) thus demonstrate an interesting linguistic shift away from the conventional militaristic, e.g. war, fight, enemy, battle, and judgmental metaphors, e.g. of plague and sin as characteristics of HIV/AIDS reporting, to a comparatively simple 'killer' metaphor.

The focus on metaphors and framing has similarly been characteristic of much of the prolific research output on media coverage of biotechnology and genetics in the last 15–20 years (Nelkin and Lindee 1995; Condit *et al.* 2002b; Hellsten 2003). A large number of American and European studies have mapped not only the general 'ups' and 'downs' in biotechnology and genetics coverage, but also the key frames/discourses/themes which have characterized such coverage (Durant *et al.* 1998; Bauer *et al.* 2002; Nisbet and Lewenstein 2002; Ten Eyck and Williment 2003; Hansen 2006). Drawing on the frame/package categories developed by Gamson and Modigliani (1989) in their analysis of nuclear issues and popular culture, studies of media and biotechnology have demonstrated how key interpretative 'packages' ('progress', 'economic prospect', 'ethical', 'Pandora's box', 'runaway', 'nature/nurture', 'public accountability' and 'globalisation': see, e.g., Durant *et al.* 1998, p. 288) have received very different prominences depending on whether the subject matter was biomedical genetics (where 'positive' frames like 'progress' and 'economic prospect' have been at the fore) or agricultural genetics/GM crops and food/animal cloning (where critical, concerned or negative packages such as 'Pandora's box', 'nature/nurture' and 'public accountability' have been prominent).

They also show the very significant shifts which have taken place in media coverage of biotechnology, cloning and the new genetics over the longer term, dating back to its origins in the 1970s. As a body of research, the literature on media coverage of biotechnology provides fascinating insights into the dynamic interplay between source-strategies, journalistic practices, media coverage, public opinion, economic interests, political manoeuvring and wider cultural resonances (Turney 1998 is particularly useful on this latter aspect; see also Toumey 1996). Thus, studies of the early phases of media coverage of recombinant DNA, cloning and the new genetics showed, for example, how key stakeholders in the debate attempted to manage relations with the news media in an effort to control public debate and curb public anxiety or resistance (Pfund and Hofstadter 1981; Altimore 1982; Goodell 1986; Nelkin 1995).

Audiences and publics in science communication

There is a wealth of research, from the last 40 years or so, on public understanding of science. In the US, surveys monitoring trends in science literacy and public understanding of science date back to the 1950s, and a regular programme of science indicators surveys have been conducted since the 1970s (Miller 1987). In the UK, a survey by Durant *et al.* (1989) marked the start of renewed intense interest in public understanding of science, as measured by levels of scientific (il)literacy. Both in the UK and in Europe generally (e.g. through the Euro-barometer surveys), numerous surveys of either general public understanding of science or of more specific aspects of science have been a prominent feature since the late 1980s.

While surveys of public understanding of science have not generally been focused on the specific role of the mass media, they have relatively consistently confirmed: (1) that there is considerable public interest in being informed about science/technology issues (e.g. MORI 2005); (2) that the mass media are a major—in some cases the major—source of public information and knowledge about science and its contribution in relation to major social issues and controversies (Hargreaves *et al.* 2004; MORI 2005), particularly where these are new and rapidly developing issues; (3) that the extent to which publics are influenced by, draw on and/or use media information/images is determined by local knowledge, first-hand experience and availability of other, more direct, sources of information (Wade and Schramm 1969; Chapman *et al.* 1997).

Combining evidence from large-scale surveys of public perception, attitudes and understanding regarding science and science-based controversies with comprehensive, often longitudinal, analyses of media coverage, numerous studies have noted intriguing degrees of co-variation. It has, however, been far more difficult to establish consensus about how public agendas and media agendas interact on science issues, let alone to establish the direction of influence. The research agenda has increasingly moved away from and beyond simplistic linear models of communication and of media influence.

Gamson and Modigliani (1989), in their longitudinal study of media coverage and public opinion on nuclear issues, for example, are a long way away from any simplistic linear cause–effect transmission model when they suggest that media coverage/images and public opinion need to be understood as 'parallel' and interacting forums of meaning creation, rather than in simple terms of media coverage 'influencing' public opinion or vice versa. In a slightly earlier study, Hilgartner and Bosk (1988) similarly point to the complex web of interactions of various 'public arenas'—including the mass media.

Very similar conclusions have been articulated in a string of studies such as Bruce Lewenstein's (1995a) study of communication about cold fusion in the early 1990s (see also Bucchi 1998), numerous studies of media coverage and agenda-building/ agenda-setting in the biotechnology debate (e.g. Durant *et al.* 1996; Durant and Lindsey 1999; Nisbet and Lewenstein 2002; Nisbet *et al.* 2003; Bauer, 2005b), and studies of media coverage and AIDS (see, e.g., the important collection by Miller *et al.* 1998 and Kitzinger 1999).

Common to all of these studies—and models, where articulated—is the recognition that media, scientists, experts, politicians and publics interact in complex ways that are anything but one-directional in the sense of the original simple transmission model of communication. What they also recognize is that to understand the role of media in science communication we need a different vocabulary with terms like interaction, engagement, information loops, neural networks, multi-directional, resonance and parallel forums of meaning creation.

At the macro-sociological level (and mainly using survey methodology and content analysis) of accounting for co-variation and interaction between 'public opinion' and 'media coverage', Bauer (2005b) offers an insightful overview of the indications from three particular 'models', namely the quantity of coverage (Gutteling 2005), knowledge gap (Bonfadelli 2005) and cultivation hypotheses (Bauer 2005a,b). These in various ways complement evidence from a long tradition of 'agenda-setting' studies which have tried to map interactions between public, media and political agendas on science, health, environment and technology issues (e.g. Ader 1995; Hertog and Fan 1995; Chan 1999; Harwood *et al.* 2005).

At a more micro-sociological level (and often based on focus-group methodology rather than surveys), studies have shown how audiences use media information, media images and media 'templates' for talking about, making sense of and understanding science or—as is more often the case—particular scientific/medical or environmental issues and controversies (Burgess and Harrison 1993; Corner and Richardson 1993; Chapman *et al.* 1997; Miller *et al.* 1998; Henderson and Kitzinger 1999; Kitzinger 2000; Shaw 2002; Holliman 2004).

Conclusion

Given the profusion of media images, the diversity of media, the blurring of boundaries between media genres and most particularly the increasingly diverse nature of public consumption of media images, it would be futile to attempt to identify a simple relation-ship between science in the media and public perceptions, attitudes or understanding. This does not, however, prevent us from recognizing that the media serve as an import-ant reservoir of readily available images, meanings and definitions, and as an important public arena, where different images and definitions—'sponsored' by different agents, groups and interested parties—compete and struggle with each other. The media, in their broad and diverse totality, provide an important cultural context from which various publics draw both vocabularies and frames of understanding for making sense of science.

Hence, rather than looking for simple linear cause–effect relationships between media coverage and public concern, I have suggested that the media are better conceived of as a continuously changing cultural reservoir of images, meanings and definitions, on which different publics will draw for the purposes of articulating, making sense of and understanding science and the politics of science-based controversies and issues.

While much communication research has tended to focus on that most visible aspect of 'science and the media', namely media content, it has long been recognized by communication researchers that research should ideally address all the major components of the communication process (production, content and audiences) in order to understand the role of media and communication in relation to science in society and associated public understanding. In practice, research has, for logistical and other reasons, often focused on only one or two of the three major components of the communication process. One point has, however, been richly and abundantly demonstrated by the last 40 years of research on science communication and the media, namely that if we wish to understand the dynamics of 'science in society' the production/building of media agendas on science is at least as important a research focus as the more conventional concern about the media's influence on public understanding and opinion.

Perhaps the most exciting development in (research on) science communication in the last 10–15 years concerns the way in which the proliferation of new communication technologies has gone hand in hand with the move toward a 'democratisation' of—or at least attempts at increasing public engagement through, for example, 'consensus conferences' (e.g. Einsiedel and Eastlick 2000)—social debates about science, expertise, risk and environmental issues. An important part of this also concerns the way in which science has lost some of its traditional monopoly on expertise, and has had to accommodate other types of knowledge or perspective including ethical, moral, philosophical and religious 'ways of knowing' and ways of determining what is right and wrong.

New communication technologies provide easy public access to a vastly increased abundance of information, opinion and debate. They have also vastly increased the scope for those (from scientists and research organizations to government, companies and pressure groups) who wish to actively contribute to and influence public debate. These developments thus pose exciting new challenges for science communication research, including questions about adapting traditional communication research approaches to new forms of communication and media, about widening/narrowing the gap between 'information rich' and 'information poor', about differential public access to/use of scientific information, about 'democratizing' science communication, and about credibility and authenticity in a communication environment where traditional (scholarly or journalistic) quality-control standards may be less evident, and where traditional notions of the epistemological superiority of science, of scientific 'certainty', and of scientific authority no longer hold sway.

REFERENCES

Ader, C.R. (1995). A longitudinal-study of agenda-setting for the issue of environmental-pollution. *Journalism and Mass Communication Quarterly*, **72**(2), 300–11.

Albaek, E., Christiansen, P.M. and Togeby, L. (2003). Experts in the mass media: researchers as sources in Danish daily newspapers, 1961–2001. *Journalism and Mass Communication Quarterly*, **80**(4), 937–48.

Allan, S. (2002). *Media, Risk and Science*. Open University Press, Maidenhead.

Allan, S., Adam, B. and Carter, C. (eds). (2000). *Environmental Risks and the Media*. Routledge, London.

Altimore, M. (1982). The social construction of a scientific controversy: comments on press coverage of the recombinant DNA debate. *Science, Technology and Human Values*, **7**(4), 24–31.

Anderson, A. (1997). *Media, Culture and the Environment*. UCL Press, London.

Antilla, L. (2005). Climate of scepticism: US newspaper coverage of the science of climate change. *Global Environmental Change: Human and Policy Dimensions*, **15**(4), 338–52.

Bauer, M. (1998). The medicalization of science news—from the 'rocket-scalpel' to the 'gene-meteorite' complex. *Social Science Information Sur Les Sciences Sociales*, **37**(4), 731–51.

Bauer, M., Durant, J., Ragnarsdottir, A. and Rudolfsdottir, A. (1995). *Science and Technology in the British Press, 1946–1990: Results (1)*. Science Museum, London.

Bauer, M.W. (2005a). Distinguishing red and green biotechnology: cultivation effects of the elite press. *International Journal of Public Opinion Research*, **17**(1), 63–89.

Bauer, M.W. (2005b). Public perceptions and mass media in the biotechnology controversy. *International Journal of Public Opinion Research*, **17**(1), 5–22.

Bauer, M.W., Gaskell, G. and Durant, J. (eds). (2002). *Biotechnology: the Making of a Global Controversy*. Cambridge University Press, Cambridge.

Bauer, M.W., Petkova, K., Boyadjieva, P. and Gornev, G. (2006). Long-term trends in the public representation of science across the 'Iron Curtain': 1946–1995. *Social Studies of Science*, **36**(1), 99–131.

Bauer, M.W., Allum, N. and Miller, S. (2007). What can we learn from 25 years of PUS survey research? Liberating and expanding the agenda. *Public Understanding of Science*, **16**(1), 79–95.

Bell, A. (1994). Climate of opinion—public and media discourse on the global environment. *Discourse and Society*, **5**(1), 33–64.

Bonfadelli, H. (2005). Mass media and biotechnology: knowledge gaps within and between European countries. *International Journal of Public Opinion Research*, **17**(1), 42–62.

Borman, S.C. (1978). Communication accuracy in magazine science reporting. *Journalism Quarterly*, **55**, 345–6.

Boykoff, M.T. and Boykoff, J.M. (2004). Balance as bias: global warming and the US prestige press. *Global Environmental Change: Human and Policy Dimensions*, **14**(2), 125–36.

Brookes, R. (1999). Newspapers and national identity: the BSE/CJD crisis and the British press. *Media Culture and Society*, **21**(2), 247–63.

Bucchi, M. (1998). *Science and the Media*. Routledge, London.

Burgess, J. and Harrison, C.M. (1993). The circulation of claims in the cultural politics of environmental change. In: *The Mass Media and Environmental Issues* (ed. A. Hansen), pp. 198–221. Leicester University Press, Leicester.

Byrne, P.F., Namuth, D.M., Harrington, J., Ward, S.M., Lee, D.J. and Hain, P. (2002). Increasing public understanding of transgenic crops through the World Wide Web. *Public Understanding of Science*, **11**(3), 293–304.

Campbell, F. (1999). *The Construction of Environmental News: a Study of Scottish Journalism*. Ashgate, Abingdon.

Carvalho, A. (2007). Ideological cultures and media discourses on scientific knowledge: re-reading news on climate change. *Public Understanding of Science*, **16**(2), 223–43.

Chan, K. (1999). The media and environmental issues in Hong Kong 1983–95. *International Journal of Public Opinion Research*, **11**(2), 135–51.

Chapman, G., Kumar, K., Fraser, C. and Gaber, I. (1997). *Environmentalism and the Mass Media: the North–South Divide*. Routledge, London.

Clarke, J.N., and Everest, M.M. (2006). Cancer in the mass print media: fear, uncertainty and the medical model. *Social Science and Medicine*, **62**(10), 2591–600.

Collins, H.M. (1987). Certainty and the public understanding of science: science on television. *Social Studies of Science*, **17**(4), 689–713.

Condit, C.M., Achter, P.J., Lauer, I. and Sefcovic, E. (2002a). The changing meanings of 'mutation': a contextualized study of public discourse. *Human Mutation*, **19**(1), 69–75.

Condit, C.M., Bates, B.R., Galloway, R., Givens, S.B., Haynie, C.K., Jordan, J.W., Stables, G. and West, H.M. (2002b). Recipes or blueprints for our genes? How contexts selectively activate the multiple meanings of metaphors. *Quarterly Journal of Speech*, **88**(3), 303–25.

Corner, J. and Richardson, K. (1993). Environmental communication and the contingency of meaning: a research note. In: *The Mass Media and Environmental Issues* (ed. A. Hansen), pp. 222–33. Leicester University Press, Leicester.

Corner, J., Richardson, K. and Fenton, N. (1989). Textualising risk: TV discourse and the issue of nuclear energy. *Media, Culture and Society*, **12**(1), 105–24.

Corner, J., Richardson, K. and Fenton, N. (1990). *Nuclear Reactions: Form and Response in Public Issue Television*. John Libbey, London.

Cox, R. (2006). *Environmental Communication and the Public Sphere*. Sage, London.

Cronholm, M., and Sandell, R. (1981). Scientific-information—a review of research. *Journal of Communication*, **31**(2), 85–96.

Davis, A. (2000). Public relations, news production and changing patterns of source access in the British national media. *Media Culture and Society*, **22**(1), 39–59.

Dornan, C. (1990). Some problems in conceptualising the issue of 'science and the media'. *Critical Studies in Mass Communication*, **7**(1), 48–71.

Downs, A. (1972). Up and down with ecology—the issue attention cycle. *The Public Interest*, **28**, 38–50.

Dunwoody, S. (1979). News-gathering behaviors of specialty reporters: a two-level comparison of mass media decision-making. *Newspaper Research Journal*, **1**(1), 29–41.

Dunwoody, S. (1980). The science writing inner club: a communication link between science and the lay public. *Science, Technology and Human Values*, **5**, 14–22.

Dunwoody, S. (1986). The scientist as source. In: *Scientists and Journalists: Reporting Science as News* (ed. S.M. Friedman, S. Dunwoody and C.L. Rogers), pp. 3–16. The Free Press, New York.

Dunwoody, S. and Peters, H.P. (1992). Mass media coverage of technological and environmental risks: a survey of research in the United States and Germany. *Public Understanding of Science*, **1**(2), 199–230.

Dunwoody, S. and Ryan, M. (1983). Public information persons as mediators between scientists and science writers. *Journalism Quarterly*, **60**, 647–56.

Dunwoody, S. and Stocking, S. H. (1985). Social scientists and journalists: confronting the stereotypes. In: *The Media, Social Science, and Social Policy for Children* (ed. E. Rubinstein and J. Brown), pp. 167–87. Ablex, Norwood, NJ.

Durant, J. and Lindsey, N. (1999). *The Great GM Food Debate*. Report to the House of Lords Select Committee on Science and Technology, Sub-Committee on Science and Society. The Science Museum, London.

Durant, J., Evans, G.A. and Thomas, G.P. (1989). The public understanding of science. *Nature*, **340**, 11–14.

Durant, J., Hansen, A. and Bauer, M. (1996). Public understanding of the new genetics. In: *The Troubled Helix* (ed. M. Marteau and J. Richards), pp. 235–48. Cambridge University Press, Cambridge.

Durant, J., Bauer, M. and Gaskell, G. (eds). (1998). *Biotechnology in the Public Sphere: a European Sourcebook*. Science Museum, London.

Einsiedel, E.F. and Eastlick, D.L. (2000). Consensus conferences as deliberative democracy—a communications perspective. *Science Communication*, **21**(4), 323–43.

Endreny, P.M. (1985). News values and science values: the editor's role in shaping news about the social sciences. Unpublished PhD Thesis, Columbia University.

Ericson, R.V., Baranek, P.M. and Chan, J.B.L. (1989). *Negotiating Control: a Study of News Sources*. Open University Press, Milton Keynes.

Fowler, R. (1991). *Language in the News: Discourse and Ideology in the Press*. Routledge, London.

Friedman, S.M. (1986). The journalist's world. In: *Scientists and Journalists: Reporting Science as News* (ed. S.M. Friedman, S. Dunwoody and C.L. Rogers), pp. 17–41. The Free Press, New York.

Friedman, S.M. (1989). TMI: the media story that will not die. In: *Bad Tidings: Communication and Catastrophe* (ed. L.M. Walters, L. Wilkins and T. Walters), pp. 63–83. Lawrence Erlbaum, Hillsdale, NJ.

Friedman, S.M., Dunwoody, S. and Rogers, C.L. (eds). (1986). *Scientists and Journalists: Reporting Science as News*. The Free Press, New York.

Friedman, S.M., Gorney, C.M. and Egolf, B.P. (1987). Reporting on radiation: a content analysis of Chernobyl coverage. *Journal of Communication*, **37**(3), 58–79.

Friedman, S.M., Gorney, C.M. and Egolf, B.P. (1992). Chernobyl coverage: how the US media treated the nuclear industry. *Public Understanding of Science*, **1**(3), 305–23.

Friedman, S.M., Dunwoody, S. and Rogers, C.L. (eds). (1999). *Communicating Uncertainty: Media Coverage of New and Controversial Science*. Lawrence Erlbaum, Mahwah, NJ.

Gamson, W.A. and Modigliani, A. (1989). Media discourse and public opinion on nuclear power: a constructionist approach. *American Journal of Sociology*, **95**(1), 1–37.

Gardner, C. and Young, R. (1981). Science on TV: a critique. In: *Popular Television and Film* (ed. T. Bennett *et al.*), pp. 171–93. British Film Institute and the Open University, London.

Geller, G., Bernhardt, B.A., Gardner, M., Rodgers, J. and Holtzman, N.A. (2005). Scientists' and science writers' experiences reporting genetic discoveries: toward an ethic of trust in science journalism. *Genetics in Medicine*, **7**(3), 198–205.

Goodell, R. (1977). *The Visible Scientists*. Little, Brown and Company, Boston, MA.

Goodell, R. (1986). How to kill a controversy: the case of recombinant DNA. In: *Scientists and Journalists: Reporting Science as News* (ed. S.M. Friedman, S. Dunwoody and C.L. Rogers), pp. 170–81. The Free Press, New York.

Goodell, R. (1987). The role of the mass media in scientific controversy. In: *Scientific Controversies* (ed. H.T. Engelhardt and A.L. Caplan), pp. 585–97. Cambridge University Press, Cambridge.

Goodfield, J. (1981). *Reflections on Science and the Media*. American Association for the Advancement of Science, Washington, DC.

Greenberg, M.R., Sachsman, D.B., Sandman, P.M. and Salomone, K.L. (1989). Network evening news coverage of environmental risk. *Risk Analysis*, **9**(1), 119–26.

Gregory, J. and Miller, S. (1998). *Science in Public: Communication, Culture, and Credibility*. Plenum Press, New York.

Gutteling, J.M. (2005). Mazur's hypothesis on technology controversy and media. *International Journal of Public Opinion Research*, **17**(1), 23–41.

Hammond, P.B. (1997). Reporting pill panic. A comparative analysis of media coverage of health scares about oral contraceptives. *British Journal of Family Planning*, **23**(2), 62–6.

Hannigan, J.A. (2006). *Environmental Sociology*, 2nd edn. Routledge, London.

Hansen, A. (1991). The media and the social construction of the environment. *Media Culture and Society*, **13**(4), 443–58.

Hansen, A. (1993a). Greenpeace and press coverage of environmental issues. In: *The Mass Media and Environmental Issues* (ed. A. Hansen), pp. 150–78. Leicester University Press, Leicester.

Hansen, A. (ed.). (1993b). *The Mass Media and Environmental Issues*. Leicester University Press, Leicester.

Hansen, A. (1994). Journalistic practices and science reporting in the British press. *Public Understanding of Science*, **3**(2), 111–34.

Hansen, A. (2000). Claimsmaking and framing in British newspaper coverage of the Brent Spar controversy. In: *Environmental Risks and the Media* (ed. S. Allan, B. Adam and C. Carter), pp. 55–72. Routledge, London.

Hansen, A. (2006). Tampering with nature: 'nature' and the 'natural' in media coverage of genetics and biotechnology. *Media, Culture and Society*, **28**(6), 811–34.

Hansen, A. (2007). Producing education coverage—a study of education correspondents and editors in the national and regional press. In: *The Status of Teachers and the Teaching Profession in England: Views from Inside and Outside the Profession: Evidence base for the Final Report of the Teacher Status Project* (ed. L. Hargreaves *et al.*), pp. 67–83. Department for Education and Skills, London.

Hansen, A. and Dickinson, R. (1992). Science coverage in the British mass media: media output and source input. *Communications*, **17**(3), 365–77.

Hargreaves, I. and Ferguson, G. (2000). *Who's Misunderstanding Whom? Bridging the Gulf of Understanding Between the Public, the Media and Science*. ESRC, London.

Hargreaves, I., Lewis, J. and Speers, T. (2004). *Towards a Better Map: Science, the Public and the Media*. ESRC, Swindon.

Harwood, E.M., Witson, J.C., Fan, D.P. and Wagenaar, A.C. (2005). Media advocacy and underage drinking policies: a study of Louisiana news media from 1994 through 2003. *Health Promotion Practice*, **6**(3), 246–57.

Hellsten, I. (2003). Promises of a healthier future. *Nordicom Review*, **24**(1), 33–40.

Henderson, L. and Kitzinger, J. (1999). The human drama of genetics: 'hard' and 'soft' media representations of inherited breast cancer. *Sociology of Health and Illness*, **21**(5), 560–78.

Hertog, J.K. and Fan, D.P. (1995). The impact of press coverage on social beliefs—the case of HIV transmission. *Communication Research*, **22**(5), 545–74.

Hilgartner, S. (1990). The dominant view of popularization—conceptual problems, political uses. *Social Studies of Science*, **20**(3), 519–39.

Hilgartner, S. and Bosk, C.L. (1988). The rise and fall of social problems: a public arenas model. *American Journal of Sociology*, **94**(1), 53–78.

Holliman, R. (2004). Media coverage of cloning: a study of media content, production and reception. *Public Understanding of Science*, **13**(2), 107–30.

Jones, G., Connell, I. and Meadows, J. (1978). *The Presentation of Science by the Media*. Primary Communications Research Centre, University of Leicester.

Jones, R. (2001). 'Why can't you scientists leave things alone?'—science questioned in British films of the post-war period (1945–1970). *Public Understanding of Science*, **10**(4), 365–82.

Kitzinger, J. (1999). A sociology of media power: key issues in audience reception research. In: *Message Received: Glasgow Media Group Research 1993–1998* (ed. G. Philo), pp. 3–20. Longman, Harlow.

Kitzinger, J. (2000). Media templates: patterns of association and the (re)construction of meaning over time. *Media Culture and Society*, **22**(1), 61–84.

Kitzinger, J. and Reilly, J. (1997). The rise and fall of risk reporting—media coverage of human genetics research, 'false memory syndrome' and 'mad cow disease'. *European Journal of Communication*, **12**(3), 319–50.

Krieghbaum, H. (1967). *Science and the Mass Media*. New York University Press, New York.

LaFollette, M.C. (1982). Science on television: influences and strategies. *Daedalus*, **111**(4),183–97.

LaFollette, M.C. (1989). Eyes on the stars: images of women scientists in popular magazines. *Science, Technology, and Human Values*, **13**(3/4), 262–75.

Lambourne, R., Shallis, M. and Shortland, M. (1990). *Close Encounters? Science and Science Fiction*. Adam Hilger, Bristol.

Leask, J. and Chapman, S. (2002). 'The cold hard facts': immunisation and vaccine preventable diseases in Australia's newsprint media 1993–1998. *Social Science and Medicine*, **54**(3), 445–457.

Lewenstein, B.V. (1989). Magazine publishing and popular science after World War II. *American Journalism*, **6**(4), 218–34.

Lewenstein, B.V. (1995a). From fax to facts—communication in the cold-fusion saga. *Social Studies of Science*, **25**(3), 403–36.

Lewenstein, B.V. (1995b). Science and the media. In: *Handbook of Science and Technology Studies* (ed. S. Jasanoff), pp. 343–60. Sage, London.

Lowe, P. and Morrison, D. (1984). Bad news or good news: environmental politics and the mass media. *The Sociological Review*, **32**(1), 75–90.

Luther, C.A. and Zhou, X. (2005). Within the boundaries of politics: news framing of SARS in China and the United States. *Journalism and Mass Communication Quarterly*, **82**(4), 857–72.

McComas, K.A. (2006). Defining moments in risk communication research: 1996–2005. *Journal of Health Communication*, **11**(1), 75–91.

McCombs, M. (2004). *Setting the Agenda: The Mass Media and Public Opinion*. Polity Press, Cambridge.

McInerney, C., Bird, N. and Nucci, M. (2004). The flow of scientific knowledge from lab to the lay public—the case of genetically modified food. *Science Communication*, **26**(1), 44–74.

McQuail, D. and Windahl, S. (1993). *Communication Models for the Study of Mass Communications*, 2nd edn. Longman, London.

Mazur, A. (1981). Media coverage and public opinion on scientific controversies. *Journal of Communication*, **31**(2), 106–15.

Mazur, A. (1990). Nuclear power, chemical hazards, and the quantity of reporting. *Minerva*, **28**, 294–323.

Miller, D. and Reilly, J. (1995). Making an issue of food safety: the media, pressure groups, and the public sphere. In: *Eating Agendas: Food and Nutrition as Social Problems* (ed. D. Mauren and J. Sobal), pp. 305–36. Aldine de Gruyter, New York.

Miller, D., Kitzinger, J., Williams, K. and Beharrell, P. (1998). *The Circuit of Mass Communication: Media Strategies, Representation and Audience Reception in the AIDS Crisis*. Sage, London.

Miller, J.D. (1987). Scientific literacy in the United States. In: *Communicating Science to the Public* (ed. D. Evered and M. O'Connor), pp. 9–37. John Wiley and Sons, Chichester.

Moore, B. and Singletary, M. (1985). Scientific sources' perceptions of network news accuracy. *Journalism Quarterly*, **62**(4), 816–23.

MORI (2005). *Information about Science And Technology*. Available at **http://www.ipsos-mori.com/ polls/2005/nesta.shtml**.

Murrell, R.K. (1987). Telling it like it isn't: representations of science in Tomorrow's World. *Theory, Culture and Society*, **4**, 89–106.

Nelkin, D. (1995). *Selling Science: How the Press Covers Science and Technology*, 2nd revised edn. W.H. Freeman, New York.

Nelkin, D. and Lindee, M.S. (1995). *The DNA Mystique: the Gene as a Cultural Icon*. W.H. Freeman, New York.

Nimmo, D. and Combs, J.E. (1985). *Nightly Horrors: Crisis Coverage in Television Network News*. University of Tennessee Press, Knoxville, TN.

Nisbet, M. (26 April 2006). *Avian Flu and the Surveillance Function of the News Media*. Available at **http://www.csicop.org/scienceandmedia/birdflu/**.

Nisbet, M.C. and Lewenstein, B.V. (2002). Biotechnology and the American media—the policy process and the elite press, 1970 to 1999. *Science Communication*, **23**(4), 359–91.

Nisbet, M.C., Brossard, D. and Kroepsch, A. (2003). Framing science—the stem cell controversy in an age of press/politics. *Harvard International Journal of Press-Politics*, **8**(2), 36–70.

Nohrstedt, S.A. (1993). Communicative action in the risk-society: public relations strategies, the media and nuclear power. In: *The Mass Media and Environmental Issues* (ed. A. Hansen), pp. 81–104. Leicester University Press, Leicester.

Nowotny, H., Scott, P. and Gibbons, M. (2001). *Re-Thinking Science: Knowledge and the Public in an Age of Uncertainty*. Polity Press, Cambridge.

Peters, H.P. (1995). The interaction of journalists and scientific experts—cooperation and conflict between 2 professional cultures. *Media Culture and Society*, **17**(1), 31–48.

Pfund, N. and Hofstadter, L. (1981). Biomedical innovation and the press. *Journal of Communication*, **31**(2), 138–54.

Priest, S. (1990). Television's NOVA and the construction of scientific truth. *Critical Studies in Mass Communication*, **7**(1), 11–23.

Priest, S., Walter, L. and Templin, J. (1991). Voices in the news: newspaper coverage of Hurricane Hugo and the Loma Prieta earthquake. *Newspaper Research Journal*, **12**(3), 32–45.

Reese, S.D., Gandy, O.H. and Grant, A.E. (eds). (2001). *Framing Public Life: Perspectives on Media and Our Understanding of the Social World*. Lawrence Erlbaum, Mahwah, NJ.

Rogers, C.L. (1986). The practitioner in the middle. In: *Scientists and Journalists* (ed. S.M. Friedman, S. Dunwoody and C.L. Rogers), pp. 42–54. The Free Press, New York.

Rogers, R. and Marres, N. (2000). Landscaping climate change: a mapping technique for understanding science and technology debates on the World Wide Web. *Public Understanding of Science*, **9**(2), 141–63.

Royal Society (1985). *The Public Understanding of Science*. Royal Society, London.

Rothman, S. (1990). Journalists, broadcasters, scientific experts and public opinion. *Minerva*, **28**(2), 117–33.

Rothman, S. and Lichter, R.S. (1982). The nuclear energy debate: scientists, the media and the public. *Public Opinion* (September), 47–52.

Rothman, S. and Lichter, S.R. (1987). Elite ideology and risk perception in nuclear energy policy. *American Political Science Review*, **81**(2), 383–404.

Rowe, G., Frewer, L. and Sjoberg, L. (2000). Newspaper reporting of hazards in the UK and Sweden. *Public Understanding of Science*, **9**(1), 59–78.

Schelde, P. (1993). *Androids, Humanoids and Other Science Fiction Monsters: Science and Soul in Science Fiction Films*. New York University Press, New York.

Schneider, J.W. (1985). Social problems theory: the constructionist view. *Annual Review of Sociology*, **11**, 209–29.

Schoenfeld, A.C., Meier, R.F. and Griffin, R.J. (1979). Constructing a social problem—the press and the environment. *Social Problems*, **27**(1), 38–61.

Schudson, M. (1989). The sociology of news production. *Media, Culture and Society*, **11**(3), 263–82.

Schudson, M. (2000). The sociology of news production revisited (again). In: *Mass Media and Society*, 3rd edn (ed. J. Curran and M. Gurevitch), pp. 175–200. Arnold, London.

de Semir, V., Ribas, C. and Revuelta, G. (1998). Press releases of science journal articles and subsequent newspaper stories on the same topic. *Journal of the American Medical Association*, **280**(3), 294–5.

Shanahan, J. and McComas, K. (1999). *Nature Stories: Depictions of the Environment and their Effects*. Hampton Press, Cresskill, NJ.

Shaw, A. (2002). 'It just goes against the grain'. Public understandings of genetically modified (GM) food in the UK. *Public Understanding of Science*, **11**(3), 273–91.

Silverstone, R. (1985). *Framing Science*. The British Film Institute, London.

Singer, E. (1990). A question of accuracy: how journalists and scientists report research on hazards. *Journal of Communication*, **40**(4), 102–16.

Snow, C.P. (1993). The Two Cultures. Cambridge University Press, Cambridge.

Solesbury, W. (1976). The environmental agenda: an illustration of how situations may become political issues and issues may demand responses from government; or how they may not. *Public Administration*, **54**, 379–97.

Spector, M. and Kitsuse, J.I. (1977/1987). *Constructing Social Problems*. Aldine de Gruyter, New York.

Steinke, J. (2005). Cultural representations of gender and science. *Science Communication*, **27**(1), 27–63.

Tankard, J.W. and Ryan, M. (1974). News source perceptions of accuracy of science coverage. *Journalism Quarterly*, **51**, 219–25.

Tannenbaum, P.H. (1963). Communication of science information. *Science*, **140**, 579–83.

Ten Eyck, T.A. and Williment, M. (2003). The national media and things genetic—coverage in the New York Times (1971–2001) and the Washington Post (1977–2001). *Science Communication*, **25**(2), 129–52.

Tian, Y. and Stewart, C.M. (2005). Framing the SARS crisis: a computer-assisted text analysis of CNN and BBC online news reports of SARS. *Asian Journal of Communication*, **15**(3), 289–301.

Tichenor, P.J., Olien, C.N., Harrison, A. and Donohue, G.A. (1970). Mass communication systems and communication accuracy in science news reporting. *Journalism Quarterly*, **47**(4), 673–683.

Tomes, N. (2000). The making of a germ panic, then and now. *American Journal of Public Health*, **90**(2), 191–8.

Toumey, C.P. (1996). *Conjuring Science: Scientific Symbols and Cultural Meanings in American Life*. Rutgers University Press, New Brunswick, NJ.

Triandafyllidou, A. (1995). The Chernobyl accident in the Italian press: a 'media story-line'. *Discourse and Society*, **6**(4), 461–94.

Trumbo, C. (1996). Constructing climate change: claims and frames in US news coverage of an environmental issue. *Public Understanding of Science*, **5**(3), 269–83.

Tudor, A. (1989). *Monsters and Mad Scientists: a Cultural History of the Horror Movie*. Blackwell, Oxford.

Tumber, H. (ed.). (1999). *News*. Oxford University Press, Oxford.

Turney, J. (1998). *Frankenstein's Footsteps: Science, Genetics and Popular Culture*. Yale University Press, London.

Wade, S. and Schramm, W. (1969). The mass media as sources of public affairs, science, and health knowledge. *Public Opinion Quarterly*, **33**, 197–209.

Wallis, P. and Nerlich, B. (2005). Disease metaphors in new epidemics: the UK media framing of the 2003 SARS epidemic. *Social Science and Medicine*, **60**(11), 2629–39.

Washer, P. (2004). Representations of SARS in the British newspapers. *Social Science and Medicine*, **59**(12), 2561–71.

Weart, S.R. (1988). *Nuclear Fear: a History of Images*. Harvard University Press, Cambridge, MA.

Weatherall, M.W. (1996). The pill scare of October 1995, the media, and the public understanding of science. *British Journal of Family Planning*, **22**(3), 151–3.

Weigold, M.F. (2001). Communicating science—a review of the literature. *Science Communication*, **23**(2), 164–93.

Weigold, M.F. and Treise, D. (2004). Attracting teen surfers to science Web sites. *Public Understanding of Science*, **13**(3), 229–48.

Weingart, P., Muhl, C. and Pansegrau, P. (2003). Of power maniacs and unethical geniuses: science and scientists in fiction film. *Public Understanding of Science*, **12**(3), 279–87.

Weiss, C.H., Singer, E. and Endreny, P. (1988). *Reporting of Social Science in the National Media*. Russell Sage Foundation, New York.

Whitelegg, E., Holliman, R., Carr, J., Scanlon, E. and Hodgson, B. (2008). *(In)visible Witnesses— Investigating Gendered Representations of Scientists, Technologists, Engineers and Mathematicians on UK Children's Television*. Research Report Series for UKRC No. 5. UKRC, Bradford.

Wilkins, L. and Patterson, P. (eds). (1991). *Risky Business: Communicating Issues of Science, Risk, and Public Policy*. Greenwood Press, Westport, CT.

Wober, J.M. (ed.). (1992). *Television and Nuclear Power*. Ablex, Norwood, NJ.

Zehr, S.C. (2000). Public representations of scientific uncertainty about global climate change. *Public Understanding of Science*, **9**(2), 85–103.

Ziman, J. (1992). Not knowing, needing to know, and wanting to know. In: *When Science Meets the Public* (ed. B.V. Lewenstein), pp. 13–20. American Association for the Advancement of Science, Washington, DC.

■ FURTHER READING

- Allan, S. (2002). *Media, Risk and Science*. Open University Press, Maidenhead. This book is an accessible and highly readable introduction to key aspects of the 'media and science' literature. Starting with an overview of debates about science, media and the public, Allan discusses science in popular culture, science journalism and how the media portray environmental risks, HIV-AIDS, food scares and human cloning.

- Gregory, J. and Miller, S. (1998). *Science in Public: Communication, Culture, and Credibility*. Plenum Press, New York. This book remains one of the best and most readable introductions to debates and research on science in society. Surveying the history and context of 'public understanding of science', Gregory and Miller provide a wide-ranging discussion of science in public culture, popular science, popularization, the role of the media, museums and science communication initiatives. They conclude with a 'protocol for science communication'.

- Nelkin, D. (1995). *Selling Science: How the Press Covers Science and Technology*, 2nd revised edn. W.H. Freeman, New York. First published over 20 years ago, Nelkin's book remains one of the most perceptive discussions, from a critical sociological perspective, of journalistic practices and media roles in the public communication of science and science-based social controversies.

■ USEFUL WEB SITES

- **The Public Communication of Science and Technology (PCST) Network: http://www.upf.edu/pcstacademy/.** The International Network on Public Communication of Science and Technology (PCST) is a network of individuals from around the world who are active in producing and studying PCST. This web site brings together information on major science communication events and conferences, such as the (now) biennial PCST Conference, with a wealth of links to research, statistics, courses, etc. on science communication.

- **The British Association (for the Advancement of Science): http://www.the-ba.net/the-ba/.** The British Association (BA) has long played a leading role in the promotion of science in society and in initiatives on all aspects of public understanding and public communication of science. The information and links under this web site's 'Science in Society' heading should be particularly relevant (**http://www.the-ba.net/the-ba/ScienceinSociety/index.html**). The BA also publishes *Science and Public Affairs*, which promotes 'awareness of the importance of science, medicine, engineering and technology to our lives'.

- **The Why Files: http://whyfiles.org.** Founded in 1996, the mission of The Why Files is to 'help explain the relationship between science and daily life' and 'to explore the science, math and technology behind the news of the day, and to present those topics in a clear, accessible and accurate manner'. Based at the USA's University of Wisconsin–Madison, The Why Files offer an interesting example of science communication, which has also been evaluated through academic research [Dunwoody, S. (2001). *Science Communication*, 22(3), 274–82].

3.2

Models of science communication

Joan Leach, Simeon Yates and Eileen Scanlon

Introduction: what does communication have to do with science?

Everything that can be thought at all can be thought clearly. Everything that can be said can be said clearly.

Ludwig Wittgenstein

Words are also actions, and actions are a kind of words.

Ralph Waldo Emerson

These two great thinkers, while writing and living on two different continents, touched on the main problems of communication and on the contradictory impulses that we have when 'talking about communication'. There is a general sentiment that Wittgenstein's aphorism, while true in a sense, is more difficult to carry out than he indicates. It is possible for us to say or write things clearly, but this takes more work than we might expect at first glance; why is it that people misunderstand what I write or say even though they may hear or read me perfectly well? And why does Wittgenstein separate 'thinking' from 'saying'? What is the relationship between what we might say and what we might think? Taken together this suggests that what we think can be different from what we say, which can be different from what is understood. Whilst this may seem self-evident, it suggests a more complex conceptualization of communication than might otherwise have initially been apparent from Wittgenstein's short quote.

Emerson's quote is also troubling. How are words also actions? Is there not a strict division between words and things, and between what we say and what we do? But most importantly for us, do these quotes from a philosopher and a poet indicate anything about the relationship between science and communication, which is our particular task at hand? There is evidence to believe that thinking about these questions may help us to think about the particular problems with communicating science and perhaps even make us better communicators. This chapter will address these issues by introducing and critiquing several models of science communication.

Cultures of communication

C.P. Snow, in his famous 1959 lecture on the 'Two Cultures', relates the following incidents on communication that he personally experienced:

There have been plenty of days when I have spent the working hours with scientists and then gone off at night with some literary colleagues . . . constantly I felt I was moving among two groups— comparable in intelligence, identical in race, not grossly different in social origin, earning about the same incomes, who had almost ceased to communicate at all, who in intellectual, moral and psychological climate have so little in common that instead of going from Burlington House or South Kensington to Chelsea, one might have crossed an ocean. In fact, one had travelled much further than across an ocean—because after a few thousand Atlantic miles, one found Greenwich Village talking precisely the same language as Chelsea, and both having about as much communication with MIT as though the scientists spoke nothing but Tibetan. . . .

I am now thinking of the pleasant story of how one of the more convivial Oxford great dons . . . came over to Cambridge to dine. The date is perhaps the 1890s . . . Smith was sitting at the right hand of the President—or Vice-Master—and he was a man who liked to include all round him in the conversation, although he was not immediately encouraged by the expressions of his neighbours. He addressed some cheerful Oxonian chit-chat at the one opposite to him, and got a grunt. He then tried the man on his right hand and got another grunt. Then, rather to his surprise, one looked at the other and said, 'Do you know what he is talking about?' 'I haven't the least idea.' At this, even Smith was getting out of his depth. But the President, acting as a social emollient, put him at his ease, by saying, 'Oh, those are mathematicians! We never talk to them'.

C.P. Snow (1959)

C.P. Snow's diagnosis that we have fallen into a 'two-culture' system in which the two cultures—broadly cast as the sciences and the humanities—do not communicate with each other has become almost a truism. Of course, whilst not accounting for the professionalization of science communication as a discipline in its own right (see Holliman *et al.* 2009), the current context for science communication retains significant echoes of Snow's now infamous concerns. The ever-increasing specialization of academic disciplines (Fuller 1997), for example, means that students of science may choose to focus on what are defined as core subjects at the expense of a more holistic education; the humanities may be dropped, and vice versa.

Moreover, one of the cultures, the scientific culture, has been blamed, as C.P. Snow indicates, for speaking 'Tibetan', a euphemism for communicating in an obscure, specialist language. As a result, it has been suggested, not least by Walter Bodmer and colleagues (Royal Society 1985), that scientists need to communicate more effectively, more often and with a wider range of audiences.

First, it might help us to be clear about what we mean by communication. It is difficult to define exactly what is meant by 'communication'; we tend to be reflective on it when it does not work instead of when it does. We say that communication is 'transparent' to us; it is something that we do and usually do not give much thought to. There is even an additional problem. When we sit down to write about communication to you, we have to imagine you as an audience and communicate to explain what communication is about. Such conundrums are common when we start to study the social world, because what might first appear to be simple becomes complex. In order to make sense of this, we need to step away from everyday situations somewhat. To do that we typically build models and make descriptions of what communication processes might be. Two such models will be discussed and critiqued in this chapter with the aim of making the processes of science communication more apparent: the transmission model and the ritual model. We will then reflect on how such critiques have been addressed in more recent models of science communication.

What is a model?

We will be going on to talk about models of communication to examine their usefulness. Philosophers of science have an extensive literature discussing differences between types of theories and models including the differences between scientific theories and other kinds of theories. The following definition will suit our purposes, but be aware that libraries are full of books debating these definitions (Box 1).

BOX 1 DEFINITION OF A MODEL

A model is an interpretation of a formal system or a representation, normally by analogy, but sometimes by metaphor, of one thing by something else. Models may be made to help us remember things, help us imagine things or interactions we cannot see, explain situations or test phenomena that are not easily or directly testable.

Models are usually descriptions, interpretations or representations of what we take to be reality. The representation of the atom that we learn in physics textbooks is a model. Notice that the model of the atom looks like the model of the solar system; we model these by analogy.

This is an example of a heuristic model; it helps us to conceive of one thing in terms of something else. Alternatively, think about wind tunnel tests of model aeroplanes. The assumption here is that we can describe the wind flow around a smaller aircraft and thus know the airflow around a bigger aircraft. This is a test purpose of a model.

In what follows we are using the word model in its weaker sense of 'view' or 'perspective'. What we hope to gain by the examination of the models that follow is a sense of what analytic categories can usefully be deployed in the study of science communication.

Two models of communication

Two alternative conceptions of communication have been alive in American culture since this term entered common discourse in the nineteenth century. Both definitions derive, as with much in secular culture, from religious origins, though they refer to somewhat different regions of religious experience. We might label these descriptions, if only to provide handy pegs upon which to hang our thought, a transmission view of communication and a ritual view of communication.

Carey (1989, pp. 14–15)

This quotation is written from the perspective of a social scientist who became interested in the discipline called 'communication' in the United States. He reviewed the research of communication and the theories of how communication works and could discern clearly two dominant ways that have been used to model communication. He goes on to argue that:

The transmission view of communication is the commonest in our culture—perhaps in all industrial cultures—and dominates contemporary dictionary entries under the term. It is defined by terms such as 'imparting', 'sending', 'transmitting', or 'giving information to others'. It is formed from a metaphor of geography or transportation. In the nineteenth century but to a lesser extent today, the movement of goods or people and the movement of information were seen as essentially identical processes and both were described by the common noun 'communication'. The centre of this idea of communication is the transmission of signals or messages over distance for the purpose of control. . . .

The ritual view of communication, though a minor thread in our national thought is by far the older of those views—old enough for dictionaries to list it under 'archaic'. In a ritual definition, communication is linked to terms such as sharing, participation, association, fellowship, and the possession of a common faith. A ritual view of communication is directed not towards the extension of messages in space but toward the maintenance of society in time; not the act of imparting information but the representation of shared beliefs. If the archetypal case of communication under a transmission view is the extension of messages across geography for the purpose of control, the archetypal case under the ritual view is the sacred ceremony that draws persons together in fellowship and communality. The indebtedness of the ritual view of communication to religion is apparent in the name chosen to label it. Moreover, it derives from a view of religion that downplays the role of the sermon, the instruction and the admonition, in order to highlight the role of the prayer, the chant and the ceremony. It sees the original or highest manifestation of communication not in the transmission of intelligent information, but in the construction and maintenance of an ordered meaningful cultural world that can serve as a control and container for human action.

Carey (1989, pp. 14–15, 18–19)

How do these models apply to communicating science?

The transmission model of communication concentrates on three things: the sender of the message (the producer), the message itself (the content) and the receiver of the message (the audience). We reflect our tacit acceptance of this model when we say things like 'I got your message'. Figure 1 is a pictorial representation of what this model might look like.

Figure 1 The transmission model—sending and receiving a message.

How might this model be helpful or harmful for thinking about the relationship between science and communication? This model emphasizes linearity; information flows in a line from the sender to the receiver(s). When we think about communication using this model, we start with the sender and 'privilege' the sender in our analysis and theories about communication, often at the expense of the receiver(s). How do we take into account the role of the audience? Furthermore, this model does not address the context of communication. Where is there a place in this model for the where, the when and the how of communication? But, before attempting to correct these oversights, we need to look closer at what this model can achieve.

The transmission model

The transmission model consists of the sender, message and receiver. Let's consider each in turn.

Sender

Remember that this model is, as Carey (1989) puts it, quite 'mechanical'. The sender is the initiator of communication. In that sense, the sender could be an individual, as when two people are talking and one person asks a question. But the sender could also be newspaper, a radio broadcast, a television show and other sorts of communication providers. What can these senders—note the plural—do to improve communication? What do we mean by improving communication? In human terms, to improve the communication role of the sender, we make the distinction between written and oral discourse. To improve written discourse, the transmission model suggests that we improve the writer first.

Traditionally, the sender can be evaluated (and hopefully improved) in five areas. These include *invention*, *disposition*, *style*, *memory* and *delivery*. *Invention* is a category that can evaluate the ways that a speaker or writer decides what subject or topic to discuss or write about. When we evaluate *disposition*, we look to how the speaker or writer has organized messages and what choices he or she has made about what material to include or exclude. *Style* is a difficult category to describe analytically. However, when we refer to the 'style' of a writer, we frequently mean word choice, poetic licence and creative input. *Memory* is a misleading word as it usually refers to the sender's (remember, writer, speaker or media source) access to information. This might be memorization in a speaker, but it may also refer to a writer's awareness of current events, access to information (e.g. see Trench, Chapter 4.2 this volume; Gartner 2009), or the ability of a radio broadcast to refer to other communications. *Delivery* is the way in which a speaker, writer or media source delivers the message; this includes issues from the use of slides, to the frequency of transmission of a radio broadcast, to access to a networked PC.

Message

Analysing a message or attempting to improve it under the transmission model again focuses on the mechanical and linear components of the communicative act. Here, the communication under scrutiny is the tangible component: what is seen, heard or read. Classically, messages have been evaluated by *genre* or type of message. There are minimally three genres of messages: *forensic, epideictic* and *deliberative. Forensic* messages are informative in scope. *Epideictic* messages are those that give praise or cast blame. *Deliberative* messages are those that make an argument or weigh different arguments on the same subject.

In science we see all three of these messages in a variety of different contexts. Laboratory reports are typically forensic as they give information about what was done in the laboratory (Medawar 1999). They can be evaluated and improved by judging the accuracy, completeness or presentation of the information. Letters to *Nature* or *Science* are deliberative. They are making an argument for the interpretation of data. They may express agreement or disagreement with other interpretations of data, but the important issue is that they make an argument. Articles about science in newspapers or magazines such as *New Scientist* are typically epideictic. They celebrate and praise scientific invention and discovery. Critics use these categories to talk about scientific communication quite often (even if they are not aware of it). One critique of popular science, for example, is that its messages are too epideictic and readers are not exposed to disagreements about the scientific process (e.g. see Nelkin 1995). These categories can be quite useful for thinking about the quality of communication in science.

Receiver

This is quite possibly the most important facet of the transmission model, although it is also the most neglected and the most difficult to analyse using this model. No communication is effective if the receiver does not get the message. Thus, this may be the most important category for analysis. Receivers of scientific information are made up of diverse publics (see Hornig Priest, Chapter 6.1 this volume). But how do they receive scientific information? Many studies suggest that the context of reception is the most determinative factor of receiving messages. Yet you will notice that nowhere does this model account for context. The receiver may not get messages if they are distorted by the communication medium; radio broadcasts may be impossible to listen to due to interference. Likewise, information about science might not be readable because it is not in a language the receiver has been trained to read.

Cognitive models of reception concentrate on the receiver's prior knowledge. This is surely important in science communication. A biologist will be a better receiver for information about local ecosystem damage than will a physicist. In some situations, non-scientists may be better receivers of scientific knowledge than scientists as their prior knowledge helps them to better evaluate the ethical or moral implications of scientific inventions or discoveries (see Holliman and Scanlon, Chapter 6.3 this volume).

The ritual model

If we think of communication in a transmission model, we must concentrate on the sender, message and receiver. In contrast to the transmission model, the ritual model emphasizes not the communication of messages in space from sender to receiver, but the communal sharing of information in a particular cultural context. Carey (1989) suggests that the ritual model is best thought of in terms of a 'theatre of communication'.

The following aspects of the ritual model help to analyse pieces of communication and make suggestions for more effective communication. It differs most from the transmission model in that these five aspects of communication can be treated equally or the analysis can centre on any one of the aspects. However, focusing on one aspect of the communication analysis changes the way in which you might analyse the other aspects.

The first thing that you will notice is that we need to have different categories to analyse the 'theatre' of communication. We have to take into account the entire stage, not just the sender, message and receiver. We will draw on the following five terms—act, agency, agent, purpose and scene (from Burke 1969)—to talk about communication under this model.

Act

What is an 'act' of communication? Communication includes many varieties of acts: arguing, persuading, pleading, even laughing can be an act of communication. Thinking about communication as an act broadens what can be considered as communication. When you enter an art museum, for example, are what you see objects of art or acts of communication, communicating beauty, relevance and design? Some acts of communication are immediate and instant. When you ask for the salt to be passed at the dinner table the communication act lasts for one moment only and is directed at an immediate audience. Others, however, are more remote and last for indeterminable durations. Literature can be seen as an act of communication. Who can be said to be the audience for Chaucer? How long will Chaucer continue to communicate to audiences of readers?

Purpose

What is the purpose of communication acts? This is a difficult question because sometimes it is impossible to know. When a friend has a bad day and answers your phone call in a cursory and even angry manner, is the purpose of his communication to let you know that he has had a bad day or it is to let you know that he is angry with you? Likewise, when reading about science, is the purpose of the communication to inform you of the latest development or to make you appreciate the 'progress of science'? It is probably best to think of communication as having multiple purposes and not just one.

Scene

This is the context of communication acts and can be one of the most important aspects to look at when analysing an act of communication. Where and when does the communication take place? How many witnesses to the act of communication are there? Think of the scientific laboratory and an interchange between two scientists. One is asking the other for information. What other elements make up this scene? There may be a laboratory apparatus present to which one scientist refers. There may be additional people working in the lab who overhear the conversation. Describing communication in this theatre-like way helps us to realize that communication never happens in a vacuum but is always in some kind of context.

Agent

This category can be thought of in an analogous fashion to the 'sender' and the 'receiver' in the transmission model. Who are the agents in any communication act? How many are there? What kind of impact can they have on the communication process? Answering these questions carefully (as some agents are not immediately obvious) is important to the understanding of the fundamental roles of communicators.

Agency

While identifying agents in communication seems somewhat easy, exploring their roles in communication acts is somewhat more difficult. It may be possible for a first-year chemistry student to explain chemical bonding to her colleagues. However, her position as student precludes her giving formal instruction to her peers and makes her communication fall under the category of 'friendly advice'. Such social roles dictate much of our communication, or at least constrain the types of things we write and say. Science journalists cannot pass judgement on the efficacy of an AIDS drug, but they may communicate their doubts about the frequency with which scientists arrive at a 'new wonder drug' for AIDS. Thinking about agency as a category constraining and motivating certain types of communication is useful for investigating the social aspects of communication.

Applying the ritual model

One possible scene for a particular scientific paper is the place where the journal or publication is read. By this, we do not mean the physical place where this article was read, but the place in the community. For example, the medical community is the place where the *British Medical Journal* is mainly read, but a journal like *Nature* prides itself on being an interdisciplinary journal so it is read by specialists in a particular field, by those working directly on similar topics and by those less directly concerned. The act of communication in this scene is both the writing and the reading of an article for a major professional journal. The agents for this act of communication are the writer, the reader and the journal. Remember, this model corrects the earlier one-way tendency of the sender–message–receiver model. But what is the agency for each of these agents? The social role of the author here is very important, as one might argue that the author must already have authority in the community to be included in an issue of any particular journal. *Nature* is one of the leading journals internationally, and of course, within the UK, which gives it a position such that its conclusions are cited in the popular press and its positions tend to inform the received view. What agency does the reader of *Nature* have in reading an article? As these readers might be part of the active specialist scientific community, do the readers have sufficient agency to make use of the scientific information that the author provides? This raises the question, what is the purpose for the author to have communicated the scientific information that he or she does in the article? As you may begin to recognize, using the ritual model to look at these issues allows us to see science communication as more than the transfer of facts from one sender to one receiver.

Scientists may well be puzzled by the use of the term model or theory in the study of communication. We use the word model in this context simply to describe the analytic categories which are evoked by the use of one model or another. Also, we do not mean to suggest by this discussion that the ritual model is necessarily always the more appropriate one to use. We simply hope to have raised awareness of ways of analysing communication beyond that of the application of the transmission model—the one which is admittedly the most widely used.

Improving communication

Now the issue becomes, how does the application of these models help us to improve communication? Models, as we have seen, are good for helping us to analyse situations and comment on the success of communication in those instances. But sometimes we want to do more than analyse. We would like to intervene and make the situation better. We might want to intervene to clarify issues, change the tenor of the debate, and focus more on the communication issues in order to make everyone's interests in the situation clear. Both models suggest ways in which to improve communication about science. Each model is also limited, however. The transmission model, in its emphasis on the sender, message and receiver of communication, suggests that to make communication better we must concentrate on these areas. The ritual model is very good for analysing our assumptions about communication and the way a particular communicative event unfolds, but it is not very good for telling us how to change our communication.

The differences between these two models, both in their ideas about how communication is best defined and how communication is best practised, should start to become clear. The transmission model of communication is assumed by the authors and teachers of most 'how-to' communication instructional handbooks and courses. They assume that if one can concentrate on making the sender and the message better, then the receiver will have an easier time understanding the message. Carey (1989) points out just how dominant this way of thinking has become and how it is deeply rooted in our ideas of trade and the history of communication technologies like the telegraph. He points out that the ritual view has been less explored by communication scholars and students alike. What might be evident at this point is the desire to combine the two models and talk about communication in a more expansive sense. We need to acknowledge that individuals need to acquire and practice skills that make them better 'senders' and make their 'messages' better. However, we also need to do some thinking about the contexts, audiences and constraints placed on the communication process.

Carey tries to find ways to talk about communication in a more expansive sense and to take account of the models that we have already studied, most importantly, by offering a revised definition:

Communication is a symbolic process whereby reality is produced, maintained, repaired, and transformed.

Carey (1989, p. 23)

In this definition, Carey places us in the middle of a debate about the relationship between language and the world in which we live. This debate, with arguments coming from scientists, philosophers, social scientists and communication specialists, has raged

---

BOX 2 ROSES, HUMAN FERTILIZATION AND COMMUNICATION THEORY

A famous line by Kenneth Burke is that 'things are the signs of words'. What could he have possibly meant by this? This also brings us back to thinking about the opening quote from Emerson that 'words are also actions, and actions are a kind of words'. What is the relationship between words and things?

Philosophers like to talk about this question and they usually use the example of a table (perhaps because philosophers spend so much of their time reading on tables). They begin by pointing out that a table does not have to be called a table. In fact, it is not a 'table' in other languages. So, they make the rather obvious point that names for things do not have any necessary relationship to the things themselves. In fact, the names for things are conventions which we have adopted over long periods of history. But this is where the problems begin. Remember the famous question from Shakespeare, 'Would a rose by any other name smell as sweet?'.

Communication theorists think that it would not. This is because the word rose has become associated with many other values and conventions in 'Western' culture like love, nobility, beauty and romance. Thus, someone might arrive home late to their partner who has been waiting for them for an hour after they have cooked a three-course meal and hand them a single rose. What the partner might discern from this statement is that this is a token of affection, gratitude and a romantic sensibility. And all this, without ever having to utter a word.

In this way, a thing is a sign of a word. This is believable enough in relation to poetry, but what about the relationship between scientific words and things? Also, how can words be actions? Can words be more powerful than things?

One example that can make this clearer is from 20th-century biology. Evelyn Fox-Keller recounts the history of research into fertilization. In the 1950s, initial research on this topic described the egg as 'passive', 'waiting' to be fertilized by the sperm which 'battled' the harsh internal environment of the female body as it 'struggled' to 'penetrate' the egg.

Such descriptions led researchers to look into the mechanisms by which the sperm was able to 'penetrate' the egg. However, they were mostly unsuccessful. Eventually, some researchers noticed that the egg itself might be 'active' instead of passive; the egg might be leading the sperm by some kind of chemical means.

Today, we see conception in terms of 'equal opportunity'. The sperm and egg are both active and passive in the process of conception. Talking about the egg as passive made a certain kind of research possible whilst also ruling out other avenues of investigation. In this way, talking about things is a form of action. From this example, we might begin to see just how complicated the relationship can be between words and things and how important the communication process is to science.

---

for nearly 2500 years. Box 2 gives some of the background to this problem. It also introduces some of the key terms in contemporary communication theory and linguistics. Some of the language used in these debates will be strange and unfamiliar. However, now you have heard or read about some of these issues for the first time, you will probably recognize them later in other contexts in your reading about communication.

At this point, some readers might want to argue with the contention that the transmission model has been shown to be inadequate at all. After all, as Carey (1989) points out, it is a simple model which has been applied in many situations. It has also been widely used in the study of communication which is mediated by a technology.

In 1949, Claude Shannon and Warren Weaver published an influential book, *The Mathematical Theory of Communication*, which is oriented to the engineering aspect of communication, and incorporates the additional elements of 'signal' and 'noise', as well as transmitting and receiving devices. This version of the transmission model arose out of the concerns of its time, in particular technological developments and the concurrent concern with the nature and effects of propaganda.

In effect, however, the problem with the transmission model arises if it is imagined that it is a complete description which covers all aspects of communication. Carey's writing about the ritual approach points out that there are other aspects to be considered in any analysis of communication. One specific shortcoming of the simple transmission model is that it ignores the role played by *mediation*. That is to say, the transmission model assumes that a point of information in the sender's mind remains the same while being transmitted and is received as the same point of information. If this is true, all that remains for science communication is to put effort into the ability to mirror messages from the sender to the receiver, and to keep the mediation as simple as possible. This assumption seriously misunderstands any process of communication by treating it in this mechanistic sense. The differences between two human beings makes the notion of a perfect copy transfer from sender to receiver almost impossible. In addition, the mediators, when they are people, add yet more human beings to the mix. For example, advertising is a powerful model of persuasive communication. How could we understand the activities of advertising agencies in a simple transmission model? They help producers in a variety of ways. For example, they help in choosing the right media mix for their campaigns. To understand such activities we would need to overlay some kind of social or institutional analysis on top of the simple transmission approach.

Mediation in science communication

The transmission model is often the common sense way in which most people understand human interaction to take place. Not only is this a model of human communication but it also represents the general starting point for most engineering approaches to communication. For example, when we talk to each other a person (the transmitter) speaks or writes (the signal) something (the message) to another person (the receiver). This model could also be applied to newspaper communication. In this case a journalist (the transmitter) writes an article (the message) which is printed in the paper (the signal) which is then read by the receiver.

Though this seems an obvious, logical and 'common sense' model of communication, we will find we need a much more sophisticated model if we are to get to grips with the workings of contemporary mass communications. We therefore need to consider the 'transmission model' critically and ask: is Figure 1 adequate to explain the full complexity and diversity of science communication in the mass media?

One of the important ways in which social science works is to question the 'common sense' understandings we have of how the social world around us operates. When something is described as being 'common sense' this tends to alert social scientists to a situation where the complexities of the social world are being ignored or even deliberately covered over. The simplicity and obviousness of the 'transmission model' hides or ignores the actual complexity of human communication—especially where the communication takes place via complex communications technology such as newspapers, television or the internet.

One possible analogy would be to see the transmission idea as a 'first go' at modelling communication. It captures some of the key ideas: a 'transmitter', a producer of the message, a 'signal', a medium for the message, a 'receiver' or a consumer of the message, and of course the message itself. From such a start, it is possible to then move towards a more sophisticated understanding of communication, for example, by critiquing this model.

In critiquing models, social scientists often try to unpack the underlying assumptions behind the model—these are not always obvious. In the case of the 'transmission model' there are three important assumptions. First, the model assumes that communication is about the transmission of information. Second, the model assumes a very specific definition of 'successful communication'. As long as the message gets through 'clearly' then the communication was successful. In other words, as long as the receiver clearly receives the message as it was originally sent then the communication has succeeded. Anything done by or to the transmitter, receiver, signal or message introduces 'interference' or 'noise' and the communication is likely to fail. These two assumptions reflect the engineering and technological basis for this model of communication, and also aspects of the deficit model that has been used to model 'science–society' relations (e.g. see Irwin, Chapter 1.1 this volume for a critique).

The most important problem with taking success to be indicated by the fact that the message was clearly and unambiguously received is this: we all know from experience that human communication is hardly ever perfect. Communications researchers concerned with spoken interactions have documented the tragic failures caused by misunderstandings between air traffic controllers and pilots. Even though all the technology worked and the words were transmitted clearly, they were not understood as they were intended to be understood. In the case of the air traffic interactions the 'noise' was not caused by the equipment (i.e. poor radio transmission) but by complex misunderstandings of the meanings and purpose of words. Though the words were transmitted clearly *the meaning was not.* When we start to explore communication between human beings we have to deal with the fact that meanings are not fixed and often depend upon the context of the communication. There are many elements to the context of human communication. First, there is obviously the 'topic'—what the communication is about. Second, there is the social context of the communication, where is the communication taking place, who is talking, what is their relationship, etc.? Third, there is the medium, is it written, spoken, on television, the internet, etc.? When human beings communicate, they take such factors into account, although often unconsciously once they have (re)learnt 'how to read' a novel form of communication. In fact, without understanding the context, most human communication would 'fail'.

The third problem with the model is that it assumes communication is just about 'messages' where in fact many acts of human communication are about 'doing' things. Consider again the air traffic controllers. They are not just passing messages, they are giving instructions to pilots about what to do—the messages should lead directly to actions. An even more overt example of 'doing' with language is that of a wedding. By saying 'I do' or some similar phrase, you become married. Saying something else, even 'I do not', would mean that you did not get married. Communications researchers call the human communication which causes things in the world to change, 'speech acts'. Speech acts can be direct, as in the marriage example, or indirect. An indirect speech act might be for someone, sitting in a meeting, to say 'it is very cold in here' at which point a colleague closes a window. The communication led to a desired event—the closing of the window. In fact, when we analyse many human interactions across different 'media' (e.g. speech, writing, print, television, computers) there are few cases where the communication is not some kind of 'doing'. These kinds of 'doing' can include persuading, ordering, apologizing and so forth; we should therefore talk about 'communicative acts' taking place during human communication.

All of these things are true of science communication. In many cases, journal articles are as much acts of persuasion as reports of results. When governments engage in the act of communicating scientific information, for example to suggest that citizens should eat at least five portions of fruit or vegetables a day, they do so in the hope that it will directly lead to changes in citizens' behaviour, in this case to eat a more balanced diet. At a deeper level, communicating science, especially amongst scientists, is as much a central part of 'doing science' as experimentation.

Contemporary models of science communication

In this section you will be introduced to some of the conceptual tools which social scientists working in the areas of communication, culture and media studies use. These have been developed in order to understand communication mainly through mass media, but they can be applied to other forms of science communication.

Given the arguments so far, it should be apparent that we need to develop Figure 1. Figure 2 does so, in part by simply replacing the terms 'transmitter', 'signal', 'message' and 'receiver' with the new terms 'producers', 'texts', 'media' and 'consumers'.

There are several reasons for these new terms, some of which will become clearer as the discussion continues. For now we can state the following:

1. These new terms are not so directly tied to engineering-based conceptions of communication—these are terms more common to the social sciences.

Figure 2 A communicative act.

2. When we communicate we create objects—letters, newspapers, photographs, television programmes, blog postings and so forth—and the term 'text' is used by communications researchers to refer in general to such objects—we shall return to why this is so later.

3. Communication is a purposeful human activity—a 'doing': we actively produce and consume texts rather than passively transmit and receive signals.

4. The plural terms indicate that communication often involves groups of people, where there are several producers, of several texts which are then used by several consumers—think of all the people who produce a hard copy newspaper, all the articles in it, all the copies of it and all the people who read it.

We can now think of groups of people actively and purposefully communicating in a more social way than that provided by Figure 1. Have we then completed the picture? The answer is, unsurprisingly, no.

If we think about the groups of people who produce and consume texts we must consider how these groups are organized. If we take newspapers and television as examples again, then the producers of the texts—media professionals, such as journalists and television producers—are part of larger organizations (e.g. the BBC or the *Guardian* newspaper group). Social scientists often refer to complex organizations as 'institutions' in order to indicate the way in which the structure and purpose of the organization has an effect on its members' behaviour, in the same way that the institutions of school and hospital have an effect on the behaviour of their members.

Think about the consumers of newspapers and television programmes—these people are also part of larger groups such as local communities, occupational groups, social classes or even distinct cultures and nations. The types of social groups which form such 'audiences' are often referred to as 'communities' to reflect that they are groups brought together by some common feature or activity, from the area in which they live, to the types of television programme they watch.

Lastly, if we consider the texts themselves, they come in a wide variety of forms, from the spoken word through to blog postings. Each of these texts is therefore partly structured and constrained by the choice of medium. Put crudely, even in the converged media landscape, hard copy newspapers do not have moving pictures. Also few texts, if any, carry only one message and do only one thing. It is this diversity of different media, messages and actions which is covered by the term 'text'. The term text derives from 'texture' which itself derives from the Latin for woven fabrics. When we communicate we produce texts, that is, objects which interweave ideas, social relationships and the communications medium into a whole. We must therefore add these elements to revise Figure 2.

Is Figure 3 then the whole answer? It does bring together the main features but it lacks a few further details. To complete the figure we need to add more elements by considering a couple of further questions. First, are the people who produce the texts always the ones who produced the information or messages contained in the texts? In the case of a scientist reporting on their work, though there is the institutional structure of the journal and review processes, the scientist is still a key participant. In the case of the newspaper report on the work of the same scientist, the link is much less clear,

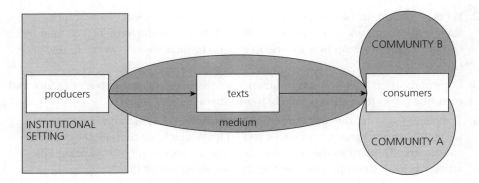

Figure 3 A communicative act with institutional settings and communities.

especially if the journalist or reporter does not interact with the scientist, e.g. by referring to a press release that describes the work (see, e.g., Trench, Chapter 4.2 this volume; Holliman 2004).

Another example would be the reporting of war or sport. The events—such as battles and world athletic records—are enacted by specific people. However, the communication of these events to others takes place via a third party—the reporter working for a media institution. Reporters, and the institutions they work for, therefore act as 'mediators' of the news. In many respects the institutions which mediate mass media communications provide information about one community, say scientists from that discipline, to other communities, say non-scientists, policy-makers or scientists working in other disciplines. Communications media are one of the ways in which communities and societies can reflect on the very events and ideas which shape them.

The second question is this. How are texts affected by other on-going communication? This is harder to make clear. Whenever we communicate, even when we speak, we are aware of the previous communications that have taken place. We are aware of previous speakers' statements, previous journalists' articles, previous scientists' publications. The constant interconnection between texts and previous texts, and the way in which this is tied to the whole process of communication, is often referred to as a 'discourse'. There are a number of definitions of 'discourse'. Put very simply, a discourse is a set of ideas, texts and ways of thinking and acting that are linked to social institutions. One can therefore think of 'the discourse of physics', or 'the discourse of sociology', or 'the discourse of UK politics'. Adding these elements to our figure just about completes the picture.

Figure 4 makes it clear that nearly all 'messages' are 'mediated' in one way or another. In fact very few messages are directly transmitted from sender to receiver in the manner implied by Figure 1. A great deal of the social and technological basis of communication could therefore be seen as adding 'noise'.

We can now draw two figures based upon Figure 4 which model two different science communication acts. In Figure 5 we have a scientist communicating to other scientists through a journal article. In Figure 6 we have the same scientific findings communicated to the public via a newspaper. Take a moment to compare the two.

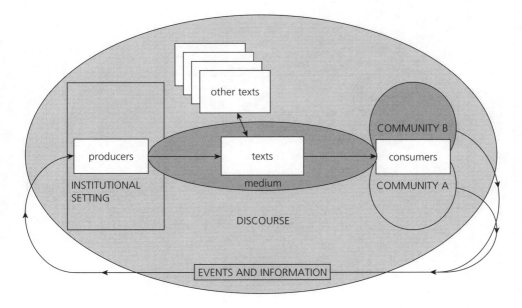

Figure 4 A communicative act in context.

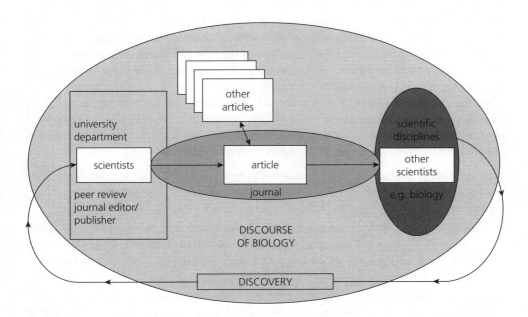

Figure 5 Publishing a peer-reviewed scientific article.

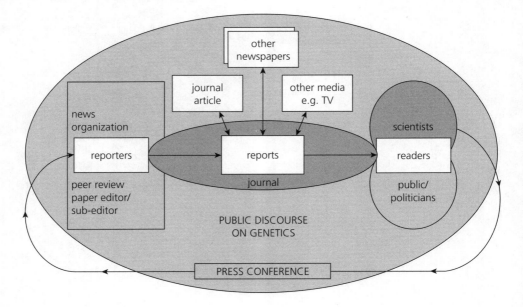

Figure 6 Reporting science in news media, promoted through a press conference.

Concluding thoughts

By asking a few key questions it should now be apparent that communication, especially mass mediated communication, is far from being the simple affair portrayed in Figure 1: the 'transmission model'. Rather, models of science communication need to take into account the many and varied agents involved in the communicative process, addressing the motivations and constraints of the institutions, discourses and communities thereof, the context of the communicative act, and so on.

Science communication and media research is an attempt to study these issues, e.g. by examining the main three aspects of the process of science communication in the context of their complex inter-relationships (see Hansen, Chapter 3.1 this volume; Thompson 1999; Kitzinger 2000; Holliman 2004). Such work addresses the:

- production of media texts, e.g. the day-to-day work of media professionals, including journalists, public relations officers, technology for making digital television programmes, structure and running of news and media companies and so on;

- consumption of media texts, e.g. audiences, increasingly seen as 'viewers' and 'users' (see Bennett, Chapter 5.1 this volume) and their preferences, the role of media in citizens' everyday lives, the social and psychological processes involved in watching or reading media texts;

- media texts themselves, e.g. the structure and content of newspaper articles, films and novel forms of science communication and the language used to describe science news events.

Any one study may focus upon one of the three areas, but the discipline of science communication and media studies as a whole attempts to bring all three elements together as in the model presented in Figure 4. Producers of texts are often based in specific institutions (e.g. newspaper offices, science laboratories, science journals) and have specific relationships to the other members of that institution (e.g. editors, other colleagues).[1] The institution itself will have relationships to other institutions (e.g. other newspapers or laboratories, government agencies). We might call the model presented in Figures 4, 5 and 6 the 'media studies model' of (science) communication. You can think of it as an amalgam of the transmission and ritual models which potentially has the features necessary to analyse the complex situations which need to be considered in contemporary science communications.

Acknowledgement

This material has been developed from that originally prepared for the Open University postgraduate science courses *Science and the Public* and *Communicating Science*.

■ REFERENCES

Burke, K. (1969). *A Grammar of Motives*. University of California, Berkeley.

Carey, J. (1989). A cultural approach to communication as culture. In: *Communication as Culture: Essays on Media and Society* (ed. J. Carey), pp. 13–36. Unwin Hyman, Boston, MA.

Fuller, S. (1997). *Science*. Open University Press, Buckingham.

Gartner, R. (2009). From print to online: developments in access to scientific information. In: *Practising Science Communication in the Information Age: Theorizing Professional Practices* (ed. R. Holliman, J. Thomas, S. Smidt, E. Scanlon and E. Whitelegg). Oxford University Press, Oxford.

Holliman, R. (2004). Media coverage of cloning: a study of media content, production and reception. *Public Understanding of Science*, **13**(2), 107–30.

Holliman, R. (2008). Communicating science in the digital age—issues and prospects for public engagement. In: *A Reader for Technical Writers* (ed. J. MacLennan), pp. 68–76. Oxford University Press, Toronto.

Holliman, R., Thomas, J., Smidt, S., Scanlon, E., and Whitelegg, E. (eds) (2009). *Practising Science Communication in the Information Age: Theorizing Professional Practices*. Oxford University Press, Oxford.

Kitzinger, J. (2000). Media templates: patterns of association and the (re)construction of meaning over time. *Media, Culture and Society*, **22**, 61–84.

Medawar, P. (1999). Is the scientific paper a fraud? In: *Communicating Science: Professional Contexts* (ed. E. Scanlon, R. Hill and K. Junker), pp. 27–31. Routledge, London.

1. Of course, the emergence of new forms of user-generated science communication means that many producers are not allied to a particular institution (see Holliman 2008).

Nelkin, D. (1995). *Selling Science: How the Press Covers Science and Technology*, 2nd revised edn. W.H. Freeman, New York.

Royal Society (1985). *The Public Understanding of Science*. Royal Society, London.

Shannon, C.E. and Weaver, W. (1998/1949). *The Mathematical Theory of Communication*. University of Illinois Press, Urbana, IL.

Snow, C.P. (1959). *The Two Cultures and the Scientific Revolution*. The Rede Lecture. Cambridge University Press, Cambridge.

Thompson, J. (1999). The media and modernity. In: *The Media Reader: Continuity and Transformation* (ed. H. Mackay and T. O'Sullivan), pp. 13–27. Sage, London.

■ **FURTHER READING**

- Bucchi, M. (1998). *Science and the Media: Alternative Routes to Science Communication*. Routledge, London. Concentrating on popular forms of science communication, this textbook provides helpful accounts of analytical models, and describes a series of case studies, including the 'cold fusion' episode.

- Kitzinger, J. (2000). Media templates: patterns of association and the (re)construction of meaning over time. *Media, Culture and Society*, **22**, 61–84. This paper introduces the analytical concept of a 'media template', one that links the elements of production, content and reception to make sense of the (re)construction of meaning over time. In this paper, the concept is applied to media reporting of child sexual abuse.

- McQuail, D. (2005). *McQuail's Mass Communication Theory*, 5th edn. Sage, London. This bestselling introductory textbook, which considers different ways to analyse communication in a wide range of forms, including new media, is currently in its fifth edition.

- Thompson, J. (1999). The media and modernity. In: *The Media Reader: Continuity and Transformation* (ed. H. Mackay and T. O'Sullivan). Sage, London. In this chapter Thompson explores a social theory of communication by examining the related elements of production, content and reception. In developing his theory he considers how new (and traditional) forms of communication are reconstituting the spatial and temporal dimensions of social life.

■ **USEFUL WEB SITES**

- **Models of science communication—how many can there be?:** **http://stadium.open.ac.uk/stadia/preview.php?s=1&whichevent=963**. This URL links to an open access 'streamed' lecture given by Brian Trench at the Open University in March 2007. The lecture explores the various communication models applied in science communication, presents a classification of these models according to various characteristics, and reviews the question in science communication circles about preferred models.

SECTION 4
Mediating science news

. . . journalists make the news . . . (Would you say that of science? the journalist might respond. Would you say that scientists 'make' science rather than 'discover' it or report it? Yes, the conscientious scholar must answer, we would say precisely that, and sociologists of science do say precisely that.)

Michael Schudson (1991). The sociology of news production revisited.
In: *Mass Media and Society* (ed. James Curran and Michael Gurevitch)

4.1

Making science newsworthy: exploring the conventions of science journalism

Stuart Allan

The desperate plight of British parents Kate and Gerry McCann trying to find their daughter Madeleine, evidently abducted from the bedroom where she was sleeping in a resort in Praia da Luz, Portugal on the evening of 3 May 2007, was front-page news around the globe. Extensive press coverage followed every step of the police investigation into the child's disappearance, with each new development subjected to intense scrutiny by journalists and commentators alike across the mediascape.

One such development transpired in early October that year, when considerable excitement was generated about a remarkable scientific breakthrough in the effort to locate the missing child. 'Forensic DNA tests "reveal traces of Madeleine's body on resort beach"', was the headline in the reputable Sunday newspaper, *The Observer* (7 October 2007). The news item continued, describing how a former detective 'renowned for locating abducted children' had performed a 'forensic analysis' which appeared to show that the child's body had either been temporarily buried in—or possibly was still beneath—the beach at Praia da Luz. The findings by the detective in question, Danie Krugel, were reported to be based on 'a combination of Madeleine's DNA sample and GPS satellite technology', a technique credited with being 'able to locate a missing person anywhere in the world using only a single strand of hair'. Following Krugel's presentation of his findings to the Portuguese police, British officers were asked to bring in sniffer dogs to extend the search. 'The subsequent reaction of the dogs to Kate's clothing—the so-called scent of death—led to the couple being declared formal suspects over the death of their daughter', the story declared. By any measure, this was an extraordinarily sensational turn in the case, one which reverberated throughout the world's press. In Britain, *The Telegraph* reaffirmed that the McCanns had used 'a scientist to help look for the missing four-year-old using a DNA-tracking device' in order to establish her 'DNA trail' from the rented apartment. Related press headlines included 'Man uses device in Maddie hunt' (*The Sun*), ' "I know where

Madeleine is buried" says body finder hired by McCanns' (the *Daily Mail*), 'I have traced her to beach' (*The People*) and 'I traced Maddie DNA to the sea' (the *News of the World*).

If it seemed that the news story about a retired detective's skill in DNA profiling was almost too good to be true, however, it was because it was precisely that—too good to be true. Within hours of the stories being published, a number of science bloggers writing on the internet were sounding the alarm. Several posted critical observations about the scientific assertions being made, especially with regard to the 'ludicrous nature' of a 'top secret' device that could perform 'forensic DNA tests' whereby an individual could be found on the basis of testing a strand of hair. Others pointed to the range of earlier news stories about Krugel and his 'bizarre machine' posted elsewhere on the web, calling into question his proclaimed scientific credibility in no uncertain terms. In light of these criticisms by bloggers, *The Observer* published a retraction of sorts the following week. Stephen Pritchard (2007), the readers' editor, conceded that the newspaper should not have reported as fact that traces of Madeleine's body had been found using Krugel's device ('the sort of thing that you expect to find in a science-fiction novel', as one blogger described it). He also admitted that the description of Krugel as being 'from the University of Bloemfontein' gave the impression that he was an academic, when in actuality he is not a scientist but rather a director of security. 'Bloggers', he wrote, 'were quick to condemn the paper for giving credence to the efforts of a man whom they said was at best a crank and whom, they claimed, may impede the search for Madeleine with his "hocus-pocus" technology'. Blame for the incident, however, was to be placed firmly on the internet in his view; reporters, under time pressure, had relied on 'several references to [Krugel's] past activity' they found there, references that 'appeared to lend credibility to his claims'. This was a disappointing state of affairs, Pritchard conceded, not least because *The Observer* deserved credit, he believed, for being the first newspaper to 'discount the wider DNA theories that swirled around the Madeleine inquiry' the month before, theories that had been 'largely discredited' since.[1]

Evident in this brief account are several issues of interest for any enquiry into science journalism, ranging from factual inaccuracies in news reports, the intense pressures constraining the reporting process, journalistic perceptions of the relative authority of informational sources and general awareness of scientific concepts and principles, amongst a host of other concerns. Even more challenging to discern, however, are the factors shaping how science is rendered newsworthy in the first place—that is, the complex ways in which journalists make their routine, everyday decisions about how to incorporate scientific claims into a story, and why. Journalists themselves will often maintain that they are simply following their 'gut feelings', 'hunches' or 'instincts' when

---

1. Ben Goldacre, on his blog 'Bad Science' (http://www.badscience.net/?p=544), remarks: 'Psychics telling your future at the fairground are fine. When it comes to newspapers printing horoscopes, I couldn't care less. But exploitative misreporting of this scale on this subject is contemptible. You're as capable as I am of reading about Krugel's work [on the internet], and so are the Observer, but still this reputable UK newspaper is presenting magic quantum box tomfoolery as serious DNA evidence on the whereabouts on a little girl who has disappeared and may well be murdered.' Goldacre, who also writes about science for *The Guardian* newspaper, provides links to other blogs commenting on the controversy.

performing this role, a process that is all the more difficult to elucidate to the extent that it is regarded as merely a matter of 'common sense'. Accordingly, this chapter's exploration of science news will aim to unravel the largely unspoken, taken-for-granted conventions which underpin these seemingly *ad hoc* judgements. In so doing, it will seek to show that an analysis of the issues involved will enhance our understanding of the challenges currently confronting science journalism.

The emergence of science journalism

Science as a newsworthy topic has consistently garnered attention from the press since its earliest days, for better or—more typically—for worse in the eyes of many scientists. It is all the more surprising to observe, therefore, that the history of science journalism has yet to attract the degree of scrutiny it deserves. Still, it is possible to identify a number of controversies involving science in historical accounts of journalism more generally, many of which revolve around a tendency to rush to judgement similar to that described above with regard to the Madeleine McCann investigation.

By any measure, one of the most noteworthy controversies occurred in August 1835, when the *New York Sun* published a brief news item about 'some astronomical discoveries of the most wonderful description' made by Sir John Herschel, son of Sir William Herschel (credited with discovering Uranus), using 'an immense telescope of an entirely new principle'. In the days to follow, the *New York Sun* proceeded to print a series of front-page stories detailing the nature of the discoveries, all of which were evidently based on facts gleaned from the pages of a 'Supplement to the *Edinburgh Journal of Science*'. To the astonishment of its readers—and the chagrin of rival newspapers hurrying to republish the claims—the *New York Sun* explained how Herschel's telescope had caught a glimpse of 'continuous herds of brown quadrupeds, having all the external characteristics of the bison, but more diminutive than any species of the *bos* genus in our natural history', amongst other signs of life on the Moon. The third instalment described the topography of the Moon's surface, before the fourth broke the most sensational revelation of all, namely that Herschel had observed 'furry, winged men resembling bats'. The *New York Sun*'s account stated:

[This lens] gave us a fine half-mile distance; and we counted three parties of these creatures, of twelve, nine, and fifteen in each, walking erect toward a small wood near the base of the eastern precipices. Certainly they were like human beings, for their wings had now disappeared, and their attitude in walking was both erect and dignified. . . . They averaged four feet in height, were covered, except on the face, with short and glossy copper-colored hair, and had wings composed of a thin membrane, without hair, laying snugly upon their backs, from the top of the shoulders to the calves of the legs. The face, which was of a yellowish flesh-color, was a slight improvement upon that of the large orang-utan, being more open and intelligent in its expression, and having a much greater expanse of forehead . . .

New York Sun, 28 August 1835 (cited in O'Brien 1928, pp. 50–51)

The ensuing frenzy engendered by the *New York Sun*'s 'scoop' spread across the country, rapidly extending to enthralled publics in Europe before reaching a rather annoyed

Herschel in Cape Town, South Africa. Soon after, in a private letter, he complained: 'I have been pestered from all quarters with that ridiculous hoax about the Moon – in English French Italian & German!!' (cited in Ruskin 2002). Evidently, the 'Great Moon Hoax', as it was later dubbed, was actually intended as a satirical intervention, at least according to Richard Adams Locke who penned the stories. His purpose, he insisted, was to demonstrate the gullibility of readers where claims about astronomy were concerned. Of his success in this regard there is little doubt, with few challenges to the authenticity of the stories having been rendered at the time. The *New York Daily Advertiser*'s avowed stance was typical: 'Sir John has added a stock of knowledge to the present age that will immortalize his name, and place it high on the page of science', it had informed its readers (cited in Levermore 1901, p. 51).

Extraordinary examples of deliberate deception aside, the emergence of science journalism as a reportorial genre has yet to be fully documented. That said, however, several researchers have sought to trace the gradual consolidation of its forms and conventions over the years—a challenging endeavour not least because of disagreements over how best to define its status as distinct from other related genres (not least science writing). In Britain, researchers have tended to examine science journalism in relation to broader trends informing the popularization of science and technology. In addition to science journals such as the *Quarterly Journal of Science* (founded in 1864), *Scientific Opinion* (1868) and *Nature* (1869), serial periodicals such as *Popular Science Monthly*, *Hardwicke's Science-Gossip*, *Pearson's Magazine*, *Tit-Bits*, *Cassell's Magazine*, *Knowledge: An Illustrated Magazine of Science* or even 'Science Jottings' in the *Illustrated London News*, amongst an array of others, promoted science and its appreciation in the interest of modernity and social progress. Each title, to varying degrees, similarly helped to configure interested 'lay' communities of readers, enthusiastic to hear the latest news about scientific developments, innovations and curiosities (see also Henson *et al.* 2004; Broks 2006; Mussell 2007).

Various analyses of these reports—together with newspapers of the day—have proven to be rich sources of insights into the inchoate features of 'science news' and the corresponding role envisaged for the science-specialist writer. In Broks' (2006) discussion of science popularization, for example, he quotes a telling bit of advice offered by English journalist W. T. Stead in 1906:

In editing a newspaper, never employ an expert to write a popular article on his own subject, better employ someone who knows nothing about it to tap the expert's brains, and write the article, sending the proof to the expert to correct. If the expert writes he will always forget that he is not writing for experts but for the public, and will assume that they need not be told things which, although familiar to him as ABC, are nevertheless totally unknown to the general reader.

cited in Broks (2006, p. 34)

In the years to follow the end of World War I, science news eventually evolved into a recognizable genre, facilitated in part by the growing professionalism of 'objective' reporting methods. It was regularly featured in newsreels in the cinema, as well as on the fledgling BBC wireless network from its start in the 1920s. By the late 1930s, an increasing number of newspaper journalists were becoming formally associated with science reporting as specialist correspondents, with individuals such as J. G. Crowther of the *Manchester Guardian* and Peter Ritchie Calder of the *Daily Herald* helping to lead the way. Growing momentum led to the formation of the Association of British Science Writers in 1947, which enriched debate about working practices and ethical standards. Television's arrival witnessed science-centred programmes emerge, not least *The Sky at Night* in April 1957, followed by the science documentary series *Horizon* in 1964, amongst others (see also Gregory and Miller 1998; Allan 2002).

Science journalism underwent a similar process of gaining legitimacy in the United States. In addition to mainstream publications, interested members of the public could turn to specialist titles, such as *Scientific American* (established in 1845), *Popular Science Monthly* (1872), *National Geographic* (1888) and *Popular Mechanics* (1902), amongst others, which were intent on helping readers to meet the challenges of 'self-instruction in science'. Relevant here is LaFollette's (1990) research, which examined the science coverage produced in the country's family magazines between 1910 and 1955. She contends that this type of reporting contributed to a 'climate of expectations' whereby science—along with scientists—was celebrated in glowing terms, with articles assuring readers that a future of endless progress beckoned (see also Burnham 1987). As in Britain, however, it was during the 1920s when especially formative developments were set in motion. Perhaps foremost amongst them was the founding of the Science Service, a syndicated news service which became the exemplar of the ideals to be emulated by major newspapers. Set up with the financial support of newspaper publisher E. W. Scripps in 1921, the service was supplying news and feature material to more than 100 newspapers by the 1940s. Reflecting on its achievements in 1958, its director, Watson Davis, suggested that in its first decade the non-profit service 'made science acceptable to the American press [while also making] science reporting acceptable and respectable in both newspaper and science circles'. Prior to this contribution, he added, 'newspapers understood little and cared less about science' (Davis 1958, p. 21; see also Foust 1995; Nelkin 1995). In the years to follow, as news organizations slowly began to add full-time science reporters to their staff (such as Waldemar Kaempffert and William Laurence, both of the *New York Times*), science journalism gradually consolidated its evolving forms and practices into a recognizable news 'assignment' or 'beat' specialism.

Interestingly, however, by the 1960s a marked shift of orientation was becoming increasingly apparent in science journalism in the UK and USA, as well as in other national contexts. Researchers examining this period have traced the emergence of sharper, more aggressive types of reporting in conjunction with its growing professionalism. If in the past science journalists had considered their role to be one of helping to promote the benefits of science and technology in the name of progress, now an increasingly sceptical, even critical, stance was being widely adopted (the publication of Rachel Carson's *Silent Spring* in 1963 being a case in point). Several studies have adopted a longitudinal approach in order to survey changes in a specific news organization's reporting, or with respect to the evolving nature of a particular topic of sciences news. Examples include Bauer *et al.*'s (2006) study of changing patterns of science news in Britain and Bulgaria from 1946 to 1995; Albaek *et al.*'s (2003) enquiry into journalists' use of researchers as experts in Danish daily newspapers from 1961 to 2001; Bucchi and Mazzolini's (2003) critique of the Italian daily press's science coverage from 1964 to 1997; Pellechia's (1997) content analysis of science reporting in three major US newspapers from 1966 to 1990;

Cho's (2006) examination of US network news coverage of breast cancer from 1974 to 2003; Nisbet and Lewenstein's (2002) content analysis of biotechnology-related coverage in the US press from 1970 to 1999; Clark and Illman's (2006) investigation of 'Science Times', the weekly science section of the *New York Times*, from 1980 to 2000; and Carvalho and Burgess's (2005) assessment of how climate change was represented in UK newspapers from 1985 to 2003, amongst others. Taken together, these studies underscore the value of a longitudinal perspective, not least with respect to discerning changing definitions of what counts as 'science news' over time, in contrast with the more typical snapshot taken in short-term analyses.

Making science news

Science, it is often said, gets a bad press. Explanations for this apparent problem, in the opinion of some journalists at least, tend to revolve around the charge that most types of science fail the test of newsworthiness. Routine science, they tend to believe, is really rather boring. It lacks the stuff of drama necessary to spark lively newspaper headlines. At the same time, some scientists maintain that on those occasions when a certain scientific development is given due prominence, it all too frequently happens for the wrong reasons. Not surprisingly, they are quick to condemn instances of sensationalist reporting—where news values have given way to entertainment values—for misrepresenting the nature of scientific inquiry, and rightly so.

Science typically appears in the press as 'an arcane and incomprehensible subject', Nelkin (1995) observed in her classic study of science journalism in the USA. Surrounding it is a certain 'mystique' that implies it is to be properly regarded as a 'superior culture' with a 'distanced and lofty image'. Far from enhancing public understanding, she argues, 'such media images create a distance between scientists and the public that, paradoxically, obscures the importance of science and its critical effect on our daily lives' (Nelkin 1995, p. 15). The processes of selection and presentation indicative of one news organization will be at variance with those of others, of course, but shared assumptions about newsworthiness recurrently underpin these daily deliberations. News coverage of science tends to favour certain areas of scientific inquiry over and above other areas, a pattern which is usually observable from one newspaper or news broadcast to the next. The process by which certain scientific developments are rendered newsworthy while others, in contrast, are deemed unworthy of attention, is the outcome of a complex array of institutional imperatives.

Journalists, together with their editors, bring to the task of making sense of the scientific world a series of 'news values'. These evaluative judgements are largely internalized by each individual in turn, effectively shaped by his or her presuppositions about the nature of a story worth pursuing. Rensberger (1997, 2002), drawing upon his own experience, points out that 'of the dozens of stories we could do on any given day, we reject most or all possibilities. In this we exercise our opinion as to what is a good story' (Rensberger 1997, p. 11). In explaining what in his opinion makes a good science story, he identifies five criteria especially valued by newspaper science journalists:

- fascination value (something that will spark the reader's amazement);
- size of the natural audience (the number of readers who are already interested in a given topic);
- importance (for society, as well as for the reader);
- reliability of the results ('Is it good science?');
- timeliness ('The newer the news', Rensberger (1997) states, 'the newsier it is').

If science items can often be the 'furthest thing from breaking news,' as Petit (1997) suggests, 'this can be their charm' (Petit 1997, p. 187). Here he draws a sharp contrast between the kinds of news stories he typically writes and the more usual kinds of items published alongside them in the same newspaper. 'Like stories in astronomy, or on fossils of prehistoric people', he observes, 'discoveries about the Earth's history and behavior provide for many people a welcome and invigorating break, a mental escape from the daily diet of human disaster, political skulduggery, and crime news' (Petit 1997, p. 187). As a result, Hotz (2002) suggests, 'science coverage today can still be more explanatory and adulatory than challenging or analytical' (Hotz 2002, p. 6).

The news values associated with sensationalised 'gee-whiz' or 'weird-and-wacky' moments differ markedly from those indicative of the more routine types of science stories likely to appear in the daily news cycle. To better understand these distinctions, researchers have attempted to explicate the means by which certain news values are embedded in the everyday procedures used by science reporters to justify the selection of certain types of events as newsworthy at the expense of alternative ones (see also Bucchi 1998; Friedman *et al.* 1999; Allan 2002). Hansen (1994) suggests, on the basis of his study of science reporting in the British press, that the 'most pronounced criterion of newsworthiness is whether science can be made recognizable to the reader in terms of human interest or in terms of something readers can relate to' (Hansen 1994, pp. 114–15). Particularly prized, as a result, are those events which illuminate the relevance of science to daily life, enabling the journalist to adopt a 'human angle' when constructing the news story. Further factors informing this process of negotiation include the efforts made by news sources or stakeholders themselves to influence journalistic judgements, as well as the relative complexity of the event itself. 'The more complex or inaccessible a piece of science news is', Hansen writes, 'the more "translatory work" it will require on the part of the journalist to make it intelligible and interesting to the readers'. Time is of the essence for journalists working to conform to a daily production schedule, especially where deadlines are concerned. That said, while a significant scientific 'breakthrough' may be judged to constitute 'hard' news, and thereby warrant immediate coverage, it is much more likely that the science involved will be, 'in news terms, a slow process of small incremental developments' (Hansen 1994, p. 115).

The norms of science journalism, it follows, seldom align comfortably with those of the science being covered. Wilcox's (2003) research identifies what she describes as the 'hype/space dilemma' shaping this alignment. Specifically, she highlights some of the contradictions encountered by journalists when scientific claims need to be hyped in order to secure the necessary space in the newspaper. '[T]he media require scientific studies to provide dramatic new findings and dramatic conflicts', she writes, 'while the conventions of scientific journalism require that the results of single studies be de-emphasized in favour of the scientific context of the research' (Wilcox 2003, p. 243). Lynch and Condit (2006), in their assessment of this dilemma, suggest that it revolves around a basic conflict, namely the journalist's struggle to make a story interesting, even sensational, in relation to other stories competing for the same space while, at the same time, reaffirming professional ideals of truth-telling and balance. Precisely how this conflict is negotiated will have significant consequences for what gets covered, but also for the standards of accuracy applied when scientific terminology is being translated into lay vocabulary. 'At best', Condit (2004) points out, 'this standard includes fidelity to sources, a balance among and inclusion of different viewpoints, and a translation that conveys some main idea from a study clearly'. What more dispassionate, thorough reporting loses with respect to excitement, it gains by helping to ensure stories avoid 'the kind of distortions that might encourage inappropriate (or even dangerous) behaviour or unrealistic expectations' (Condit 2004, p. 1415).

News values are thus context-specific, and as such are never fixed once and for all—instead they are always evolving over time, being inflected differently from one news organization to the next. Still, the ostensibly 'common sense' criteria informing definitions of newsworthiness in science journalism have proven to be surprisingly consistent over the years. 'For all of modern science's sophisticated concepts and technology', Young (1997, p. 114) observes, 'journalism's traditional five *W*s and an *H* – who, what, when, where, why and how – still form the core of science reporting'. Toner (1997) agrees, but with an important caveat:

Editors insist – and many believe – that busy readers have no time for the 'rest of the story.' The five *W*s that journalists once revered are often reduced to four. Yes, we can fit the 'who', 'what', 'where' and 'when' into the little space on the front page. But skip the 'why'. That fifth *W*, after all, may only raise more questions than it answers.

Toner (1997, p. 130)

This is the key issue, in Toner's view. Science journalists who fail to ask the question 'why' are too often providing information that satiates people's curiosity as opposed to stimulating it. 'Like a weed', he writes, 'curiosity has a habit of popping up in the wrong place. It can be unruly and hard to control. It is robust and tenacious. And where one question sprouts, many more are bound to follow' (Toner 1997, p. 128). It is this contagious aspect of curiosity, in his opinion, that renders it 'such a powerful tool for journalism'. That is to say, he adds, 'by planting the seeds of curiosity, we make our audience accomplices in the pursuit of knowledge' (Toner 1997, p. 129).

Complicating these efforts to attract and hold the public's attention is a wide array of difficulties, however. A growing number of commentators are expressing their alarm about how economic pressures on news organizations are forcing some to scale back their commitment to science reporting, cutting jobs and refocusing priorities around non-specialist topics. This decline is particularly apparent where investigative reporting of science is concerned (see Bucchi 1998; Gregory and Miller 1998; Reed 2001; Allan 2002; Hotz 2002; Anderson *et al.* 2005; Hargreaves *et al.* 2003; Priest 2005; Kiernan 2006). Dean (2002), former science editor of the *New York Times*, identifies three further difficulties. First, she notes, science journalism's reach has to be remarkably broad: 'We cover everything from anthropology to astrophysics to atherosclerosis. We advise other departments when a ballplayer is injured or a court overturns a pollution regulation'. Second, given that science is becoming increasingly specialized, 'it is hard for journalists, even journalists with advanced training, to know what is important and what is not important'. Third, she points out, scientific research is becoming increasingly commercialized, creating widespread conflicts of interest. As more and more researchers are turning 'their labs into test beds for their own companies, or have grants from major commercial concerns, or seek venture capital', she writes, 'they have powerful motives for making the most of their results and playing down anything that might challenge them' (Dean 2002, p. 25; see also Thomas, 2009).

The relationship between journalists and scientists, some suggest, is becoming increasingly fraught. 'More than ever', Hotz (2002) contends, 'scientists aggressively court media attention, even as – paradoxically – unprecedented commercial secrecy comes to shroud so much of what scientists do today and financial conflicts of interest among researchers have become so common' (Hotz 2002, p. 6). Scientific publications, including several of the leading peer-reviewed journals, have become caught-up in this commercialization, which leaves journalists wondering which sources they can trust. This problem can be further compounded at times by what is described as the embargo system, whereby journals send out advance details about scientific research to journalists on condition that they withhold reporting on them until the embargoed time elapses. Kiernan's (2006) analysis of its impact demonstrates how these journals succeed in exercising considerable control over the range and diversity of science stories, with their editorial judgements having an especially strong influence on non-specialist journalists struggling to cope with scientific complexity. Nonetheless, it is important to recognize that this occurs with the full cooperation of those journalists who prefer a 'level playing field' with rival news organizations. Consequently, he writes, 'if journalists are in a stranglehold, it is a self-inflicted stranglehold – and one that does not serve the public interest' (Kiernan 2006, p. 122). Whitehouse (2007) concurs, contending that this 'cosy and secretive arrangement' acts as a 'marketing tool' for the journals while it simultaneously 'encourages lazy reporting and props up poor correspondents'. Genuine scoops, as a result, 'are few and far between in science as the embargo system militates against them'.

Several researchers investigating the ways in which science journalists put together news stories have underscored the importance of examining the actual language used to make complex scientific findings meaningful in the 'lay press'. Such a process will necessarily entail moving beyond direct dependence on scientific discourse so as to find more suitable modes of expression, sometimes requiring the journalist—as Friedhoff (2002) maintains —'to think about communicating ideas and concepts that in many cases were previously unthinkable'. Findings, she adds, 'need to be embedded into a commonly understood context' (Friedhoff 2002, p. 23). In order to unravel this process of embedding, some researchers have utilized the concept of 'framing' to help specify how science stories tend to respect certain 'principles of organisation' (Goffman 1974). That is to say, framing is often described as a strategy employed by journalists to define the nature of an occurrence as a meaningful event. In the words of Nisbet and Mooney (2007):

Frames organize central ideas, defining a controversy to resonate with core values and assumptions. Frames pare down complex issues by giving some aspects greater emphasis. They allow citizens to rapidly identify why an issue matters, who might be responsible, and what should be done.

Nisbet and Mooney (2007, p. 56)

Framing, it follows, is the subject of intense negotiation, and as such frequently proves to be a contested process—not least between journalists and their sources (as well as between journalists and their editors within a news organization, where decisions about placement, headlines, images, captions and the like, become important: see Mellor, Chapter 5.2 this volume). Indeed, in addition to helping to define what is significant to know, frames have far-reaching implications for how claims made by sources are selected (or not) as newsworthy, the narrative conventions guiding the ways in which they are reported and the possible consequences for influencing public perceptions. Sources themselves are often acutely aware of the importance of the framing process, so will make every effort to try and ensure that their preferred definition of the issue or event is placed in a positive light.

To clarify the nature of the tensions which can arise in this framing process it is worth considering it from the vantage points of the science journalist, on the one hand, and the scientist as a prospective source, on the other. Of the myriad of concerns confronting both of them, respectively, several are particularly pertinent. Turning first to the journalist, he or she will probably want to determine the answers to questions such as the following:

- Who is a credible, trustworthy and legitimate scientific source for me to contact? Who possesses sufficient expertise to interpret the significance of possible risks, dangers or hazards? Whose scientific authority will reinforce the impartiality that I am striving to achieve in my news account?

- Wherein lies the news value of a story? Which aspects of it will be of interest to my editor, fellow journalists and audience? Does the story have a clear narrative structure, a straightforward beginning, middle and end, which will allow facts to be expressed quickly and easily?

- What is potentially controversial about this story? If a possible risk is at issue, is it safe? In the event scientific sources refuse to offer a 'yes' or 'no' to my question, how best to communicate possible implications in simple—but not simplistic—terms? How do I put a human face on scientific principles, let alone risk calculations?

- What happens when my sources disagree? If balanced reporting, by definition, means that there are always at least two sides to every story, should alternative accounts be given equal weight in the story? How do I determine what counts as sufficient evidence to sustain a claim in the absence of scientific consensus about its significance?

- How best to differentiate the public interest from what interests the public?

Looking at this process of mediation from the opposite point of view, namely that of the prospective source, throws an alternative range of possible questions into sharp relief. Scientists, for example, may ask themselves questions such as:

- Why does this journalist want to talk to me? Where will I be made to fit within the larger structure of the news story? That is to say, how will my views be aligned in relation to alternative views, how will the facts of the matter be framed or contextualized?

- Can journalists be trusted to explain complex findings to the public in a manner that will be accurate? Will cooperating with the journalist enhance my profile in the field, or will it diminish my credibility, upset my colleagues, antagonize my rivals?

- Will there be benefits for my research in having it turned into a news story? Is this an opportunity to be exploited, especially where attracting potential funding is concerned, or is it best to play safe and pass it up? Do scientists have a particular obligation to explain their work to members of the public?

- When discussing controversial science, which may entail possible risks or hazards, how can I acknowledge that it is impossible to eliminate problems entirely without, at the same time, calling into question my authority? In the absence of guarantees, how do I explain which risks should be avoided, which ones may be worth taking, and which ones are insignificant?

- Is scientific uncertainty an inevitable price to pay for progress? Who is entitled to criticize science, and who is not? Wherein lies responsibility for how science is used in society? Is it possible to separate out science from the moral or ethical implications of its application?

Needless to say, the symbiotic nature of the scientist–journalist relationship can quickly become hotly contested under certain circumstances, especially when trust breaks down. It is certainly the case that answers to the questions above will demand attention, as well as careful negotiation, during each and every encounter.

Shaping public perceptions

In rounding out this discussion, it is worth pausing to return to the 'Great Moon Hoax' discussed above. 'Science entered the popular press with the moon hoax', Hughes (1940) maintained in her study of news and human interest journalism. The reason why the story attracted such an extraordinary response from members of the public, she believed, was because of the way the *New York Sun* reports inflected the language of science. 'The discoveries had an air of authenticity because of the technical jargon', she observed. Consequently, 'when the excited readers learned "these are doubtless innocent and happy creatures . . . notwithstanding some of their amusements would but ill comport with our terrestrial notions of decorum," they took it for the inhuman and tantalizing restraint of the man of science, and were doubly convinced' (Hughes 1940, p. 185).

Journalistic conceptions of news values, it goes almost without saying, must necessarily resonate with those of the readers, listeners or viewers. Hughes' turn of phrase 'the inhuman and tantalizing restraint of the man of science' is itself interesting in this regard. From the vantage point of today, these words sound anachronistic (not least the use of the male pronoun to describe scientists) and yet, rather tellingly, all too familiar. At issue is the implied authority of ostensibly dispassionate scientific discourse, but also—crucially—the question of its very legitimacy in the eyes of diverse publics. The *New York Sun*'s readers, it is being assumed, were likely to be mystified about what was being communicated by news reports relying upon 'technical jargon', but at the same time arguably more inclined to accept the claims being made nonetheless. 'It was probably the first news story to win so heterogeneous an audience', Hughes contends, 'and it showed that what enthrals the plain man will also interest the sophisticate' (Hughes 1940, p. 185). The fact that the story itself was an instance of fakery, it follows, is almost beside the point. 'The moon hoax represented the new use of the newspaper as something to amuse', and so to the extent that science—fabricated or otherwise—generated good copy about the 'unanswered questions' concerning the 'curiosities and mysteries of everyday life', it was likely to feature prominently in any calculation of newsworthiness. The hoax, Hughes concluded, 'is thus one of many stories which may be said to define a universal interest in the half-explained, the unknown, and the uncontrollable' (Hughes 1940, p. 187).

It would appear that this assertion about public fascination with the 'curiosities and mysteries of everyday life' proves equally valid in the brave new world of the internet today. While examples of various hoaxes perpetrated there are plentiful, there is nonetheless a growing recognition of its importance for science journalism. In the USA, a study conducted by the Pew Internet Project and the Exploratorium (2006) found that the internet was a primary source for science news and information (second only to television) for 40 million American adults, and that the use of online science resources was 'linked to better attitudes about the role science plays in society and higher assessments of how well they understand science'. More specifically, the findings of the November 2006 report, based on a survey of 2000 people in January 2006, suggested that:

- Nearly 9 in 10 (87%) online users have used the internet to look up the meaning of a scientific concept, answer a specific science question, learn more about a scientific breakthrough, help complete a school assignment, check the accuracy of a scientific fact, download scientific data, or compare different or opposing scientific theories.

- Most Americans say they would turn to the internet if they needed more information on specific scientific topics. Two-thirds of respondents asked about stem cell research said they would first turn to the internet and 59% asked about climate change said they would first go to the internet. Most of those searches would begin with search engines.

- Nearly three-quarters (71%) of internet users say they turn to the internet for science news and information because it is convenient.

- Two-thirds (65%) say they have encountered news and information about science when they have gone online for a different reason in mind (Pew Internet Project and the Exploratorium 2006).

Even bearing in mind the usual sorts of qualifications where opinion surveys are concerned (margins of sampling error—in this study said to be plus or minus three percentage points—as well as interpretations of question wording, practical difficulties and so forth), these results would appear to indicate that the internet is creating important, and increasingly globalized, spaces for science news and information to circulate. 'People's use of the internet for science information has a lot to do with the internet's convenience as a research tool, but it also connected to people's growing dependence on the internet for information of all types', stated John B. Horrigan, the study's principal author. 'Many think of the internet as a gigantic encyclopedia on all subjects and this certainly applies to scientific information' (cited in Pew Internet Project and the Exploratorium 2006; press release 20 November 2006). For science journalists, the report similarly made for interesting reading, not least with regard to its finding that it is a widespread practice amongst users to go online in order to double-check the reliability of science news reports they encounter in more traditional media.

Academic studies into how science news is being interpreted by internet users are at a relatively early stage of development (see also Hansen, Chapter 3.1 this volume). Important here is a context-sensitive approach to any bold claim, whether positive or negative, about how people are being influenced (or not) as a result. Using a variety of methods, researchers have sought to discern the subtle, frequently contradictory ways in which public perceptions of science are negotiated as part of everyday life. Rather than invoking a language of causative 'effects' or 'impacts', they strive to understand how people draw upon science news on the internet to help them make sense of different scientific controversies. Relevant examples of research undertaken to date include Eveland and Dunwoody's (1998) examination of internet users' engagement with 'The Why Files' website, which aims to provide 'the science behind the news' in story narratives; Richardson's (2001) critique of the dynamics of news exchange via internet newsgroups discussing possible health risks of BSE, or 'mad cow disease'; Eisend's (2002) investigation of the effects of internet use on traditional scientific communication media among German social scientists; Treise *et al.*'s (2003) research into the factors which appear to influence the perceived credibility of a science website in the USA; Koolstra *et al.*'s (2006) comparison of Dutch television with the internet in terms of public perceptions of their relative reliability as sources of information for science communication; and Weigold and Treise's (2004) study of how teenagers use the web to find science information, which revealed that participants 'regularly link from news stories to science sites while reading interesting science-related news stories' (Weigold and Treise 2004, p. 237).

Studies such as these invite a reconsideration of certain longstanding assumptions about science news in the internet age, as well as a heightened degree of self-reflexivity amongst science journalists themselves. Many of the latter, not surprisingly, continue to be sceptical. The internet—especially in conjunction with the embargo system discussed above—makes it 'easy to churn out story after story, usually without leaving your desk', states Whitehouse (2007), a former BBC science correspondent. 'The result of this', he continues, 'is that science coverage can be indistinguishable across outlets. The quick communication and comparison made possible by the internet has resulted in a uniform blandness of science reporting'. Pertinent here is Trumbo *et al.*'s (2001) investigation of e-mail and web use by science journalists in the USA, suggesting that the rapid diffusion of these technologies has been due, in the main, to a positive orientation toward the quality of web information, trust in its sources, and advantages gained through connectivity; Massoli's (2007) examination of the 'journalistic information approach' adopted on the web sites of various European public research institutions; as well as Pinholster and O'Malley's (2006) study of how EurekAlert!, the science news web service of the American Association for the Advancement of Science (AAAS), helps reporters to cover the fast-breaking science beat around the globe (see also Trench, Chapter 4.2 this volume).

Speaking in more conceptual terms, Hermida (2007) similarly strikes an upbeat note, expressing his conviction that the internet 'offers us new ways of rethinking how science is reported and explained'. As he explains:

In our traditional print model, we expect audiences to come to us – to pick up the *Vancouver Sun* in the morning and read about the research at UBC [University of British Columbia] into leatherbacks [turtles]. But in an online world, you cannot simply expect people to come to you out of habit. The emphasis is on reaching out to audiences – to create more opportunities for people to stumble across your content. . . . Increasingly more and more news organizations are taking this approach

– providing their audiences with ways of taking their content [to] share it with friends via [the social networking sites] Facebook, YouTube or their own blog.

Hermida (2007)

Hermida believes that it is vitally important that science journalists adopt a 'digital mindset', one that retains the values of accuracy and fairness while seeking to capitalize on the non-linear, interactive and participatory nature of social networks in online environments. Chalmers (2007, 2009) offers an assessment of the ways in which the internet is transforming how scientists—in this case physicists—report their findings and communicate with one another. The internet, he contends, is 'tailor-made' for the 'democratization of science, and is quickly becoming a much more social medium than the one-way "click and download model" of the Web as it was originally conceived'. Blogs such as *Cosmic Variance*, for example, are proving remarkably popular amongst physicists—'3000 readers per day and [it] hosts discussion "threads" ranging from the latest preprints in theoretical particle physics to how to mix the perfect cocktail'—as well as with a wider public of non-scientists (Chalmers 2007, p. 20).

To close, it is becoming increasingly apparent that tectonic shifts are reshaping the landscape of science journalism, not least with respect to the growing convergence of 'old' media with 'new', internet-based media. Today's science journalist is increasingly expected to be multi-skilled in order to work comfortably across a range of digital plat-forms while, at the same time, warding off the challenges posed by ordinary citizens—not least the bloggers and YouTubers of 'the iPod generation'—threatening to storm the ramparts of the journalistic profession. Precisely how, and to what extent, the internet is changing the characteristics of science news is deserving of our close attention. Even those who are dismissive of the celebratory claims being made about its potential for creating new spaces for 'public engagement with science' need to recognize that it is here to stay, and that it promises to dramatically recast science journalism's familiar norms and values in unanticipated ways.

■ **REFERENCES**

Albaek, E., Christiansen, P.M. and Togeby, L. (2003). Experts in the Mass Media: Researchers as Sources in Danish Daily Newspapers, 1961–2001. *Journalism and Mass Communication Quarterly*, **80**(4), 937–48.

Allan, S. (2002). *Media, Risk and Science*. Open University Press, Buckingham.

Anderson, A., Peterson, A. and David, M. (2005). Communication or spin? Source–media relations in science. In: *Journalism: Critical Issues* (ed. S. Allan), pp. 188–98. Open University Press, Maidenhead.

Bauer, M., Petkova, K., Boyadjieva, P. and Gornev, G. (2006). Long-term trends in the representations of science across the Iron Curtain: Britain and Bulgaria, 1946–95. *Social Studies of Science*, **36**, 97–129.

Broks, P. (2006). *Understanding Popular Science*. Open University Press, Maidenhead.

Bucchi, M. (1998). *Science and the Media: Alterative Routes in Science Communication*. Routledge, London.

Bucchi, M. and Mazzolini, G.R. (2003). Big science, little news: science coverage in the Italian daily press, 1946–1997. *Public Understanding of Science*, **12**, 7–24.

Burnham, J.C. (1987). *How Superstition Won and Science Lost.* Rutgers University Press, New Brunswick, NJ.

Carson, R. (1963). *Silent Spring.* Hamish Hamilton, London.

Carvalho, A. and Burgess, J. (2005). Cultural circuits of climate change in UK broadsheet newspapers, 1985–2003. *Risk Analysis*, **25**(6), 1457–69.

Chalmers, M. (2007). A revolution in bits. *Physics World*, **20**(1), 18–21.

Chalmers, M. (2009). Communicating physics in the information age. In: *Practising Science Communication in the Information Age: Theorizing Professional Practices* (ed. R. Holliman, J. Thomas, S. Smidt, E. Scanlon and E. Whitelegg). Oxford University Press, Oxford.

Cho, S. (2006). Network news coverage of breast cancer, 1974–2003. *Journalism and Mass Communication Quarterly*, **83**(1), 116–30.

Clark, F. and Illman, D.L. (2006). A longitudinal study of the *New York Times* Science Times section. *Science Communication*, **27**(4), 496–513.

Condit, C.M. (2004). Science reporting to the public. *Canadian Medical Association Journal*, **170**(9), 1415–16.

Davis, W. (1958). Science service and the dissemination of science. *AIBS Bulletin*, **8**(5), 21–3.

Dean, C. (2002). New complications in reporting on science. *Nieman Reports*, **56**(3), 25–6.

Eisend, M. (2002). The Internet as a new medium for the sciences? *Online Information Review*, **26**(5), 307–17.

Eveland, W.P. and Dunwoody, S. (1998). Users and navigation patterns of a science World Wide Web site for the public. *Public Understanding of Science*, **7**, 285–311.

Foust, J.C. (1995). E. W. Scripps and the Science Service. *Journalism History*, **21**(2), 58–64.

Friedhoff, S. (2002). Rethinking the science beat. *Nieman Reports*, **56**(3), 23–5.

Friedman, S.M., Dunwoody, S. and Rogers, C.L. (eds) (1999). *Communicating Uncertainty.* Lawrence Erlbaum, Mahwah, NJ.

Goffman, E. (1974). *Frame Analysis.* Harper and Row, New York.

Gregory, J. and Miller, S. (1998). *Science in Public.* Perseus, Cambridge.

Hansen, A. (1994). Journalistic practices and science reporting in the British press. *Public Understanding of Science*, **3**, 111–34.

Hargreaves, I., Lewis, J. and Speers, T. (2003). *Towards a Better Map: Science, the Public and the Media.* ESRC, Swindon.

Henson, L., Cantor, G., Dawson, G., Noakes, R., Shuttleworth, S. and Topham, J.R. (eds) (2004). *Culture and Science in the Nineteenth-Century Media.* Ashgate, Aldershot.

Hermida, A. (2007). Reimagining science journalism. *Future Directions in Science Journalism Conference*, University of British Columbia, 9–10 November. Available at: **http://reportr.net/2007/12/05/future-directions-in-science-journalism-video**.

Hotz, R.L. (2002). The difficulty of finding impartial sources in science. *Nieman Reports*, **56**(3), 6–7.

Hughes, H.M. (1940). *News and the Human Interest Story.* Transaction, New Brunswick.

Kiernan, V. (2006). *Embargoed Science.* University of Illinois Press, Urbana.

Koolstra, C.M., Bos, M.J.W. and Vermuelen, I.E. (2006). Through which medium should science information professionals communicate with the public: television or the internet? *Journal of Science Communication*, **5**(3), 1–8.

LaFollette, M.C. (1990). *Making Science Our Own: Public Images of Science, 1910–1955.* University of Chicago Press, Chicago.

Levermore, C.H. (1901). The rise of metropolitan journalism, 1800–1840. *The American Historical Review*, **6**(3), 446–65.

Lynch, J. and Condit, C.M. (2006). Genes and race in the news. *American Journal of Health Behavior*, **30**(2), 125–35.

Massoli, L. (2007). Science on the net: an analysis of the websites of the European public research institutions. *Journal of Science Communication*, **6**(3), 1–16.

Mussell, J. (2007). Nineteenth-century popular science magazines. *Journalism Studies*, **8**(4), 656–66.

Nelkin, D. (1995). *Selling Science: How the Press Covers Science and Technology*, 2nd edn. W.H. Freeman, New York.

Nisbet, M.C. and Lewenstein, B.V. (2002). Biotechnology and the American media. *Science Communication*, **23**(4), 359–91.

Nisbet, M.C. and Mooney, C. (2007). Framing science. *Science*, 6 April, 56–7.

O'Brien, F.M. (1928). *The Story of the Sun*. Greenwood, New York.

Pellechia, M.G. (1997). Trends in science coverage: a content analysis of three US newspapers. *Public Understanding of Science*, **6**, 49–68.

Petit, C. (1997). Covering earth sciences. In: *A Field Guide for Science Writers* (ed. D. Blum and M. Knudson), pp. 180–7. Oxford University Press, New York.

Pew Internet Project and the Exploratorium (2006). *The Internet as a Resource for News and Information about Science* **http://www.pewinternet.org/**

Pinholster, G. and O'Malley, C. (2006). EurekAlert! Survey confirms challenges for science communicators in the post-print era. *Journal of Science Communication*, **5**(3), 1–12.

Priest, S.H. (2005). Risk reporting: why can't they ever get it right? In: *Journalism: Critical Issues* (ed. S. Allan), pp. 199–209. Open University Press, Maidenhead.

Pritchard, S. (14 October 2007). The readers' editor on . . . DNA and the hunt for Madeleine. *The Observer* [comment pages], p. 38.

Reed, R. (2001). (Un-)professional discourse? Journalists' and scientists' stories about science in the media. *Journalism*, **2**(3), 279–98.

Rensberger, B. (1997). Covering science for newspapers. In: *A Field Guide for Science Writers* (ed. D. Blum and M. Knudson), pp. 7–16. Oxford University Press, New York.

Rensberger, B. (2002). Reporting science means looking for cautionary signals. *Nieman Reports*, **56**(3), 11–13.

Richardson, K. (2001). Risk news in the world of internet newsgroups. *Journal of Sociolinguistics*, **5**(1), 50–72.

Ruskin, S.W. (2002). A newly-discovered letter of J.F.W. Herschel concerning the 'Great Moon Hoax'. *Journal for the History of Astronomy*, **33**, 71–4.

Thomas, J. (2009). Controversy and consensus. In: *Practising Science Communication in the Information Age: Theorizing Professional Practices* (ed. R. Holliman, E. Whitelegg, E. Scanlon, S. Smidt and J. Thomas). Oxford University Press, Oxford.

Toner, M. (1997). Introduction. In: *A Field Guide for Science Writers* (ed. D. Blum and M. Knudson), pp. 127–30. Oxford University Press, New York.

Treise, D., Walsh-Childers, K., Weigold, M.F. and Friedman, M. (2003). Cultivating the science internet audience. *Science Communication*, **24**(3), 309–32.

Trumbo, C.W., Sprecker, K.J., Dumlao, R.J., Yun, G.W. and Duke, S. (2001). Use of e-mail and the web by science writers. *Science Communication*, **22**(4), 347–78.

Weigold, M.F. and Treise, D. (2004). Attracting teen surfers to science web sites. *Public Understanding of Science*, **13**, 229–48.

Whitehouse, D. (23 July 2007). Science reporting's dark secret. *The Independent* [Media Weekly], p. 13. Available at: **http://www.independent.co.uk/news/media/science-reportings-dark-secret-458300.html.**

Wilcox, S.A. (2003). Cultural context and the conventions of science journalism: drama and contradiction in media coverage of biological ideas about sexuality. *Critical Studies in Media Communication*, **20**(3), 225–47.

Young, P. (1997). Writing articles from science journals. In: *A Field Guide for Science Writers* (ed. D. Blum and M. Knudson), pp. 110–16. Oxford University Press, New York.

■ FURTHER READING

- Friedman, S.M., Dunwoody, S. and Rogers, C.L. (eds) (1999). *Communicating Uncertainty*. Lawrence Erlbaum, Mahwah, NJ. This edited collection explores the news reporting of scientific uncertainty, addressing topics such as biotechnology, dioxins and global warming, amongst others. A variety of perspectives are offered into how scientists represent uncertainty, and the implications for journalists striving to report on it in an accurate manner. Public responses to media portrayals are also examined, providing insights into how attendant controversies are understood.

- Kiernan, V. (2006). *Embargoed Science*. University of Illinois Press, Urbana. This book focuses on the processes behind science news, paying particular attention to the embargo system. Specifically, it documents how a small number of elite scholarly journals endeavour to effect control over news media coverage of scientific developments, namely by distributing advance copies of their articles on the condition that reporters agree to wait until the embargo lifts before releasing news stories. It is argued that this system fosters uncritical, shallow forms of 'pack journalism'.

■ USEFUL WEB SITES

- **Association of British Science Writers (ABSW): http://www.absw.org.uk/**. This UK web site aims to contribute to the improvement of science journalism, primarily by helping individuals interested in writing about science and technology. A range of informational resources are available (including briefing documents, a 'find an expert' service, as well as advice about training and career guidance). Also available on the web site is a blog called ABSW, and a regular newsletter, entitled '*The Science Reporter*'.

- **National Association of Science Writers (NASW): http://www.nasw.org/**. This US web site represents an extension of NASW's remit, which is to promote 'the dissemination of accurate information regarding science and technology, through all media normally devoted to informing the public', as well as to foster 'the interpretation of science and its meaning to society, in keeping with the highest standards of journalism'. The web site offers online resources, such as links to web sites for the latest science news, teaching materials for educational institutions as well as tools and tips for freelance science journalists, amongst other features.

4.2

Science reporting in the electronic embrace of the internet

Brian Trench

Science reporting and the internet

Scientists and journalists were among the professional groups most thoroughly affected by developments in information and communication technologies in the latter part of the 20th century. The need to share information between researchers drove each successive phase of the development of the internet from the 1970s up to the emergence of the World Wide Web to wider publics from its birth in a European particle physics laboratory, CERN. For at least two decades, internet communication in various forms has become naturalised in science.

Over a similar period, journalists have experienced rapid and disruptive technological change. In the lifetimes of many individual journalists, newspaper production moved from technologies largely unchanged for almost a century to technologies that facilitate global information-sharing and simultaneous publication in several countries or regions. A wide range of facilities previously provided on separate platforms and sometimes mediated through other groups of professionals have become available on journalists' desktop computers.

From around 2000, e-mail and web-searching became effectively universalized in professional journalism. The use of online resources by US newspaper journalists in news-gathering in general grew from 57 per cent of survey respondents in 1994 to 92 per cent in 1999 and reported daily use of online resources grew from 27 per cent to 63 per cent (Garrison 2001). Over the same period, the internet and web moved from sixth position among online resources used—behind data bases such as Dialog and Nexis UK, and bulletin boards—to clear first position.

The internet is the primary means by which scientific information is shared within scientific communities. It connects to repositories of information accumulated over several decades. It is the means by which projects are established, conferences are organized, relationships are struck up and, increasingly, scientific findings are formally disseminated.

For researchers, web-mediated data bases of journals and other publications are a primary resource. For research centres, a web presence is essential—without it, the centre in some sense does not exist. For anybody regularly looking for information about science in a professional capacity or otherwise the internet is likely to be the first recourse.

The technological shifts in the professional communities of science and of journalism have created overlapping information and communication spaces in which both communities have a significant presence. Very few other areas of media interest inhabit electronic spaces that enclose the worlds of both media sources and media producers so completely. In politics, sport, arts, law, maybe business, media are required to a greater extent to observe events and interact directly with sources in order to perform their routine tasks.

Journalists specializing in science were earlier than most in becoming comfortable and competent in the use of internet technologies, in part reflecting their closeness to their source communities and in part down to their curiosity about and disposition to new developments. Science journalism was becoming established as a specialism in the same period (the 1970s onwards) as journalism and science were experiencing successive waves of technological change, and science journalists were at the leading edge of these developments. Between 1994 and 1999 e-mail use increased nearly fourfold among journalist members of the National Association of Science Writers (NASW) in the United States; in 1994, many science journalists had contacted scientific sources by e-mail and by 1999, nearly all had done so and nearly all had used information from the web in their reports (Trumbo *et al.* 2001).

Among journalists beyond science and beyond the United States such widespread use was slower in coming. But a 2006 survey of Italian journalists found that 60 per cent connected online more than ten times a day (Fortunati *et al.* 2006). Search engines, personal e-mail, news web sites and web sites of sources were ranked third to sixth among the most important means of communication in newsrooms, behind face-to-face and phone conversations. The surveyed journalists agreed strongly that the use of the internet was making journalism more of a 'desktop job', and slightly less strongly with the propositions that internet use was getting more information and a wider range of sources into stories. A related survey of Greek journalists found that over two-thirds of Greek journalists use the internet as a research medium and information medium to a very large or large extent (Panagiotarea and Dimitrakopoulou 2006).

The increasing attention being paid from the 1980s in many scientific communities to promoting scientific literacy influenced how information and communication technologies were applied in communicating science. Across the developed world—and, increasingly, in the developing world—the concern with scientific literacy, awareness and understanding spawned government initiatives, actions by professional societies, schools' programmes and media campaigns, and from the mid 1990s the web and e-mail emerged as important platforms for these efforts.

Internet technologies have simultaneously provided a means for scientific institutions and communities to communicate more effectively with media but also directly with publics, without the mediation of journalists. Also, many of the services introduced on the web with media as primary targets are accessible to all internet users. A former press officer with the Royal Society notes that:

. . . web sites are becoming an increasingly important public relations tool for disseminating messages, and are competing with and affecting media coverage.

Ward (2007, p. 171)

The coexistence of competition and influence is just one of the several paradoxes that have characterized the web's development in science communication generally and in the reporting of science in the media. I have developed this observation elsewhere (Trench 2007a, 2008), but here I shall outline some of the resources and strategies put in place with the aid of internet technologies to support or 'subsidise' media reporting of science.

Source strategies and subsidies

As already indicated, the use of internet technologies in communicating science and communicating between science and media, in particular, is shaped by social as much as technological factors. From the 1980s, universities, research institutes, research funding agencies, professional societies and scientific publishers have increasingly adopted corporate public relations strategies, as they competed for students, researchers, subscribers, political support, research funding, media attention and commercial relations. In this, they followed other institutions, where:

. . . the employment of public relations practitioners (PRPs) . . . has recently expanded at a significantly higher rate than most studies have acknowledged.

Davis (2000, p. 39)

Among non-government and voluntary organisations too 'a much larger range . . . have begun employing PR strategies and personnel' (Davis 2000). Those with a stake in science, particularly environmental activists (Hansen 1993; Anderson 2003), became notable exponents of various facets of PR, including technology-assisted PR, from the 1980s onwards.

Whereas some accounts described the developing role of science information officers in relatively benign terms (Rogers 1986), Nelkin tellingly titled her seminal work on science and media *Selling Science—How the Press Covers Science and Technology* (Nelkin 1987, 1995) and between the two editions increased the emphasis on the place of science public relations. By the early 1990s those working in science PR in Britain were plentiful enough to warrant forming a professional network, Stempra (science, technology, engineering and medicine public relations association), to 'share information and expertise'.[1]

Driven by the changing social contexts of knowledge production and facilitated by online and other digital media technologies, many higher education and research organizations have employed teams of information officers, science writers, webmasters and others, to boost their media (and broader public) profile. Following the lead of American universities, higher education institutions in Europe are putting together communication units of eight, ten and more staff, several of these focused on media relations and generally one or two on development and maintenance of web sites.

---

1. http://www.stempra.org.uk.

This adoption of corporate PR approaches in science also acknowledges what became widely accepted in institutional and corporate public relations in recent years, namely that there is greater value in providing information that is easy and inexpensive to use, and credible, than in seeking explicitly or covertly (e.g. through 'spin') to persuade media and publics. Whereas the internet can be used to provide information in timely and user-friendly fashion, it cannot readily be used for spin: there is always another source just a click or two away that will reveal the subterfuge.

Within public relations the practical uses of the internet have been much discussed in the professional literature (e.g. Witmer 2000; Phillips 2001; Kelleher 2007) and the internet has been seen as contributing to a 'revolution' (Philips 2001) or 'transformation' in PR (Gregory 2004). The web is, for example, 'the first controlled mass medium', in that those providing information have control over how it reaches end-users (White and Raman 2000). Similarly:

. . . the press officer is no longer the only means for access by a journalist. Corporate information is also available on company, government, media, academic and personal web sites.

Phillips (2001, p. 3)

Rethinking the corporate communication strategy has not come naturally, however. Several studies have been critical of the web sites of higher education and research institutions, saying, for example, that information about research at these institutions was not prominently featured in home pages and:

. . . the institutions are more interested in obtaining money from alumni than attracting new students or providing services to current students, faculty and staff, parents and family, or visitors.

Will and Callison (2006, p. 182)

A study of US university web sites found that some were weak in 'relational communication' (Kang and Norton 2006). Yet others demonstrated how universities in European countries and European public research institutions were failing to realize the research communication possibilities of the web (Lederbogen and Trebbe 2003; Jaskowska 2004; Massoli 2007; Trench 2007b).

The application of e-mail and e-mail–web combinations to media relations has been more consistent. Technological changes have facilitated the supply of information directly to the desktops of individual news editors, section editors, picture editors and reporters in forms that allow its direct transfer to the media production process. Callison (2003), noting the ways in which US journalists use companies' web sites, looking for personnel information, PR contacts, press releases and corporate profiles, underlines to PR professionals the value to journalists of materials that can be easily found, can be downloaded and 'can be quickly edited and type-set' without needing to be re-keyed.

A survey in 2000 of PR practitioners among the membership of the NASW showed that nearly three-quarters considered e-mail 'essential' in their media relations work, with 40 per cent 'routinely' sending releases out by e-mail (Duke 2002). Over two-thirds thought e-mail use and web use—with very similar numbers in both cases—had 'somewhat increased' or 'greatly increased' media coverage of their organization; 91 per cent were 'routinely' posting their releases to the organization's web site.

Interviews with journalist members of the NASW on their uses of e-mail and web established that these journalists found communication easier with sources, including for checking-back on information or interpretation, and helped expand the range of possible sources (Dumlao and Duke 2003). Several reporters referred to the convenience of having releases or alerts sent directly, one interviewee commending some sources as 'amazing at getting coverage of their stuff, simply because they have such a great e-mail alert system' (Dumlao and Duke 2003, p. 293). A prominently mentioned use of the web was that of tracking down individuals for contact by e-mail; the interviewees were generally more sceptical about the value of this technology for information-gathering, except where the sources are known and trusted, in which case they may be checked routinely.

A survey of media professionals published by the European Commission (2007) showed that the sources most used for information about science and research were 'scientific and peer-reviewed journals' (62 per cent of respondents), 'internet, including search engines' (54 per cent), 'news agencies' (40 per cent) and 'personal contacts with researchers' (37 per cent). Assuming that some, even most, of the contact with journals, and with other named sources such as newspapers, public research organization and European institutions, is mediated through the internet, the real result for the internet, more broadly defined, could be the highest, and by a wide margin.

The reported impact of supply of information by e-mail on the levels of coverage may be taken to demonstrate the increased value of PR practitioners' 'information subsidies' in communicating science. This influential concept refers to the supports given to media to facilitate transfer of information into media spaces. The defining elaboration of the concept (Gandy 1982) was preceded by a study that focused on the supply of information subsidies by interests in health and medical sciences (Gandy 1980) and which suggested that:

. . . success in providing information subsidies to one's chosen targets is closely tied to the resources available to the subsidy giver, since considerable resources are necessary for the creation of pseudo-events.

(Gandy 1980, p. 106)

Internet technologies have given subsidy-givers more resources while diminished resources on the part of subsidy-receivers have further tilted the balance. The ability of PR practitioners to influence news production 'has been given added impetus by a rapid decline in editorial resources and a growing media dependency on sources' (Davis, 2000, p. 39). Journalists interviewed about their relations with scientist-sources referred to the impact of organizational and technological factors, specifically shortening deadlines, on their information-gathering (Reed 2001). In these circumstances the market for information subsidies has expanded 'and sources are increasingly employing PRPs to supply this market with their own individualised brands of subsidy' (Davis 2000, p. 44).

It has been claimed, independently of technological considerations, that 'science journalism tends to be source driven and source framed', though few studies have looked at 'processes of contestation and negotiation among news sources that impact upon science coverage' (Anderson *et al.* 2005). One of the few such detailed studies, on a case in psychology, charted the 'natural history' of a news item on false memory syndrome, focusing on a framing contest between the British Psychological Society and the British

False Memory Society (Deacon *et al.* 1999). That study's authors concluded that it demonstrated the influence on the media of sources that were given privileged access.

Evidence for the source-driven character of science journalism also comes from professionals in the field. Wilkie (1996) describes journalists covering medical science as 'trawling regularly' the top general-science and medical science journals and refers to press releases and news agency stories as generating 'an electronic deluge of information every day'. The electronic deluge has become increasingly targeted as more recent technologies provided the means to improve the availability and accessibility of scientific and higher education organizations, to provide information in many formats and in media-friendly, even ready-to-go, forms and to filter information according to the needs of media professionals.

Internet services for media science

A vast array of services is offered via the internet to subsidise media reporting of science. Some of these represent modifications of services previously offered by other means (phone, fax, post, courier, etc.) and others are specific to the internet. The universal usage of e-mail by journalists reporting science makes e-mail a primary means of bringing science 'news' to the media's attention. HTML-formatted e-mails that mimic the on-screen or on-paper look of printed or electronically published magazines and journals make the bridge between e-mail and other platforms.

Scientific journals and popular science magazines provide to media and, in many cases, also to prospective subscribers, e-mail alerts on forthcoming stories and reports; online 'specials' outside of the normal publishing cycle; news services on web sites updated several times daily in order to keep visitors returning to the sites; e-zines sent by e-mail that present some of the print publication's content in magazine layout, some of it available in full text; electronic tables of content (eTOCs), presenting the content of the next edition in a series of headlines, some or all of these—depending on the publication, and with variations in the levels of immediate access—linked to the full text of the items.

Many journals and press release services supply updates from their web sites by RSS (really simple syndication), meaning that updates matching the registered users' stated interests will display directly on the users' screens, rather than having to be sought in the e-mail inbox. Some journals offer short audio clips in podcast form and short video packages through the web site *YouTube*.

The most 'traditional' of such information subsidies is the press release sent under embargo several days in advance of publication, giving notice of selected items in the next edition. The *European Respiratory Journal* (ERJ), for example, provides by e-mail a press release with journalistic versions of a half-dozen selected items from the latest edition of the journal. These summaries are ready-to-use but also include all the contact details for nominated authors to facilitate media interviews. The full article can be requested by e-mail. In the ERJ 'press pack' for the December 2007 edition of the journal, a summary of an article was provided under the heading, 'Vaccinate everyone against influenza?' This referred to a journal article entitled, 'Influenza- and respiratory syncytial virus-associated mortality and hospitalisations'. The press release for the January 2008 edition offered 'Improved outcome of experimental lung transplantation' as a version of the journal article, 'Keratinocyte growth factor prevents intra-alveolar oedema in experimental lung isografts'.

What these two examples among many thousands of possible examples from a wide range of journals illustrate is that one of the important information subsidies is in the provision of vernacular terms and everyday references to ease the passage of the formal scientific information into the general media. Science journalists interviewed on their interactions with scientists (Reed 2001) referred to the value of metaphors and analogies in improving accessibility to scientific information. *Nature*'s press releases often provide masterly demonstrations of this service; that journal's press releases will be considered in more detail in the following section.

Common to the practices of many journals is the prominent inclusion of an embargo date and/or time. The mere presence of the embargo represents a kind of information subsidy: it provides a 'news peg', it creates an 'event' where none previously existed, the publication 'today' of material that may have been several years in gestation (Kiernan 2003). But the embargoes of *Nature* and *Science* also contribute to:

. . . a remarkably predictable rhythm in the weekly science news cycle. Every Thursday and Friday, newspaper and web sites nearly always feature articles about scientific research published in those journals—news coverage that's carefully orchestrated by the journals' publishers.

Peterson (2001, p. 252)

Kiernan (2003, p. 917) concludes his study by commenting that:

. . . when it comes to breaking news about scientific research, newspapers try to make sure they cover the stories that other newspapers cover. The goal is not to be different, but to be the same.

The web sites of scientific institutions, including journals, also provide services that are particularly targeted at the media. These include 'press room' services, sometimes for registered media representatives only, and sometimes open to all web users, where press releases may be posted at the same time as they are distributed directly to media, and an archive of releases is maintained. Images may be made available for download in such press room facilities—the NASA web sites, for example, provide a range of multimedia content for use in print and broadcast media.[2] Other scientific institutions provide 'briefings' on selected topics that are in the news, such as the Royal Society's guide to 'Facts and fictions about climate change'.[3] The unlimited storage space on the web means that these briefings can be kept accessible for later use when the topic comes back into the news.

'News' pages with journalistic accounts of current or recently completed research and selections of stories culled from the mass media about such research are common features on the web sites of scientific institutions. An increasing number of such institutions provide audio or video interviews with the researchers involved. These may not be of a

2. http://www.nasa.gov/multimedia/index.html.
3. http://royalsociety.org/page.asp?id=4761.

standard to make them directly transferable to the broadcast media but may alert these media to the ability of the researchers to undertake an interview on camera.

Other media services that have grown up on the internet provide intermediation between journalists and scientists and their institutions. The AAAS established EurekAlert! as a means for researchers to make their media releases very widely available. Among the other web-based services channelling press releases from (mainly US) universities and research centres to the media and wider publics are Science Daily[4] and Newswise.[5]

The Royal Institution in Britain took the initiative in 2001 of setting up the Science Media Centre, intended as a resource to support better coverage of science in the media, or, in its own terms, 'first and foremost a press office for science when science hits the headlines'.[6] This initiative corresponded to the suggestion made by Reed's (2001) inter-viewees for a central 'meeting point' for scientists and journalists, enabled by technological developments and supplementing the services of public relations professionals. But the Science Media Centre also sees itself as 'pro-active' in facilitating 'more scientists to engage with the media when their subjects hit the headlines'.[7] Much of that engagement is direct, e.g. through face-to-face briefings, but the centre's guides to scientific subjects 'in a nutshell' and its press releases are published over the web.

Professional societies that are also journal publishers use their web sites not only to support their members and to build public awareness of and support for their area of scientific interest but also to promote their journals by encouraging media reporting of their content. The Institute of Physics (IOP) in Britain maintains a news service[8] that often features items based on material in IOP-published journals. According to Peterson (2001, p. 252):

. . . much [US] media coverage of physics concerns papers in the journal *Physical Review Letters* that were highlighted [on the web] by public affairs staff at the American Institute of Physics.

The cases of *Nature* and AlphaGalileo

The *Nature* press service and AlphaGalileo, a multilingual distributor of releases from European scientific institutions to the media, are among a small number of services high-lighted as 'resources' on the web sites of several professional groups, such as the NASW, the Association of British Science Writers (ABSW), the World Federation of Science Journalists and the European Union of Science Journalists' Associations.

The *Nature* press release is perhaps one of the best-known examples of a science media service. It long pre-dates the web but its operation has become more sophisticated still through the use of e-mail and the web. In the 1990s, when fax was a principal medium

4. http://www.sciencedaily.com.

5. http://www.newswise.com.

6. http://www.sciencemediacentre.org/.

7. http://www.sciencemediacentre.org/.

8. http://physicsworld.com.

of communicating releases to the media, *Nature*'s system allowed journalists receiving their releases to retrieve selected items of journal content through a fax-connected data base, using discrete codes for individual items.

The *Nature* release is sent by e-mail to 4000 registered press service subscribers around the world on the Friday before publication of the following week's edition of the journal. It refers the recipient to a subscriber-restricted web site where the papers and articles to which the release refers (plus those from a further 20 *Nature* subject-specific journals) may be accessed.

The *Nature* release states prominently, and repeatedly, the restrictions on usage of the journal material to which the registered users have access. Information on the embargo hour—18:00 GMT on the evening (Wednesday) before formal publication of the print edition—is given in relation to several global time zones. There is detailed information on the times from which the journal's contents may be accessed from the *Nature* press site. The release itself is declared to be *Nature*'s copyright, and there is a prominent plea that it should not be redistributed. All of these injunctions underline the privilege that the journalist enjoys in receiving the release.

The release presents selected items of the journal's content in 100–200 words each, written in a journalism style with headlines and with vernacular phrasing that suggest ways in which the content might be made more accessible to general readers. Thus, an item on a study of North Sea and North Atlantic fish stocks was summarized under the heading, 'Herring buoyant, but cod still in its plaice' (for edition of 24 August 2000), and, in the same release an item on the 'noble' gas, argon, was headed, 'A noble cause'. For many years, the *Nature* release has often introduced its last item with the phrase, 'and finally', mimicking a feature of British television news that marks the last item as lighter or quirkier.

The release also crucially contains details on how to contact the author(s), often with mobile phone numbers, and may, in cases where an item is considered worthy of special attention, give details of a press conference by the author(s). Here too internet technologies have strengthened *Nature*'s offering: press conferences may be relayed over the internet, allowing journalists in several continents to participate without having to leave the office.

Science, the main rival to *Nature* as a general-science reference journal, operates a similar media service. Both journals also run web-based news services—*News@Nature* and *Science News*—with frequent daily updates. These also provide a resource to general media in their coverage of science.

AlphaGalileo was established initially with the support of British scientific institutions and subsequently of institutions across Europe to provide a means for higher education and research institutes in the continent to distribute their releases to media across the world. It was conceived as a complement to EurekAlert! and is seen by its backers as:

. . . an essential ingredient not only in the dissemination of European research news but also in enabling wider engagement and dialogue between researchers and civil society—a process that rests, at least in part, on the provision of high quality information through the popular media.

SIRC (2006, p. 6)

As well as media releases from research organizations—the core of the service—AlphaGalileo provides a calendar of events, book announcements, expert data base and image library, as well as other resources to support media reporting of research. At the time of writing (January 2008), the service has 6000 registered users who can contribute or download media releases; these invariably contain relevant contact details and sometimes include hyperlinks to further material such as the reports or papers on which the release is based. In the final quarter of 2007, the average monthly figure for media releases posted on the service was 650. Thus, using no other means of gathering information for reporting science but AlphaGalileo, a journalist could have over 20 potential story leads daily. Notice of releases posted on topics for which journalist users have registered comes directly through e-mail alerts. AlphaGalileo sends out 60,000 e-mails per week, meaning that most users are getting several e-mails daily drawing their attention to something on the web site. A minority of users also receive notice of updates through RSS feeds.

AlphaGalileo provides a high level of technical and other supports to its users, reporting in periodic newsletters[9] on additions to its staff and services, including the increased number of languages in which news releases are provided.

A survey of service users showed that e-mail alerts and RSS feeds, on the one hand, and browsing of the site for research news, on the other, were used at approximately equal levels—nearly 90 per cent of users in both cases (SIRC 2006). Ninety per cent of survey respondents, of whom 60 per cent were journalists, subscribed to the site's science section, over three times or four times, respectively, the subscription levels for society and humanities. Reflecting the synergistic use of internet-mediated services, AlphaGalileo users reported that they also used EurekaAlert! (47 per cent), *Nature* services (37 per cent), the *Reuters* news agency services (28 per cent) and Cordis, the European Commission's web-based information service on funded research (18 per cent), with a range of other internet-mediated services named as having lower levels of usage. Ease of use, ease of understanding and timeliness were among the positive attributes of AlphaGalileo's information provision mentioned most by journalist and editor users. Over 80 per cent of all survey respondents rated the releases as 'excellent' or 'good' (SIRC 2006).

Implications for science journalism

We still do not have precise accounts of how, in the making of individual stories, journalists reporting science use the internet as a resource. In relation to a single story, such use might include accessing a press release, following a link to sending an e-mail seeking contact or clarification, searching the web or journal data bases for others working on the same topic, using the staff directory at a university site to get a phone number, and more.

This assumed procedure draws attention to one significant implication of internet usage: the increasingly desk-bound character of journalism in general, and of science journalism in particular (Holliman 2000). At the European Forum on Science Journalism, held in Barcelona in December 2007, Tim Radford, former science editor of the elite UK newspaper *The Guardian*, reflected on this phenomenon by recounting a story of a

---

9. http://www.alphagalileo.org/en/about-us/AlphaGalileo-eNews.

visit he made to the Natural History Museum in London to interview an individual on a particular topic, but that led indirectly to encounters with several other researchers at the museum and thus to further stories.[10]

Nothing can trump the value—so the implied moral of the story—of the face-to-face meeting. On the other hand, being desk-bound, but with access to a wide range of sources on and through the internet, can add another kind of value to the journalistic enterprise. Already in the early 1990s, before the web caused the internet to become ubiquitous, the available resources for electronic information-gathering—including scientific, techno-logical and business data bases, bulletin boards and news groups—were considered the possible basis for a rebalancing of relations between reporters and sources and the provision of news in a 'critical and more balanced context' (Koch 1991, p. 310).

More recent reflections on the impact of such practices on professional roles of the journalist have included the argument that:

. . . journalists [will] be less gatekeepers and more cartographers pointing out interesting news paths online rather than filtering and packing a closed news product.

Santamaria (2004)

Greek journalists surveyed considered that the roles of 'information specialist' and 'critical analyst' would be much more important than that of 'neutral information broker', or the traditional reporting role (Panagiotarea and Dimitrakopoulou 2006).

Perhaps rather more than with other subject specialisms in journalism, the practice of science journalism has conformed to a transmission model, that is, a faithful relaying of information from privileged sources to diverse publics. In politics and arts, for example, the specialist journalist is expected to be a critic and interpreter, to put new develop-ments into relevant contexts and to assess how well stated or presumed aims have been achieved. In science, with the pervasive use of the internet in all aspects of commun-ication within scientific communities, between these communities and between the scientific world and other worlds, journalists have materials at their desktops that allow them to perform these roles: on more or less any item that passes a 'newsworthiness test' (see Allan, Chapter 4.1 this volume), they may be able to find further scientific publications, conference presentations or research reports beyond the one at hand, details of the research programme of which it is part, biographies of the people involved, personal web pages of some of these and discussions of the topic among interested parties, including dissenting parties.

Crucially, these possibilities are also open to other web users who have the compet-ence and the motivation to explore behind the news contained in a media release or news item posted on an institutional web site. Access to the web has opened up many aspects of scientific research previously hidden from the general public (Peterson 2001). Members of interested, but non-specialist, publics have access to information prepared by professional organizations primarily for consumption by professionals. Parts or all of sites maintained by scholarly societies and scientific journals are open-access or require only that users register by name. Online discussion forums, personal web pages, blogs,

---

10. Author's notes of the panel discussion: 'Privileged? Brutalised? Beleaguered? Are science journalists needed any more?' At the European Forum on Science Journalism, Barcelona, 3 December 2007.

open-access publications, pre-print servers, are just some of the generally accessible means by which we can retrace some of the steps scientists take in their backstage preparation, and explore how uncertainties are negotiated in doing science.

The continuing proliferation of these information sources is changing the environment for science reporting in the mainstream media and for reception of such coverage. But science journalism's adaptation to these conditions has been very limited. Tim Radford, in the discussion mentioned above, described it as 'the curse of science journalism [to] spend all day recycling material that is already available'. The success of the privileged institutional sources in expanding their electronic embrace of the media may have reinforced the deference of science journalism to those sources. Meanwhile, reflexive scientists have for some time argued the need for a more critical press (e.g. Goldsmith 1986; Lévy-Leblond 1996; Rose 2004), and the internet provides important means to elaborate that critique.

In relatively early days of the discussion about public understanding of science and scientific literacy in Britain, John Durant (1993), one of the founding figures in the academic study of these issues, distinguished between three kinds of knowledge members of the public may or may not have about science: knowledge about essential facts; knowledge of how science works; and knowledge of how science really works. Using strategies of story-telling, rather than fact-downloading, and using the resources available to them, journalists reporting science could help throw light on how science really works, showing the continuing struggle with the uncertainties of science and, in this way, giving media audiences a stronger sense of the limitations and of the achievements of science.

■ REFERENCES

Anderson, A. (2003). Environmental activism and news media. In: *News, Public Relations and Power* (ed. S. Cottle), pp. 117–32. Sage, London.

Anderson, A., Petersen, A. and David, M. (2005). Communication or spin? Source–media relations in science journalism. In: *Journalism: Critical Issues* (ed. S. Allan), pp. 188–98. Open University Press, Maidenhead.

Callison, C. (2003). Media relations and the internet: how Fortune 500 company web sites assist journalists in news gathering. *Public Relations Review*, **29**, 29–41.

Davis, A. (2000). Public relations, news production and changing patterns of source access in the British national media. *Media, Culture and Society*, **22**(1), 39–59.

Deacon, D., Fenton, F. and Bryman, A. (1999). From inception to reception: the natural history of a news item. *Media, Culture and Society*, **21**, 5–31.

Duke, S. (2002). Wired science: use of World Wide Web and e-mail in science public relations. *Public Relations Review*, **28**, 311–24.

Dumlao, R. and Duke, S. (2003). The Web and e-mail in science communication. *Science Communication*, **24**(3), 283–308.

Durant, J. (1993). What is scientific literacy? In: *Science and Culture in Europe* (ed. J. Durant and J. Gregory), pp. 129–37. Science Museum, London.

European Commission (2007). *European Research in the Media: What do Media Professionals Think?* European Commission Directorate General for Research, Brussels.

Fortunati, L., Sarrica, M. and de Luca, F. (2006). The influence of the Internet on the practices and routines of Italian journalists. In: *The Impact of Internet on Mass Media in Europe* (ed. N. Leandros), pp. 267–80. Arima, Bury St Edmunds.

Gandy, O. (1980). Information in health: subsidised news. *Media, Culture and Society*, **2**, 103–15.

Gandy, O. (1982). *Beyond Agenda-setting: Information Subsidies and Public Policy*. Ablex, Norwood, NJ.

Garrison, B. (2001). Diffusion of online information technologies in newspaper newsrooms. *Journalism*, **2**(2), 221–39.

Goldsmith, M. (1986). *The Science Critic: a Critical Analysis of the Popular Presentation of Science*. Routledge and Kegan Paul, London.

Gregory, A. (2004). Scope and structure of public relations: a technology driven view. *Public Relations Review*, **30**(3), 245–54.

Hansen, A. (1993). Greenpeace and press coverage of environmental issues. In: *The Mass Media and Environmental Issues* (ed. A. Hansen), pp. 150–78. Leicester University Press, Leicester.

Holliman, R. (2000). Representing science in the UK news media: 'Life on Mars?', cell nucleus replacement and Gulf War syndrome. Unpublished PhD thesis. Faculty of Social Sciences, The Open University, Milton Keynes.

Jaskowska, M. (2004). Science, society and internet in Poland. In: *Scientific Knowledge and Cultural Diversity: Proceedings of Eighth International Conference on Public Communication of Science and Technology, Barcelona*, pp. 263–7. Rubes Editorial, Barcelona.

Kang, S. and Norton, H.E. (2006). Colleges and universities' use of the World Wide Web: a public relations tool for the digital age. *Public Relations Review*, **32**, 426–8.

Kelleher, T. (2007). *Public Relations Online: Lasting Concepts for Changing Media*. Sage, London.

Kiernan, V. (2003). Embargoes and science news. *Journalism and Mass Communication Quarterly*, **80**(4), 903–20.

Koch, T. (1991). *Journalism for the 21st Century: Online Information, Electronic Databases, and the News*. Adamantine Press, Twickenham.

Lederbogen, U. and Trebbe, J. (2003). Promoting science on the Web: public relations for scientific organizations—results of a content analysis. *Science Communication*, **24**(3), 333–52.

Lévy-Leblond, J.-M. (1996). *La Pierre de Touche: la Science a l'Épreuve*. Gallimard, Paris.

Massoli, L. (2007). Science on the Net: an analysis of the websites of the European public research institutions. *Journal of Science Communication*, **6**(3) (**http://jcom.sissa.it/archive/06/03**).

Nelkin, D. (1987). *Selling Science: How the Press Covers Science and Technology*. W.H. Freeman, New York.

Nelkin, D. (1995). *Selling Science: How the Press Covers Science and Technology*, 2nd revised edn. W.H. Freeman, New York.

Panagiotarea, A. and Dimitrakopoulou, D. (2006). Greek journalists in the digital era: innovators or laggards? In: *The Impact of Internet on Mass Media in Europe* (ed. N. Leandros), pp. 337–50. Arima, Bury St Edmunds.

Peterson, I. (2001). Touring the scientific web. *Science Communication*, **22**(3), 246–55.

Phillips, D. (2001). *Online Public Relations*. Kogan Page, London.

Reed, R. (2001). (Un-)professional discourse? Journalists' and scientists' stories about science in the media. *Journalism*, **2**(3), 279–98.

Rogers, C.L. (1986). The practitioner in the middle. In: *Scientists and Journalists: Reporting Science as News* (ed. S.M. Friedman, S. Dunwoody and C.L. Rogers), pp. 42–54. American Association for the Advancement of Science, Washington, DC.

Rose, S. (1 April 2004). Stop pandering to the 'experts'. *The Guardian*.

Santamaria, D. (2004). *Professional Routines in Catalan Online Newsrooms—Online Journalism in Real Contexts*. Posted at **http://www.cibersociedad.net/congres2004/grups/ fitxacom_publica2.php?group=89&id=112&idioma=es**.

SIRC (Social Issues Research Centre) (2006). *AlphaGalileo: Review of Performance Against Strategic Goals 2005–2006 (Phase Two)*. Social Issues Research Centre, Oxford.

Trench, B. (2007a). How the Internet changed science journalism. In: *Journalism, Science and Society: Science Communication Between News and Public Relations* (ed. M. Bauer and M. Bucchi), pp. 133–41. Routledge, London.

Trench, B. (2007b). Irish research looks inwards—Irish research organisations on the web. Unpublished presentation to European Research Public Relations Seminar, Dublin, October 2007.

Trench, B. (2008). Internet: turning science communication inside-out? In: *Handbook of Public Communication of Science and Technology* (ed. M. Bucchi and B. Trench), pp. 185–98. Routledge, London.

Trumbo, C., Sprecker, K.J., Dumlao, R.J., Yun, G.W. and Duke, S. (2001). Use of e-mail and the Web by science writers. *Science Communication*, **22**(4), 347–78.

Ward, B. (2007). The Royal Society and the debate on climate change. In: *Journalism, Science and Society: Science Communication Between News and Public Relations* (ed. M. Bauer and M. Bucchi), pp. 159–72. Routledge, London.

White, C. and Raman, N. (2000). The World Wide Web as a public relations medium: the use of research, planning and evaluation in web site development. *Public Relations Review*, **25**(4), 405–19.

Wilkie, T. (1996). Sources in science: who can we trust. *Lancet*, **347**, 1308–11.

Will, E. and Callison, C. (2006). Web presence of universities: is higher education sending the right message online? *Public Relations Review*, **32**, 180–3.

Witmer, D. (2000). *Spinning the Web: a Handbook for Public Relations on the Internet*. Longman, New York.

▓ FURTHER READING

• Allan, S. (2006). *Online News—Journalism and the Internet*. Open University Press, Maidenhead. Although not specifically focused on the reporting of science, this book provides a wide-ranging analysis of online news, and the factors affecting contemporary journalism. The book includes a discussion of citizen journalism, drawing on evidence from reporting of the recent South Asian tsunami and Hurricane Katrina.

• Bauer, M. and Bucchi, M. (eds) (2007). *Journalism, Science and Society: Science Communication Between News and Public Relations*. Routledge, London. This collection includes contributions from media professionals, both in science reporting and in science public relations, and from media analysts. Some journalists recount their difficulties in telling certain kinds of science stories; they and several of the academic contributors reflect on the source dependence of science reporting and on the historical shift of the balance between journalists and PR.

• Cottle, S. (ed.) (2003). *News, Public Relations and Power*. Sage, London. This edited collection examines a number of issues relevant to the strategic management of news, in particular the promotion of source material to newsrooms and how this is affecting news production. The chapter by Anderson examines the strategies used by environmental activists to gain access to newsrooms with the aim of securing media representation.

■ **USEFUL WEB SITES**

• **AlphaGalileo: http://www.alphagalileo.org/; EurekaAlert!: http://www.eurekalert.org/.** These two web sites represent the two foremost online news centres/hubs, connecting research institutions, government agencies, journals, news media, journalists, individual researchers and the public. AlphaGalileo, operated by the independent not-for-profit AlphaGalileo Foundation, is the online news centre for European research in science, medicine, technology, the arts, humanities and social sciences. EurekAlert! is an online, global news service operated by the American Association for the Advancement of Science (AAAS).

• **Science and Development Network (SciDev.Net): http://www.scidev.net/.** This multilingual web site seeks to provide reliable information resources 'to help both individuals and organisations in developing countries make informed decisions about how science and technology can improve economic and social development'. Separate web pages are devoted to news, editorials, features and opinions, as well as regional gateways and dossiers. The 'E-guide to Science Communication' provides links to original material, including practical guidance on topics such as 'reporting on science', 'dealing with the media', 'interacting with policymakers', and so forth. Extensive contacts, as well as e-mail lists, are also available.

SECTION 5
Communicating science in popular media

Representation *is* an essential part of the process by which meaning is produced and exchanged between members of a culture. It *does* involve the use of language, of signs and images which stand for and represent things. But it is a far from simple process, as you will soon discover.

Stuart Hall (1997) *Representation: Cultural Representations and Signifying Practices* (emphasis in original)

5.1

From flow to user flows: understanding 'good science' programming in the UK digital television landscape

James Bennett

Introduction

During Ofcom's (the Office of Communications) 2004 review of the role of public service broadcasting in the emerging digital television landscape, the regulator opined that whilst the major free-to-air channels had made some 'significant achievements' in fulfilling public service obligations, there were also 'important short comings' (Ofcom 2004a, p. 4). In particular, across the course of this review Ofcom highlighted the importance of 'challenging' programming, as that which makes 'viewers think' (Ofcom 2004b, p. 7), in defining the key characteristics of public service broadcasting that may come under threat in the digital landscape. However, Ofcom's analysis did not lay the blame solely at the feet of the broadcasters; rather it recognized that as digitalization led to a proliferation of channels, let alone platforms on which to view 'TV', audiences may begin to 'drift [*sic*] away from the more challenging types of programming, traditionally thought to be at the heart of UK television' (Ofcom 2004a, p. 4). Writing about recent trends in science programming, Michael Jeffries (2003, pp. 528–32) has argued that good science programming is 'characterized by change, crisis and challenge', placing the genre squarely within those most likely to experience a decline both in terms of audiences and production in the digital television landscape. Indeed, this seems evidenced by Ofcom's own report, which went on to detail that it has been 'some of the more serious and challenging programme types' which are 'most affected by multichannel competition. *Horizon*, *Newsnight* and the *South Bank Show* all had a viewing share more than 50% lower in multichannel homes compared with homes with only the main terrestrial TV channels' (Ofcom 2004a, p. 6).

For Ofcom, science programming, along with other 'serious factual' genres, therefore becomes part of the central 'citizen rationale' for public service broadcasting in the digital landscape. In such a view public service broadcasting should, in addressing likely market failures in the digital age, provide programming which, amongst other things, aims 'to stimulate our interest in and knowledge of arts, science, history and other topics through content that is accessible, encourages personal development and promotes participation in society' (Ofcom 2004a, p. 9). Indeed, despite the move away from tying public service broadcasting to specific genres in the previous year's Communications Act (DCMS, 2003), such a view of the importance of—and threat to—science programming is evident in the government's regulation of the emerging digital television landscape which enumerates specific genres 'to be supported' by public service broadcasters; for example, science, education, children's programming, religion, news and current affairs. At the same time, as Ofcom notes

It is widely recognised in the industry . . . that, at a certain point in digital take-up, the current balance of privilege and obligation, particularly for ITV, will so have eroded that, absent other measures, their public service broadcasting obligations will become commercially unviable.

Ofcom (2004a, p. 3).

There is not space here to examine the way in which each individual broadcaster is dealing with the role of science programming in fulfilling, or relieving itself of, these public service obligations. As a result, the discussion below concentrates on the British Broadcasting Corporation's (BBC's) science programming in relationship to the development of interactive television and the importance of science presenters, making reference to trends on commercial channels where appropriate. This is, however, not merely a question of the constraints of space. Rather, as the DCMS has suggested, in the digital television landscape the BBC's role will be to act as the 'cornerstone of public service broadcasting', increasingly becoming the central, if not the sole, site of public service broadcasting obligations (DCMS 2005, 2006).

However, even before Ofcom's 2004 report on the changing nature of the television landscape and the difficult place for challenging programming within this, academic accounts of science programming, such as those by Silverstone (1984), Darley (2003), Jeffries (2003) and Gardner and Young (1981), have continually asserted that whatever television science programming does achieve, when it consistently reinforces the perception of science as a simplistic, linear, consensus-making process, it fails to present science as 'fundamentally or irremovably in dispute' (Silverstone quoted in Darley 2003, p. 237). In such accounts, whatever its failings, *Horizon* (BBC, 1964–) has come to be established as the criterion against which 'good' science programming is measured. Jeffries' (2003) work is exemplary here, unpicking what he sees as a 'schism' that has developed between the BBC's production of natural history programming and other sciences. He argues that the success and investment in the former has been to the neglect of other sciences, such as cosmology, geology and non-wildlife biology. Moreover, Jeffries argues that natural history programming has become 'dominated by a paradigm of nature as essentially balanced and ordered', which sits in stark contrast to the 'good science' epitomized by *Horizon* and its presentation of science 'as changeable, challenging, contingent' (Jeffries 2003, pp. 527–43). However, as I've suggested, such programming is increasingly under threat, with the industry magazine *Broadcast* recently reporting that *Horizon* was among a slew of serious factual programmes likely to face a 20% cut in production budgets as part of the BBC's larger restructuring and cost-saving initiatives in the face of the Corporation's lower than expected licence fee increase in 2006 (Thompson 2007). The problem of production cuts is further reinforced by fears of 'dumbing down', evident not only within academic discussions of other science programming but within the industry's own understanding of the place of *Horizon* in the digital landscape. As the head of specialist factual programming for Channel 4 noted recently, programming strands like *Horizon* 'have special difficulties. The *Horizon* brand is about being the best science around, but if it's too serious people won't watch it' (quoted in Rouse 2007). It would therefore seem obvious to conclude that the problems facing good science programming are only likely to become more difficult in the digital television landscape.

However, by examining the place of science programming in relation to a series of emergent production strategies and technologies in the digital television landscape, as well as larger debates within television and media studies, I want to suggest that we should perhaps best not judge the presentation of science simply against individual programmes or, indeed, series. Rather, if we look across the 'flow' and management of 'user flows' of television (Box 1), we find a diversity of opinion represented and an

---

BOX 1 FLOW AND USER FLOW

Writing in 1975, Raymond Williams suggested that television's cultural and textual form was not the individual programme—as in cinema's filmic form—but rather amounted to an aggregation of a variety of texts, planned by the channel or broadcaster in the form of the schedule and experienced by the viewer. Williams argued that this 'phenomenon, of planned flow, is . . . perhaps the defining characteristic of broadcasting, simultaneously as a technology and as a cultural form' (Williams 1992 [1975], p. 86). The term flow, therefore, indicates the way we might speak of 'watching television', 'picking up on the general rather than the specific experience' of an individual programme (Williams 1992 [1975], p. 89). The concept of flow, and its usefulness for studying television, has been a consistent debate within television and media studies, with the most enduring criticisms of Williams' work being for the way in which it diverts attention from the individual text or programme. That is, for Williams the individual programme's meaning is only ever apparent through its relationship with the other texts it appears with: part of a rather messy and indistinct flow of other programmes, advert breaks, sponsored messages, news flashes, etc. that make up the experience of television.

However, together with the criticism that such an approach moves our attention away from detailed critical textual and aesthetic analysis, recent work has begun to suggest that Williams' model of flow is of ever more limited use in a television landscape where interactivity and multi-platform programming are increasingly common. John Caldwell usefully suggests that flow, and its relationship to the programming practices of scheduling in the broadcast or network era of television, might best be understood as a 'first shift aesthetic practice' of the television industry. As such, flow is being supplemented, but not totally replaced, by the 'second shift aesthetic' practices that are emerging with the rise of TV/dot-com synergies (Caldwell 2003).

BOX 1 CONTINUED

Caldwell outlines a number of second-shift practices of the industry, such as niche-ing, dispersal, aggregating, tiering, branding and marshalling user flows, which all accept that the television audience will inevitably move away from the programmed flow of the schedule in a digital landscape where interactivity, user-generated content and convergence are the norm. The importance of Caldwell's work therefore lies in the way in which it enables us to understand the seemingly newly empowered audience—who can interact, download, watch on-demand and even create their own media content —as still subject to the programming practices of the industry: that is, it is the 'flow' of users' own movements, interactions and creations that is now subject to the industry's programming practices. This need not be, as I show here, a pessimistic account of digital media forms, but does enable a critical attention to the texts of digital television as well as to the economics and production practices that shape them.

---

engaging, participatory and changing, if not always challenging, presentation of science. Thus, my argument below is two-fold. Firstly, I suggest that by looking at new production strategies of 'interactive' and 'multi-platform' television, we must examine how audience engagement, now re-imagined as 'interactive', is managed and structured. In so doing, my second concern is to suggest that such an approach to science is as concerned with issues of universality and the creation of engaged audiences as it is with 'challenging' science. The criterion of 'challenging' programming does not simply disappear in such a view, but rather must be understood in relationship to notions of interactivity, flow, user flows and participation. For a number of reasons that I set out below, I examine the sub-genre of natural history programming in this discussion of TV science. I deal first with the highly controversial *Walking with . . .* series and their use of interactive TV in relationship to debates about science programming by looking at the third instalment, *Walking with Cavemen* (*WWC*) (BBC/Discovery, 2003), in some detail. The chapter concludes with a discussion of the BBC's coverage of the issue of climate change in relation to the use of particular television presenters, new technologies and the role of the 'interactive audience'.

From flow to user flows: public service broadcasting and digital programming strategies

Before going on to examine these programmes in more detail, it is worth dealing briefly here with the digital programming strategies I am discussing and their relationship to debates within television and media studies more broadly. The BBC commenced delivering interactive TV across all digital television platforms—terrestrial, cable and satellite —with *Walking with Beasts* (*WWB*) (BBC/Discovery, 2001). This second instalment of the *Walking with . . .* series marked the flagship programme for the arrival of interactive television (iTV) on the BBC. 'Red button' iTV, a term which refers to the way audiences access iTV through the colour-coded buttons on their remote control previously used to

Figure 1 *Walking with Cavemen* title sequence. The interactive features are laid out on screen for the viewser to select between using the appropriately coded colour keys on their remote control: red for 'on now'; green for 'connect' and so on (see also Plate 1). © BBC/Discovery, 2003.

access Teletext features, allows the audience to 'do other things with your [*sic*] television', including vote, access news stories and video, as well as extra coverage and materials (http://www.bbc.co.uk/digital/, site visited 2 August 2006).

In the case of *WWB*, this allowed audiences to access a range of videostreams in addition to the programme's 'Mainstream', including 'Making of', 'Evidence' and 'Facts' as well as the provision of a more in-depth 'Alternate commentary' voiced by *Horizon*'s Dilly Barlow. By allowing a choice of content, delivered simultaneously, such 'applications' therefore offered manifold experiences of the same programme, differing not only from the broadcast programme but also according to the individual's personal choices (Box 2). These additional applications are illustrated in Figure 1 (see also Plate 1) by the title sequence from *WWC*, the third series in the *Walking with . . .* trilogy.

As Bennett and Strange (2008) note, the BBC's introduction of iTV in 2001 was part of a wider strategy by the Corporation to adopt a multi-platform television strategy in order to exploit the opportunities of convergence afforded by digitalization. In particular, the BBC aimed to take advantage of (and to deal with the problems that Ofcom noted in my introduction) television and web synergies that have resulted in content and users migrating across platforms in the digital era. As such, we need to understand this change as not simply a technological, digital revolution; but an industrial and cultural event that, as I show below and elsewhere, is a slow process with links to earlier forms of display. Previously termed 'projects' by Ashley Highfield (BBC Director of New Media and Technology) in 2001, the BBC have recently rebranded this approach as a '360 degree commissioning and production [strategy to] ensure creative coherence and editorial leadership across all platforms and media' (quoted in

BOX 2 DIGITAL TELEVISION: FROM PROGRAMME TO INTERACTIVE APPLICATIONS, MULTI-PLATFORMING AND THEIR AUDIENCES

Red button iTV began in the UK in 1999 with the launch of BSkyB's service for its coverage of Premier League football (soccer) matches on its digital satellite platform 'Sky'. The BBC began experimenting with iTV a year later, trialling a largely text-based service on the now defunct ITV Digital platform for its coverage of the annual tennis tournament at Wimbledon, before rolling out a much larger investment in iTV in 2001, when it began running 'applications' across all digital television platforms: cable, satellite and digital terrestrial.[1] I use the term 'application' to refer to the iTV text, which through foregrounding various elements of choice and access to extra and different materials combines television's programming forms with those of computing software and data base structures. In contrast, the term 'programme' is used throughout to denote the linear, broadcast text that appears as part of a channel's scheduled flow. Accessing interactive applications therefore removes audiences from the linear flow of broadcast television, allowing them to 'do more things with your television' as the BBC promised, including see extra news and sports coverage, play games, go shopping or, as here, 'join in with programmes' and access additional materials.

It's important to note at this point that the 'interactivity' of red button applications is extremely limited: restricted to a series of choices between different forms of content on offer. In my discussion here I focus on how these choices are structured within individual programmes' associated applications: that is, as I explain below, how choices are structured so as to manage user flows by marshalling the user's navigation of the application. Whilst I am therefore not concerned with interactivity as 'a measure of a media's potential ability to let users exert an influence on the content and/or form of the mediated communication' (Jensen and Toscan 1999, p. 201) my discussion of multi-platform television examines how the marshalling of these user flows creates engaged users, interested and involved in the outcome of scientific debates and experiments. As such, a crucial element of the programming practices of interactive and multi-platform television can be understood as the 'call to action', which notifies the audience of the availability of interactive applications or multi-platform elements that will extend their experience of the text. As such, interactive and multi-platform strategies must invoke the audience to take up a different position from the traditionally 'passive' state of television watching; that is, interactive and multi-platform television invite the audience to leave the broadcast flow of television behind and press the red button or go online. Whist this might not necessarily require the audience to 'interact' immediately, as both interactive applications and multi-platform elements are usually available beyond the scheduled transmission of a programme, the 'call to action' informs the audience of the presence of these options. Calls to action might take the form of direct invitations from presenters, as I discuss below, or simple on-screen graphics called 'DOGs' (digital on-screen graphics).

At this point it is worth making a final brief point about terminology. I use the term 'audience' to refer to the position of traditional broadcast television viewers, whilst Dan Harries' (2002) term 'viewser' is adopted to denote the 'in-between-ness' of interactivity and watching demanded by iTV applications. In turn, the term 'user' defines the audiences engaged with the interactive text of multi-platform projects and campaigns. The distinction between one position and the other is not immutable or absolute; indeed, one group is always necessarily the other—that is, audiences are also viewers and users. As such it's important to note that the television audience is not necessarily more passive than the viewser or user; ethnographic work on the television audience has consistently demonstrated that in myriad ways the TV viewer is always 'active' in their watching of television. (For more on the 'active audience' of television, see the Further Reading section at the end of this chapter.)

1. For a more detailed discussion of the development of interactive TV in the UK see Bennett (2008a).

Bennett and Strange 2008). As Bennett and Strange's exposition of the BBC's projects suggests, these diffuse and dispersed multi-platform elements are held together by threads of cross promotions and 'calls to action' for the audience to move from television screen to internet, to mobile device, interactive application and back again. More importantly, Bennett and Strange argue that the BBC use the 'real world' as another platform to engage users, which encourages participation in national or local events such as 'building a digital picture of Britain' or, as I detail below, engaging with the issue of climate change.

As I noted in my introduction, the proliferation of channels and platforms for watching 'television' on is part of the wider context within which we must understand the place of science programming. In such a landscape, traditional programming strategies, including the scheduling of 'challenging' programming in prime time or to follow on from popular programmes, can no longer be assured of attracting an audience merely by dint of their place in the television schedule. Discussing the changing industrial practices in this landscape, John Caldwell differentiates between these 'first shift' practices and understandings of broadcast television—such as scheduling and Raymond Williams' infamous account of television's cultural and industrial form as one of 'flow' (Williams 1992 [1975])—and what he terms the 'second shift aesthetics' of the digital landscape. His work demonstrates how the practices of TV/dot-com synergies, such as that enunciated by Highfield, must now attempt to 'master textual dispersals and user navigations that can and will inevitably migrate across brand boundaries . . .' (Caldwell 2003, p. 136). Caldwell convincingly demonstrates that the personalization of the text across dispersed sites is not without restriction of the user's movements. Rather users' movements are 'herded' in particular directions or large audiences and user groups are aggregated under umbrella brands that provide multiple offerings. That is, Caldwell's work charts a shift from the industrial strategy of flow in the network era of television to a strategy whereby TV/dot-com synergies attempt to rationalize 'user flows' in order to keep users on their proprietary content or those of their content affiliates for as long as possible.

As I've suggested elsewhere, we can usefully draw on Caldwell's work to examine the way in which the BBC use such second-shift practices to fulfil public service remits (Bennett 2006). In this view, whilst commercial second-shift practices are simply concerned with 'monetizing user flows', the practices of public service broadcasting in the digital television landscape are as much about keeping audiences engaged with scientific debates (or other genres) across a range of platforms, sites and encounters with television as they are about directly 'challenging' audiences in individual programmes. Thus, in drawing on Caldwell's work here, I want to suggest that we can't simply look at the individual programme as a stable entity against which to judge the delivery of 'good science' programming. Rather, we must pay attention to the way in which the expanded experience of the programme, through interactive applications and multi-platform elements, structure user/viewer choices, engagement and participation. Moreover, drawing on Helen Wheatley's (2004) work on landmark natural history programming, I argue that such an approach shouldn't come at the expense of understanding the continuing role of the scheduled flow of television in providing audiences with an understanding of science that is challenging, changing and in crisis.

Walking with . . . good science?

I want to now turn to the *Walking with* . . . series to explore the issue of 'good science' programming in relation to changing production strategies in the digital television landscape (Box 3). Discussions of the *Walking with* . . . programmes have predominantly focused on their use of computer-generated-imagery (CGI) to re-create extinct life in a format Scott and White (2003) have termed 'un-natural' history programmes. In terms of Jeffries' discussion of science and natural history programming, the series' un-natural history is therefore importantly produced by the BBC's science department, rather than the renowned Natural History Unit Jeffries is so critical of. As such, its place in debates about science programming is interesting for the fact that it lies at the inter-section of natural history's relationship with other sciences, with the programmes relying on, and including, scientific experts from fields such as palaeontology, meteorology and archaeology. As a digital production strategy itself, the use of CGI functioned as a form of spectacle which, as Andrew Darley's account of *Walking with Dinosaurs* (WWD) (BBC/Discovery, 1999) suggests, represented 'a new approach to televising science' that

---

BOX 3 *WALKING WITH* . . . CO-PRODUCTIONS

The *Walking with* . . . trilogy began in 1999 with *Walking with Dinosaurs* (*WWD*), continuing with 2001's *Walking with Beasts* (*WWB*) and 2003's *Walking with Cavemen* (*WWC*). As part of a larger co-production deal between the BBC and the American channel Discovery, the three series in the trilogy share a great deal in terms of production value, the use of scientific expertise, animatronics, CGI and, between *WWB* and *WWC*, the development of iTV applications. The success of these series saw the BBC effectively sell the *Walking with* . . . brand as a franchise to Impossible Pictures, who pro-duced a further three instalments: *Swimming with Seamonsters* (2003) (bizarrely subtitled 'A Walking with Dinosaurs Trilogy' in an attempt to exploit the successful *Walking with* . . . brand), *Walking with Monsters* (2005) and *Space Odyssey: Voyage to the Planets*, which was originally titled *Walking with Spacemen* (2004). As Helen Wheatley's (2004) work on the BBC–Discovery co-production series the *Blue Planet* (BBC/Discovery, 2001) elucidates, co-production deals intertwine the imperatives of public service broadcasting and commercial objectives, requiring the careful balancing of each. Drawing on John Caughie's (1986) work on heritage drama, she suggests how the economic context of this co-production deal infuses the *Blue Planet* with an aesthetic of 'visual pleasure', which made the series highly saleable in the international market. Thus, whilst co-production deals reduce the risk of investing in high-production dramas or documentaries for public service broadcasters by spreading the economic investment between companies, one must also question—as Wheatley and Caughie do—to what extent the 'requirements of the international market [are] compatible with the interests of the national public which public service broadcasting is there to serve' (quoted in Wheatley 2004, p. 333). On the face of it, the tacky 'whistles and bells' of the CGI-led *Walking with* . . . series suggests that the commercial imperatives of Discovery's involvement overshadow the public service requirements of 'good science' programming. However, as I'll go on to suggest, the issue of balancing commerce and public service might be more complex than this.

was 'in keeping with . . . "postmodern" culture' (Darley 2003, p. 228). In contrast Jeffries is cautiously optimistic that the series, together with its successor *WWB*, did create new and engaging modes of scientific display on television that 'at least [moved] some of the audience . . . beyond celebratory awe and into scientific debate via the series' message boards'. However, the use of CGI in the series has largely been denounced as detracting from the presentation of both natural history and science, failing to balance the spectacle of such display with the didactic aims of the genre. Indeed, Jeffries own conclusions as to the 'new and engaging modes of scientific display' found in the series is rather contradictory, basing the evidence for his claims on the complaints of message board users about the use of CGI and speculation in the series. As Darley suggests, despite the novelty of CGI offering new possibilities for the presentation of science on TV, the series arguably fell 'prey to contemporary aesthetic strategies that tend to negate representation and meaning (content), promoting instead the fascinations of spectacle and form (style)' (Darley 2003, p. 229). As a result, through a comparison with Roger Silverstone's 1984 exposition of the similarly themed *Horizon* programme *Death of the Dinosaur*, Darley suggests that earlier 'rhetorical' strategies of science programming's use of the visual image might be better suited to acknowledging the diversity of views in scientific debates, 'demanding' more of its viewers and providing 'opportunities for reflection' (Darley 2003, p. 244). By this, Darley seems in favour of a more staid aesthetic style, whereby scientific exposition is achieved by a balance of visual styles—from the use of scientists as talking heads, to models, on-screen graphics and the general avoidance of dramatic framing and lighting techniques, let alone the use of CGI to create an artificial diegesis that is dominated by spectacular encounters between predator and prey.

Ultimately, therefore, it is the 'sheer spectacle of the unfilmable' that tips the quotidian in favour of entertainment for Darley in *WWD*; its 'hyper-realist' imagery and 'stock stereotypes' mask the 'extraordinary level of speculation' in the programme and its failure to engage with the current 'situation of scientific research in the field' (Darley 2003, p. 246). In this view, science is only enlisted insofar as it is present in the 'Making of' documentaries that appended the series, where it is 'used as legitimization for what is being attempted at the aesthetic level' (Darley 2003, p. 247). Darley's critique of *WWD* might aptly be applied to the series' later instalments where the display of CGI spectacle remains paramount in the attempts to attract an audience to science programming. However, the problem Darley notes here is arguably not only much older than the introduction of spectacular forms of CGI in *WWD* but also complicated by the introduction of interactive applications in the series' subsequent instalments *WWB* and *WWC*. As I've suggested, we need to think about how such interactive applications manage 'user flows' in understanding their relationship to the delivering of 'good science' programming as a form of public service broadcasting, and it is to this issue I want to turn now through a discussion of *WWC*.

As Alison Griffiths' work has demonstrated, the display of natural history and anthropology has always been a 'site [of] complex negotiations . . . between anthropology, popular culture and commerce in attempting to strike the right balance between education, spectacle and profit' (Griffiths 2002, p. 47). Her work and that of Tony Bennett elucidates how the development of the museum, in contradistinction to the 'vulgar' display of anthropology and natural history in amusement parks, cabinets of curiosity and fairs, attempted to focus on more didactic regimes of display (Bennett 1995). As I've argued elsewhere, in tracing the links between the organization of the audience's movement through the spaces of natural history display in the museum and interactive television, the use of CGI in *WWD* and *WWB* simply represents the incarnation of a form of spectacle related to earlier forms of human-made verisimilitude, such as the use of taxidermic displays in natural history museums (Bennett 2007). However, as I also noted in that discussion, the *WWB* interactive application's structuring of user flows also privileged such forms of spectacle over and above the didactic aims of the natural history genre. Nevertheless, in understanding the BBC's role in negotiating this balance, it is important to recall the Corporation's public service obligations to provide services that are universally available and accessible to all—a particularly challenging remit in the digital landscape where audiences are fragmenting across channels and new technologies. In such a view the interactive application for *WWB* enabled the BBC to cater for different audiences at the same time: that is, by offering multiple experiences of the same text, the *WWB* application aggregated a larger audience under the 'textual dispersal' of one programme. As Mark Goodchild, the series' Interactive Executive Producer remarked, the scheduling of *WWB* (prime-time on the BBC's most popular channel, BBC One) was 'meant to show *Walking with Beasts* wasn't meant for a science audience. So the interactive application catered for the core science audience by offering something more substantial', signalled by both the 'trusted voice' of Dilly Barlow providing the more scientifically in-depth alternative commentary and the presence of the 'Evidence' and 'Making of' videostreams (interview conducted 25 January 2003).

It is, however, this issue of extending the appeal, presentation and address of science programming through interactivity that concerns me here. Whilst the interactive application for *WWC* continued to rehearse debates that I have outlined above in relation to the balancing of spectacle and education, it did so in a context where the use of iTV had become increasingly '*de rigueur*' in television production practices (cf. Bennett 2008a). In particular, apart from the criticisms of *WWB* for its failings as scientifically 'challenging' programming, the interactive application had been lambasted by industry critics for being far too complex and 'not working as a piece of interactive television'.[2] As Tom Williams (BBC's Creative Director of Interactive Television) expressed in an interview conducted in 2004, the BBC's interactive production design works under a rationale of simplicity: 'if one menu is bad [in that it requires viewser to understand and navigate it], two menus is even worse' (interview 26 August 2004). Thus in *WWC* we find a much more 'pared down' interactive application, which fitted within the BBC's broader remit of designing interactive applications to be 'comfortably exciting' (BBC 2004).

WWC's premise was to send the presenter, Professor Robert Winston, back in time for a closer look at our evolutionary ancestors. Moving away from the use of CGI in *WWD* and *WWB*, the series was predominantly populated by a large number of extras in prosthetics and ape-suits, who would make appropriate 'ooh-ooh' noises and clamber enthusiastically across the set. Their behaviours and evolution were explained by a

---

2. *WWB*'s application was a controversial winner of a BAFTA award for 'enhancement of a linear media'. Jonathan Webb, then Flextech's director of interactive media, claimed that *WWB* simply 'did not work as a piece of enhanced TV' and the application received its BAFTA award before it had actually aired on television (quoted in *Broadcast*, 25 November 2002).

Figure 2 The type of 'pop-up' labelling that punctuates *Walking with Cavemen* appears at the bottom of the screen and simply labels the display that is on screen: species, time of life and other basic facts. © BBC/Discovery 2003.

mixture of Winston's voiceover and his actual appearance in the diegesis alongside the 'cavemen', turning the 'heroic field naturalist' of Outram's (1996) analysis of natural history presenters into a 'time-traveller'. As Roger Silverstone's analysis of science programming in the early 1980s suggested, whatever the precise register of the appearance of scientists in such programming—from scientist as thinker, to scientist as interpreter, demonstrator, labourer or technician—all go towards constructing 'the scientist as hero' (Silverstone 1984, p. 400). Similarly to the other *Walking with . . .* trilogy instalments, these re-created extinct animals were given narratives, personalities and spectacular encounters, such as inter-ape-clan wars and being chased by a variety of predators. Away from *WWB*'s delivery of multiple video and audiostreams for viewers to choose from, *WWC*'s interactivity came in the form of simple 'pop-up' text boxes, which explained, interpreted and provided scientific evidence for events on screen, and the transmission of a 'Making of'/evidence programme at the end of the main feature's schedule slot. Both of these elements of the interactive application sought to provide the viewer with a greater degree of scientific veracity than Robert Winston's rather whimsical presentation and commentary, which I shall return to below. As the 'Making of'/evidence element of the interactive application was effectively comprised of the kind of interviews with scientists that Darley critiques *WWD*'s 'Making of' featurette for, I want to concentrate on the pop-up labels that featured throughout the main programme itself.

The pop-up text boxes, or labels, can be divided into two categories: firstly, those that are automatic, occupying a section in a bottom third of the viewer's screen (in Figure 2); and secondly, a series of more in-depth 'fact-files' (Figure 3),[3] which the viewer could

3. These fact-files are available online via the BBC's Press Office web site, where users can download a press pack for the series.

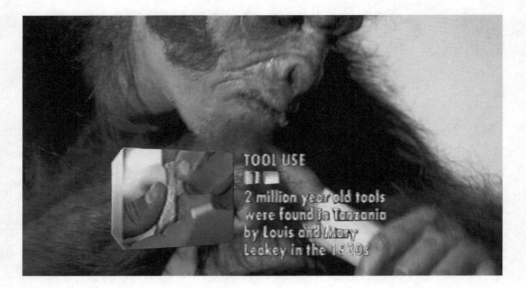

Figure 3 The more in-depth fact-files of *Walking with Cavemen* were called up by the viewser in a now-view-on-demand model, which indicated to the viewser when these fact-files were available. © BBC/Discovery 2003.

choose to call up or ignore at particular moments in the programme. The more in-depth labelling practices depicted in Figure 3 obscure almost the entire screen, leaving the viewser with an impression of what the visual content might be, but retaining the audio-stream of the main feature. In relationship to the BBC's status as a public institution with a remit of universal access, education, information and entertainment, one can profitably link the role of these labels to the practices of the museum as an earlier public institution similarly concerned with such issues.

Just as the BBC's management of user flows must distinguish the Corporation from the practices of its commercial rivals, Tony Bennett shows how the 'evolutionary museum', distancing itself from the cabinet of curiosity and the 'enlightenment museum', developed a system of labelling that was, for the first time, explanatory. Written in English, labelling as conceived for the 'new museum idea' by George Brown Goode of the Smithsonian Institute, was intimately related to the new liberalism of early 20th century museum display that fixed the museum's position as a public educator. Prior to written English accompanying the artefact, museum labelling practices had served the purpose of reinforcing the visual spectacular of the objects on display by, in the cabinets of curiosity, operating 'on the same plane rather than serving as a key to their meaning' by themselves, being 'sources of visual delight' in display (Jardine quoted in Bennett 2004, p. 168). Or, alternatively, in the enlightenment museum they acted as an index of material, written in Latin that privileged the object's visual appeal and allowing access to knowledge only if one was of a certain class (Bennett 2004, p. 168). Thought of in this way, labelling in public service broadcasting natural history programming should continue the course of public education— elucidating and explaining the object on display in a way that is accessible to all. *WWC*'s interactive application is therefore arguably a continuation of such practices—its pop-up labelling functioning like an immediate object label in the museum, whilst the more detailed fact-files can be read as summative museum displays, explaining where the viewser/visitor is at in relation to the display's overall meaning/evolution or purpose.

However, as I've suggested, if we pay attention to how the choices of the interactive viewser are structured we do still find a privileging of the forms of spectacle and display that Jeffries (2003) and Darley (2003) find problematic. Thus, whilst the automatic pop-up labels may appear at virtually any time, regardless of the presence of spectacle in the mainstream, this is because they are both relatively unobtrusive and uninformative. So during episode 2 of the series, 'Blood Brothers', a group of *Homo habilis* ape-men work together to drive off a lion from a carcass in order to obtain food. The footage amounts to one of the series' more obvious attempts at spectacle, ape-men and women clambering over rocks, whooping and charging with a series of quick edits and mobile camera framing giving the scene a feeling of pace and excitement. During the build up to the *habilis'* driving-off of the lion, automatic pop-up labels appear but do not distract or obscure the narrative and visual spectacle. In contrast, because the summative labelling of the actively called up fact-file obscures the visual image, the opportunity for the viewser to explore these is limited to moments where the main feature's visual splendour and spectacle are at a minimum. Thus it is not until the lion has been driven off by the *Homo habilis* that a fact-file becomes accessible. The emphasis on spectacle is further reinforced by the absence of presenter Robert Winston's narration, allowing spectacle to speak for itself, rather than be interpreted by either scientist or label.

Nevertheless, the introduction of pop-up labelling practices serves important public service educative goals that relate to the use of labelling technology in the early 20th century new museum idea. George Brown Goode's account of principles of labelling museum displays can easily be applied to this digital address: 'formerly accessible only to the wealthy, and seen by a very small number of people each year, exhibits [through clear labelling practices] are now held in common ownership and enjoyed by hundreds of thousands' (Goode, quoted in Bennett 2004, p. 169). Placing the BBC as public institution in this discourse of accessibility, Goode's ideas relate specifically to television's ability to pervade the everyday and reach people from all walks of life. This may not be the 'challenging' science demanded by Jeffries (2003) and Darley (2003), but the interactive application promised the possibility of engaging an audience with more explanatory and detailed discussions of what was occurring on screen: that is, managing their 'user flows' in a didactic regime akin to that of the museum.

The television personality, technology and the shifting paradigms of the BBC's (participation) science

Despite these problems, *WWC* is also interesting for the way in which it was exemplary of a longer-term shift to the use of high-profile presenters in science programming. Discussing the cyclical nature of science programming, whereby significant trends emerge and prove briefly successful in attaining very large audience numbers, Phillip

Dolling, executive producer of BBC Science, suggests that 'typically by the fourth or fifth iteration, the approach begins to diminish' (quoted in Rouse 2007). Thus *WWC* represented a move away from the spectacle of CGI created un-natural history and a return to the importance of the scientific expert as presenter. Within the *Walking with . . .* format, *WWC* was followed by *Swimming with Seamonsters* (BBC/Discovery, 2004), which again featured the presence of a presenter, this time zoologist Nigel Marvin, in the diegesis of the un-natural history's extinct world. This shift is important for two final reasons in terms of the concluding arguments about interactivity, public service broadcasting and audience participation that I want to make here. Firstly, the use of presenters as a marketable commodity or brand in themselves represents another tactic for attracting larger audiences to science programming. Secondly, I want to suggest that the presence of such presenters, who often bring qualities of 'trust', 'intimacy' and 'credibility' to programming through their intertextual appearance in other programmes or publicized skills in real life, is intimately connected to driving audience use of new forms of interactive and multi-platform television: that is, making more engaged users and viewers out of audiences.[4]

The shift to a focus on the presenter is far from a new trend. As Gardner's discussion of science programming in the 1970s attests, the introduction of ITV and competition for viewers had led to a shift from the formal 'scientific lecture-style' of presentation towards an attempt to be 'less patronizing and more engaging, to be entertaining at the same time as it is informing' (Gardner, quoted in Darley 2003, p. 230). It is hardly surprising, therefore, to find that the increased competition for audience numbers brought about by the development of a multi-channel, multi-platform digital landscape has been similarly met by an importance placed on presenters. As Dolling suggests, in the process of developing new science programmes or series, 'Often you can only go forward when you find the right presenter, as with James Burke 30 years ago and Robert Winston more recently' (quoted in Rouse 2007). Dolling is discussing the BBC's use of James May, who is best-known for fronting car-lovers' show *Top Gear* with Jeremy Clarkson, to present *James May's 20th Century* (BBC/Open University, 2007) as a successful strategy to 'bring an audience in'. May's series attracted an audience of 2.4 million, or an 11.5% share of the audience, whilst SkyOne opted for a similar tactic in signing co-*Top Gear* presenter Richard Hammond to front the first four series of its 'lads'' science programme *Braniac: Science Abuse* before opting for an even more high-profile presenter in Vic Reeves to present season five (Granada for SkyOne, 2003–). Whilst both programmes are outside of the natural history genre I focus on here, their use of such high-profile presenters with appeals to a larger market than that of the niche science market are demonstrative of the importance of presenters in such production strategies.[5] They are also, as counter examples to the

4. For a full account of the importance of televisual and vocational skill in relation to the concepts of authenticity and credibility see my work on television personalities in the Further Reading section below (Bennett 2008).

5. The replacement of Hammond in series 5 of *Braniac* by popular comedian and television personality Vic Reeves underscores both this importance of and the increasing reliance on the presenter's 'televisual skill' rather than 'vocational skill'. However, *Braniac* arguably represents the far end of this move away from the scientist as presenter and it is more instructive to remain within our example of the BBC and its natural history programming.

interactive and multi-platform television I'm discussing here, prime examples of the fear of 'dumbing down' evident in my introduction.

Of course, in relation to the presentation of natural history the most important and high-profile presenter remains Sir David Attenborough whose 'blue chip' wildlife pro- grammes (Bouse 2000) attract both the most prestigious and largest funding and audiences in science television (cf. Wheatley 2004). Presenters such as Winston and Attenborough, through their intertextual appearance across programmes and in raising awareness of particular charities or other work, have developed a 'televisual image' that is built on forg- ing a bond of 'trust' with audiences: their widely circulated background as vocationally skilled scientist and zoologist, respectively, add both credibility and authority to the information presented in their programmes. Returning to the presence of Robert Winston in *WWC* we can, as I've suggested, understand his role within the programme as part of a longer lineage of scientists-as-hero or adventurer on TV science. However, the pro- gramme's status as an un-natural history programme belies such a simple interpretation. Although well-known for his work in fertility and genetics, Winston's expertise does not extend to natural history and his presence across a raft of BBC scientific programming is predicated on his ability to make information and complex ideas accessible. In this way, Winston has cultivated a televisual image that extends the credibility of the knowledge he imparts across a wide array of scientific discourses. Arguably the value of this televisual image extends to encouraging audiences to use the new, and at first unfamiliar, techno- logies of digital television. Thus during the opening credit sequence for *WWC*, Winston appears against a video-wall of screens filled with images from the series and, with a remote control in hand, explains the features on offer in the interactive application, the ease of using them and structure of the application before finally inviting viewers to 'press red now, and come walking with cavemen' (Figure 4).

Figure 4 Professor Robert Winston's invitation to use the 'red button' at the start of *Walking with Cavemen*. (The inset has been added to illustrate the point.) © BBC/Discovery 2003.

Winston's introduction provides a reassuring presence to viewers new to the inter-active spaces of digital television through the trust and credibility of his, by now, familiar televisual image. (Winston had by this stage been appearing regularly on BBC programming since 1998.) Whilst statistics aren't available for the number of people who actually accessed the interactive application for *WWC*, the success of Winston's association with the technology was continued for his presentation of *How to Sleep Better* (BBC, 2005), which featured an interactive application that gave viewers personalized information on their own sleeping patterns and behaviours and was accessed by 2 million of a possible 6 million viewers (Willis 2006). As Marc Goodchild (2003) has argued, the BBC emphasize presenter endorsements to let audiences know about interactive applications with David Attenborough's similar call to action in the *Life of Mammals* (BBC, 2002–2003) represent-ing a particularly successful example of this tactic. Indeed, the promotion of interactivity via familiar and trusted presenters extends beyond the genre of natural history or science programming and includes the use of other trusted television personalities, such as Alan Titchmarsh and Andrew Marr, to introduce a plethora of interactive applications with didactic elements that might otherwise seem unappealing or inaccessible to audiences.

However, it is David Attenborough's association with these new technologies that remains of interest here. Examining his mode of address and call to action is particularly revealing in the context of understanding how science programming in the digital broad-casting era might not only fulfil public service obligations but do so through audience participation and, in turn, present a view of science and the world 'as changeable, chal-lenging, contingent' across the flow and user flows of digital television (Jeffries 2003, p. 543). Just as I have suggested that the credibility and expertise of Attenborough's scientific knowledge are lent to new television technologies, Jeffries has argued that it is Attenborough's televisual image that is central to the rise and success of the 'natural history paradigm' on television. Jeffries is, of course, not alone in this criticism of Attenborough. In particular, the 'balanced and ordered' nature of his programmes has increasingly come under attack in relation to debates about climate change and his failure to engage with the issue. However, Attenborough, as part of a wider move in BBC natural history pro-gramming, has begun to acknowledge the importance of climate change and, in so doing, a series of multi-platform television strategies have been developed to manage user flows so as to productively involve users in climate change as an issue of challenge, change and, most evidently, crisis.

Most notable amongst these was *Climate Change – Britain Under Threat* (BBC/Open University, 2007), a multi-platform television and web event that relied on Attenborough's televisual image to manage user flows between television and web. Throughout 2006, via the presence of a short clip of David Attenborough, the BBC's web site made an appeal for users to take part in an experiment on climate change by downloading a model that would help calculate climate change. By giving over unused space in their computer's virtual memory, and being able to keep in touch with the progress of the results, the model worked as a form of 'distributed computing' that engaged users, however passively, in scientific discovery (see the Useful Web Site section at the end of this chapter for more information). More importantly, across the duration of the experiment a thread of cross-promotion for the upcoming 'results' television programme worked as a way of managing user flows, keeping users aware of related television events and, hopefully, returning them to their television screen to watch these. Over 200,000 people, across 171 countries ran the model with 4.8 million people watching the related television programme in January 2007 (nearly 20% of the audience share). *Climate Change – Britain Under Threat* was then tied, via online and television promotion, to a series of *Saving Planet Earth* (BBC, 2007) programmes.

Saving Planet Earth represented a series of celebrity-led nightly documentaries that highlighted the plight of the world's most endangered species and were aired during late June and early July 2007, leading up to a live final event gala on the last weekend of the series. This again featured Attenborough to the fore, whose presence not only drove multi-platform user flows, imbuing the new technologies with the credibility and trust of his televisual image, but also lent these qualities to the various television personalities and celebrities involved in the series. This was underscored by the playing of a direct appeal by Attenborough for people to donate money to the *Saving Planet Earth* fund, either directly through iTV, online or via the phone, at the end of each night's documentary. The programmes thus represented the natural world in crisis, rather than simply a balanced and ordered world of spectacle and visual pleasure as in other natural history programming fronted by Attenborough directly. As a whole, therefore, the *Saving Planet Earth* season, as an extension of the climate change programming at the start of the year, underscored the entwined importance of presenters and the use of new technologies and programming strategies for the presentation of science in the digital landscape. Most importantly, it was successful in doing so, with the associated appeal for donations raising £15 million and over 25 million people tuning in to at least one of the documentaries or live finale. In relation to the debates about science programming with which I opened this chapter, I am not suggesting that these celebrity-led documentaries, nor the live finale, were 'good science' in and of themselves. Rather, as a way of managing user flows they created a sustained engagement with a crucial debate within science across a wide audience, aggregating it under one umbrella campaign and managing it across a series of screens, sites and moments to encourage a kind of participation.[6]

Conclusion: from user flows to flow—the continuing importance of the linear broadcast schedule

If one of the themes of my discussion above has been to demonstrate how new programming strategies have emerged to allow the BBC to produce new modes of engagement with viewers/users, I have also been concerned to trace historical links to older technologies, debates and regimes of display that these new digital technologies and strategies evoke. By way of conclusion, in evaluating whether all of this adds up to 'good science', I want to return to the importance of Raymond Williams' notion of flow and the way in

---

6. As I discuss elsewhere, in public service debates, such 'herding' of user flows provides an important rebuke to those market-led attacks on the BBC, which accuse the Corporation of failing to make appropriate use of new digital programming strategies to encourage participation (cf. Bennett 2008b; Bennett and Strange 2008).

which the linear 'old media' broadcast schedule continues to structure our experience of television. That is, if we look across both the flow and user flows of the BBC's *Saving Planet Earth/Climate Change* programming we find a consistent message of the planet in crisis, understood through a mode of address that adopts climate change as a scientific certainty in the threads of BBC promotion. As such, the BBC's science programming could be said to be guilty of failing the charge that television cannot present science as 'fundamentally or irremovably in dispute', which I laid out at the start of this chapter. Indeed, the Corporation seemed to admit as much itself when it bowed to perceived public pressure in pulling a planned live *Planet Relief* concert event for 2008. In part, this was because doubts were raised amongst senior BBC executives that *Planet Relief* would expose the Corporation 'to the charge of bias' (Black 2007). As Peter Barron, BBC2's *Newsnight* editor, stated in relation to *Planet Relief*: 'It is absolutely not the BBC's job to save the planet' (quoted in Black 2007). However, we need to examine not only how the new digital strategies of television manage user flows, but how individual programmes, or even multi-platform events, sit within the scheduled flow of television. As Wheatley argues, rather than 'relying on single examples of "good television" to either vindicate or vilify public service broadcasters, it would seem that a more holistic picture of provision across the whole schedule needs to be given' (Wheatley 2004, p. 336).

In this context, it is important to note that the Sunday following the *Saving Planet Earth: Live* finale the BBC also broadcast *Top Gear: Polar Special*. In a specially designed 4×4, the programme's main host, Jeremy Clarkson, together with James May, sets out to prove that polar exploration is easy, taking a rather different attitude to climate change from the season of programming I have detailed thus far. Clarkson is, of course, famous for his 'industrial', 'petrol-head' attitude and 'anti-environmentalist' stance, having previously raced cars across untouched landscapes in Britain and Africa. The programme deals with science as both technology, in the form of other cars and other technologies to help them deal with the expedition, and nature, in the form of human biology, the environmental conditions of the Arctic and climate change. The juxtaposition of *Top Gear* and Clarkson's stance on climate change is not simply about damaging the immediate environment through his driving across the North Pole, but the programme's attitude to the issue. This is best encapsulated in the short space here by the programme's final credit sequence. The special closes with a montage of shots of Clarkson's rugged 4×4 crashing through snow and ice, triumphantly driving off into the sunset of a polar wind storm as Clarkson reflects on his journey:

The fact is that two middle aged men, deeply unfit and mostly drunk had made it, thanks entirely to the incredible machine that took us there. They said we'd never get to the pole because of the damage the car has already done to the icecap. Perhaps, then, that's what we've proved most of all: really, the inconvenient truth is [laughs] it doesn't even appeared to have scratched the surface.[7]

---

7. Clarkson's description of the findings of his polar adventure as an 'inconvenient truth' of course allude to the Davis Guggenheim/Al Gore environmental preservation film of the same name, *An Inconvenient Truth* (Participant/Lawrence Bender Productions, 2006), which provided a slightly more scientifically veracious account of the state of global warming.

The *Top Gear* special achieved similar rating figures to the Attenborough climate change special with 4.5 million people watching the programme, but was met with outrage by environmentalists and the left-leaning press and its readers. Whilst *The Guardian* (Dean 2007) suggested the programme was only worth watching to 'see if Clarkson gets attacked by polar bears', Greenpeace called for the BBC not to show the programme at all. Of course, it is possible to argue that such indignant reactions to Clarkson's stance are precisely the response he would hope to provoke from the left-leaning intelligentsia and environmentalists. But it does challenge them and us, as an audience, to find counter-examples, other scientific opinion, to perhaps lend our computers to the BBC's future climate change experiments, follow the hints on green living or donate to appeals from *Saving Planet Earth* or perhaps participate in other projects, whether digitally interactive, or more simply, such as *Springwatch*'s (BBC, 2004–) amateur science of monitoring the seasons through changes in nature. I'm not suggesting that *Top Gear*'s 'lads'' science is therefore better science because of this challenge it makes, or the debate it provokes—a categorization that seems inevitable if we follow Jeffries' approach to placing 'challenge' at the heart of understanding 'good science'. Rather, I'm suggesting that if we think across the 'flow' of the BBC's output, we do find challenge, change and crisis. More importantly, as people respond with donations—financial, physical and virtual—to the BBC's multiplatform strategies we find that science in the digital broadcasting era should involve and engage the participation of the users, viewers and audiences of digital television as much as challenge them.

■ REFERENCES

BBC (2004). *Building Public Value*. BBC, London.

BBC (2005). *BBCi Brand Guidelines*. BBC, London.

BBC (2006). *Creative Futures*. BBC, London.

Bennett, J. (2006). The public service value of interactive television. *New Review of Film and Television Studies*, **4**(3), 263–85.

Bennett, J. (2007). From museum to interactive television: organizing the navigable space of natural history display. In: *Multimedia Histories: From the Magic Lantern to the Internet* (ed. J. Lyons and J. Plunkett), pp. 148–67. Exeter University Press, London.

Bennett, J. (2008a). Your window-on-the-world: the emergence of 'red button' interactive TV in the UK. *Convergence*, **14**(2), 161–82.

Bennett, J. (2008b). Interfacing the nation: remediating public service broadcasting in the digital television age. *Convergence*, **14**(2), forthcoming.

Bennett, J. and Strange, N. (2008). The BBC's second shift aesthetics: interactive television, multiplatform 'projects' and public service 'content' for the digital era. *Media International Australia*, Special Issue: Beyond Broadcasting, no. 126(February), 106–19.

Bennett, T. (1995). *The Birth of the Museum: History, Theory, Politics*. Routledge, London.

Bennett, T. (2004). *Pasts Beyond Memory: Evolution, Museums, Colonialism*. Routledge, London.

Black, R. (5 September 2007). BBC switches off climate special. *BBC News Online*. Available at: **http://news.bbc.co.uk/1/hi/sci/tech/6979596.stm**.

Bouse, D. (2000). *Wildlife Films*. University of Pennsylvania Press, Philadelphia.

Caldwell, J.T. (2003). Second shift media aesthetics: programming, interactivity and user flows. In: *New Media: Theories and Practices of Digitextuality* (ed. J.T. Caldwell and A. Everett), pp. 127–44. Routledge, London.

Caughie, J. *et al.* (1986). Co-production in the next decade: towards an international public service. In: *The BBC and Public Service Broadcasting* (ed. C. MacCabe and O. Stewart), pp. 92–105. Manchester University Press, Manchester.

Darley, A. (2003). Simulating natural history: Walking with Dinosaurs as hyper-real entertainment. *Science as Culture*, **12**(3), 227–56.

Dean, W. (21 July 2007) Television: Wednesday 25: pick of the day, *Top Gear: Polar Special*. *The Guardian: Guide*, p. 81.

DCMS (Department of Culture Media and Sport) (2003). *Communications Act*. HMSO, London.

DCMS (Department of Culture Media and Sport) (2005). *Green Paper—Review of the BBC's Royal Charter: a Strong BBC, Independent of Government*. HMSO, London.

DCMS (Department of Culture Media and Sport) (2006). *White Paper—A Public Service For All: the BBC in the Digital Age*. HMSO, London.

Gardner, C. and Young, R. (1981). Science on TV: a critique. In: *Popular Television and Film* (ed. T. Bennett *et al.*), pp. 171–93. BFI Publishing, London.

Goodchild, M. (2003). Buttons and Boards Across Borders: Community and User Participation Panel. Speech given at the American Film Institute's Interaction '03: a Trans-Atlantic Producers' Forum Conference, 18 September 2003. Available at: **http://www.afi.com/education/dcl/interaction03/panelc.aspx**.

Griffiths, A. (2002). *Wondrous Difference: Cinema, Anthropology and Turn-of-the-century Visual Culture*. Columbia University Press, New York.

Harries, D. (2002). Watching the Internet. In: *The New Media Book* (ed. D. Harries), pp. 171–82. BFI Publishing, London.

Jeffries, M. (2003). BBC natural history versus science paradigms. *Science and Culture*, **12**(4), 527–45.

Jensen, J. and Toscan, C. (1999). *Interactive Television: TV of the Future or the Future of TV?* Aalborg University Press, Aalborg.

Ofcom (2004a). *Looking to the Future of Public Service Television Broadcasting*. HMSO, London.

Ofcom (2004b). *Phase 2: Ofcom Review of Public Service Television Broadcasting*. HMSO, London.

Outram, D. (1996). New spaces in natural history. In: *Cultures of Natural History* (ed. N. Jardine, J.A. Secord and E.C. Spray), pp. 249–65. Cambridge University Press, Cambridge.

Rouse, L. (19 July 2007). Commissioning focus—BBC science. *Broadcast*.

Scott, K.D. and White, A.M. (2003). Unnatural history? Deconstructing the Walking with Dinosaurs phenomenon. *Media, Culture and Society*, **25**(3), 315–32.

Silverstone, R. (1984). Narrative strategies in television science: a case study, *Media, Culture and Society*, **6**, 337–410.

Thompson, S. (9 August 2007). BBC plans cuts to flagship factual shows. *Broadcast*. Available online at: **http://www.broadcastnow.co.uk/news/bbc_plans_cuts_to_flagship_factual_shows.html**.

Wheatley, H. (2004). The limits of television? Natural history programming and the transformation of public service broadcasting. *European Journal of Cultural Studies*, **7**(5), 325–39.

Williams, R. (1992). *Television: Technology and Cultural Form* (originally published 1975). Routledge, London.

Willis, J. (5 May 2006). Speech given to the Toronto Documentary Forum. Available at http://www.bbc.co.uk/pressoffice/speeches/stories/willis_toronto.shtml.

■ **FURTHER READING**

- Bennett, J. (2008). The television personality system: televisual stardom revisited after film theory. *Screen*, **49**(1), 35–50. This article offers further insight into the way in which television personalities function as a site of meaning in television.

- Darley, A. (2000). *Visual Digital Culture: Surface Play and Spectacle in New Media Genres*. Routledge, London. Darley deals with the way in which new digital media technologies are implicated in the continuing importance of 'spectacle' as an attraction and aesthetic form in media practices.

- Gillespie, M. (ed.) (2006). *Media Audiences*. McGraw Hill/Open University Press, London. The entries in Gillespie's edited collection provide a useful introductory discussion of key debates about the 'activity' of media audiences, situating the TV 'audience' in relation to the computer 'user'.

- Haraway, D. (1989). *Primate Visions: Gender, Race and Nature in the World of Modern Science*. Verso, London. Haraway's work provides a critical insight into the way in which anthropology is marked by discourses of race and gender at the point of display, discussing the exhibition of natural history and anthropology in the museum and other public institutions.

- Jardine, N., Secord, J.A. and Spray, E.C. (eds) (1996). *Cultures of Natural History*. Cambridge University Press, Cambridge. Entries in Jardine *et al.*'s anthology give students an opportunity to follow up on debates about the display of natural history in film and television that are touched on in this chapter.

- Silverstone, R. (1986). The agnostic narratives of television science. In: *Documentary and the Mass Media* (ed. J. Corner), pp. 81–106. Edward Arnold, London. Silverstone builds on his 1984 study of science programming in this article on the place of science television within debates about the form and style of documentary television.

■ **USEFUL WEB SITES**

- Ofcom: http://www.ofcom.org.uk/. Ofcom is an independent regulatory body that has responsibility for television, radio, telecommunications and wireless communications. The web site has useful documentation on the changing shape of the communications industries as they move towards digitalization, particularly the three-part review of public service broadcasting conducted in 2004, as well as continuing quarterly updates that provide detailed statistics on the increasing penetration of digital television and radio into UK homes.

- BBC Science and Nature, Prehistoric Life: http://www.bbc.co.uk/sn/prehistoric_life. This page gives users access to information on the BBC's *Walking with . . .* series, including production details and history. Through the 'Caveman Profiles', users can access the same information that was presented to viewers of *Walking with Cavemen*'s interactive application (although largely reformatted and slightly more extensive). The site also offers a myriad option of games and puzzles for younger users that evidences the series' attempts to address to a broader, non-science audience.

- **BBC Climate Change page: http://www.bbc.co.uk/sn/climateexperiment/**. This page gives access to the results, methodology, technology and data from the BBC–Open University climate change experiment. The main page features a video address from David Attenborough, which recalls his earlier 'call to action' in setting up the experiment. The navigation of the site is marshalled by the authority of Attenborough's voice, which directs user flows to further results, clips from the BBC One television programme and ways to get involved.

- **Flow TV: http://flowtv.org/**. A critical resource for students interested in following up debates within television, media and new media studies about the changing production practices, texts, aesthetics, audiences and the cultural role of television in society.

5.2

Image–music–text of popular science

Felicity Mellor

This chapter examines the ways in which the form and structure of a piece of communication, as well as its explicit content, play an important part in the making of meaning. It looks at two examples of media representations of science—a newspaper article and a TV documentary—to explore how visual images, sounds and the written word carry meanings in addition to their overt literal meaning and how these different modes of communication function together within media texts.[1] The examples presented here are not particularly significant in their own right, nor are they necessarily representative of all media science, but they do illustrate the ways in which the form of a text can shape its meaning. The chapter draws on concepts from semiotics—the study of signs—but it does so selectively, introducing concepts by using them rather than by explaining them in the abstract. Rather than provide a systematic introduction to a particular theoretical approach, the aim of this chapter is to encourage a critical orientation towards media science and to show what sort of things such analyses can reveal. The chapter concludes with a brief discussion of what semiotic analyses can bring to the study of popular science.

The two little pigs

It is worth beginning with a very obvious observation: what we encounter in the media is mediated. Even for factual content, layers of representation lie between us, the readers or viewers, and the subject being presented. To re-present something entails making choices about *how* to represent it and each possible choice will carry slightly different connotations. The choice of this word rather than that one, this picture rather than

---

1. Throughout this chapter I will use the word 'text' in a general sense to refer to any representational product, including images and sound as well as written text.

The Guardian Saturday October 13 2001 **15**

Beacons of hope Coloured pigs could aid transplant research

Tania Branigan

US researchers who genetically modified a litter of pigs say that the fluorescent snout and trotters represent a major advance towards growing animal organs suitable for transplant to humans.

The piglets were produced by a team led by Randy Prather at the University of Missouri, who inserted jellyfish genes into pig cells grown in a laboratory.

The nuclei from these cells were placed inside donor pig eggs whose own DNA had been removed, and the eggs were implanted into a surrogate mother. All but one of the five resulting piglets, which were born in March this year, have yellow snouts and hooves which glow under ultraviolet light.

Professor Prather said the study, to be published in the Animal Biotechnology journal, showed it was possible to modify pig clones to express desired traits. That was a step towards modifying genes to produce organs acceptable to the human body.

Vicky Robinson, scientific officer of the RSPCA, said: "There is no evidence it will help to solve the problem of organ shortages."

Piglets cloned using jellyfish genes. The yellow one is called NT-2, for nuclear transfer, the scientific name for the cloning process

Figure 1 Newspaper article published in *The Guardian*, 13 October, 2001, p. 15. © Guardian News & Media Ltd 2001.

that one, sparks off different sets of associated meanings which influence how we think about the thing being represented.

To show how this happens, I will take as my first example a photograph that accompanied a news report in *The Guardian* newspaper (Figure 1). The photograph occupied four times as much space as the body of the written text. It shows two piglets, each held in the arms of a person standing behind. The piglets are shown face-on and side-by-side. The photograph is close-cropped so all that can be seen of the people holding the piglets are their fingers and part of the chest of one and the shoulder of the other. Clearly, it is the piglets, not the people, who are the subject of this photograph.

Semioticians would say that the photograph *denotes* two piglets—this is what the photograph describes. In other words, the overt message of the photograph is: here are two piglets. Already, though, the choice of subject matter carries further levels of meaning. Whenever I show this photograph to a class of students, the general reaction is: 'Aaaah, cute!' Why is that? Partly, it's because these are the young of an unthreatening domestic animal. Think how my students would react differently if the picture showed two snakes face-on in the same composition. Even two adult pigs wouldn't evoke quite that same 'cute' factor. But it is also partly because in British culture piglets are associated with childhood; for example, through the story of *The Three Little Pigs* or the nursery rhyme *This Little Piggy Went to Market*. Thus those round shapes of ink on the page of the newspaper not only denote piglets, they also *connote* feelings of nostalgia, cuteness, domesticity, and hence, perhaps, of safety. The choice of subject matter signifies a cluster of meanings beyond the literal, denoted meaning.

There are, then, two levels at which meaning is generated: the level of denotation which concerns the literal content of the text; and the level of connotation which concerns the feelings and ideas evoked by the form of the text interpreted through the values of a culture. As this suggests, the connotative meaning of the photograph is culturally dependent. On one occasion when I showed this picture to a class of students, one Muslim student had a very different reaction. This student, drawing on different cultural associations of pigs as unclean, found the picture repellent. But for the majority of *The Guardian*'s readership, piglets are more likely to signify cuteness than religious proscription.

The composition of the picture helps reinforce this reading. The face-on portrait style shot endows the animals with a certain human-ness—we gaze into their eyes and connect with them. The way they are cradled in the arms and hands of the people holding them makes one think of a pair of human twins cradled by their parents. And the way they are held side-by-side with ears and trotters touching adds to the cosiness of the scene. How lovely. The close-up shot also helps keep our attention on the piglets. Think how the picture might carry rather different meanings if it were a long shot showing the piglets in their environment—a pig sty, perhaps, with lots of other pigs, or a muddy field or farmyard. That might provoke thoughts about agriculture and animal husbandry. The close-up helps cut out such thoughts. The choice of shot thus directs the reader's gaze towards a particular interpretation of the image.

The close-up also means that we see very little of the people holding the piglets. Being able to see at least something of them signifies contact between human and piglet and helps further humanize the animals, but the close-up on the piglets means that the people don't hold our attention. However, we do see just enough of them to note that they are both dressed in white, and the way the clothing of the person on the left is buttoned up the front with a V-shaped collar at the top is suggestive of a white coat. Could these be vets? Are the poor piglets ill? Or maybe these are scientists about to experiment on the piglets? Or are they butchers about to slaughter them? The glimpse of white coat hints at such terrors but does not emphasize them. Again consider how different that would be if a close-up had not been used; then it would be harder to ignore questions about what these people were doing with these piglets.

The photograph accompanied a short report on the inside pages of *The Guardian* (Branigan 2001). The headline of the report narrows down the uncertainty over who the white coats might belong to, but it does so in such a way as to retain the focus on the piglets: Beacons of hope Coloured pigs could aid transplant research. With the reference to scientific research here and elsewhere in the written text, the white coats come to signify scientists even though, in reality, they may actually have been the overalls of the butchers. It follows that the photographic image of the white coats is 'polysemous'—it has several possible meanings. The French theorist Roland Barthes (1977, p. 39) talks of photographs having 'a floating chain of signifieds', a multitude of meanings, which can be fixed or 'anchored' by the written text. In this example, the written text anchors the denoted meaning of the white coats as signifiers of scientific research.

The headline also, however, anchors the positive connotations found in the overall composition of the image. The emboldened 'Beacons of hope' suggests that the piglets are showing the way to a 'brighter' future. But this headline is also a pun, for, as the rest of the written text explains, the trotters and snout of one of these pigs glows under ultraviolet light. This leads on to another aspect of the image which I have so far ignored. The photograph was printed in colour and the most prominent meaning of the image is encoded in this use of colour: whilst the piglet on the right has a normal pink-coloured snout, the snout of the piglet on the left is yellow (Plate 2). The piglets were from a litter of five clones which had been modified with a fluorescent gene from a jellyfish. All but one of the piglets expressed the gene, hence the glowing yellow nose of the piglet on the left. It is a black-and-white version of the image that evokes the 'Aaaah, cute!' response from my students. The colour image evokes more ambivalent feelings, more like a puzzled 'Ooo?'

The use of colour and the positioning of the piglets side-by-side draws attention to the difference in the piglets' snouts. Without the normal pink-nosed piglet beside it, we might interpret the yellowness of the other piglet as a printing error, but compared with the pink we understand the yellow to denote the actual colour of the piglet's snout. The headline—'Beacons', 'coloured'—again helps here by anchoring the meaning of the image: the colour is indeed significant. The caption of the photograph also anchors the meaning of the image, as well as adding further information not available from the image alone: 'Piglets cloned using jellyfish genes. The yellow one is called NT-2, for nuclear transfer, the scientific name for the cloning process'.

So these piglets are not so touchingly familiar as they first appeared. They are clones. They incorporate the DNA of an entirely different species. One has a glowing nose. Cloning, transgenics, experimenting on domestic animals—these are controversial issues. Scientists and other commentators often complain that the press reports such issues in a sensationalist way, hyping the controversy and stressing negativity.[2] However, the analysis of this image shows that the opposite is the case here, as has been found to be the case in other analyses of science reporting (Nelkin 1995). Although the difference between the piglets' colouring is emphasized—as it must be since this is the reason why the piglets are newsworthy—in all other respects the image is reassuring and comforting. The composition highlights difference yet also emphasizes sameness—the piglets are held in similar poses, they are given equal space, they appear similar in all respects save for the colour of their snouts and trotters. Their difference thus becomes configured as no more than a difference in eye colour or hair colour in a portrait of two human babies. Arguably, the bright pink and yellow noses simply reinforce the childhood nostalgia already noted—more like a choice of cheerfully coloured sweets at the confectioners rather than the controversial outcome of genetic experimentation.

Considering the picture first without colour has revealed the subtle ways in which the form of this picture minimizes a negative or disturbing reading. The written text does

---

2. For instance, Ian Wilmut, one of the scientists who created Dolly the cloned sheep, has said of the media coverage of cloning: 'There seems to be a pressure in the media to come up with sensational stories. That is disturbing' (quoted in Ahlstrom 2000). For an analysis of the actual UK media coverage of cloning, including the Dolly episode, see Neresini (2000), Priest (2001) and Holliman (2004).

the same. Most obviously, the references in the headline to 'hope' and 'aid' focus on the positive potential of transgenic research rather than any ethical or safety concerns it might raise. The matter-of-fact phrasing of the caption serves to downplay any anxieties the picture might evoke, whilst the first paragraph of the body copy refers to the fluorescent snouts and trotters as 'a major advance'. The story is framed as a medical breakthrough, preparing the way for growing animal organs which can be used for transplants in humans. It so happens that this was also the way the story was framed in the associated press release produced and issued by the University of Missouri where the research was conducted (Jenkins 2001).

As with the picture and headline, the rest of the written text also keeps the focus firmly on the piglets rather than the researchers. The researchers only become the grammatical subject of sentences for what they say rather than for what they did. What they did is presented in subclauses: 'US researchers who genetically modified a litter of pigs say . . .'. Despite journalists normally avoiding the use of the passive voice in order to make their writing seem more dynamic and active, in this short story two sentences describing what the researchers did are in the passive: 'The piglets were produced by a team . . .', 'The nuclei from these cells were placed . . .'. The piglets, by contrast, are presented as subjects: they 'could aid transplant research', they 'represent a major advance', they 'have yellow snouts and hooves'. This serves to minimize the sense of active intervention on the part of the scientists by rendering them as relatively passive and placing the piglets in the foreground of the action.

Finally, the written text also includes two quotes. The first is a 28-word indirect quote in the fourth paragraph from the lead researcher. This quote is paraphrased from the press release (Jenkins 2001) and repeats the claim that the research is a step towards providing organs for human transplants. The second quote is a 14-word direct quote in the fifth (and final) paragraph from a scientific officer at the RSPCA questioning whether the research would solve organ shortages. The literal meaning of this last quote is to record what this person said, but again it also carries meaning at another level. Its presence makes the report appear impartial—it signifies balance, the journalistic norm that requires a report to give both sides of the story (see Allan, Chapter 4.1 this volume for discussion). Yet this is a very limited form of balance, as can be seen by again considering how it would seem different if another choice had been made here. What if the quote addressed the ethics of the research, challenging the research itself rather than the narrower question of whether it has long-term potential to solve a particular problem? What if the quote came at the start of the story, not the end of it? And what if it had come from a prominent authoritative figure rather than an unknown RSPCA officer? As Conrad (1999) has found, journalists tend to draw on cautionary quotes for balance, rather than on quotes which undermine the validity of the research itself. In this case, the use of such a quote enables balance and impartiality to be signified without disrupting the overall positive framing of the story.

To summarize: this example has shown that the meaning of a text does not arise just from its subject matter, but also from more subtle elements such as the composition of an image, the use of colour, the structure of the written text and the choice, format and order of particular words. By imagining how the meaning of the text would change if one of its elements were altered (snakes instead of piglets; black-and-white instead of colour; a different order to the written text), it is possible to uncover the meaning of some of these elements. This technique is often called a commutation test.

This analysis has also shown that both words and images signify meaning at two different levels: the level of denotation and the level of connotation. Where denotation concerns *what* the image shows, connotation concerns *how* it is shown. The connotations of words and image can reinforce each other. In particular, written text can serve to anchor the meaning of a photograph which, on its own, may be polysemous or open to different interpretations. In this example, words and image worked in such a way as to produce a 'preferred reading'—the interpretation that is most strongly determined by the text—that was positive, reassuring and hopeful, despite the potentially controversial and disturbing subject matter.

This example has also revealed how the inclusion of two different voices in the written text can work to signify balance and impartiality even whilst the arrangement and sourcing of that text promotes one view (that of the University of Missouri press office) over any others. To put it bluntly, the text appears to be impartial whilst in fact it is not; it (re)constructs a preferred reading of a complex issue. In fact, not only did the news report draw heavily on one press release, but the photograph was also provided by the same press office.

Overall, then, *The Guardian*'s treatment of this potentially controversial story served to downplay possible negative connotations (e.g. about ethical and safety issues) and to offer a preferred reading that was accepting of the interpretation offered by the researchers and their institution's press officers. Whatever one feels about this sort of research, one might expect a newspaper—perhaps particularly a quality paper like *The Guardian*—to question news sources and seek out alternative sources and framings of a story rather than what appears to be an uncritical, single-sourced reproduction of the messages of a press office. (Of course, in many cases national newspapers do fulfil this 'fourth estate' function. Whether newsroom production processes for science journalists are now more reliant on promotional source materials is an issue addressed elsewhere, e.g. see Allan, Chapter 4.1 this volume; Trench, Chapter 4.2 this volume; Bauer and Bucchi 2007.)

What is clear is that analysing the form of a news report using these analytical techniques—examining what it connotes as well as what it denotes, in relation to the use of source materials such as press releases—can help reveal the ways in which the construction of news may serve the interests of the sources of the news rather than the wider public interest. Further research would be required, of course, to investigate how members of the public interpret and contextualize a complex issue, such as the cloning of non-human animals for therapeutic reasons.

Science as other

As in newspaper journalism, TV producers must also make choices about the form of a programme which can influence its overall meaning. The moving image can be deconstructed in just the same ways as can a still image, but in addition it is also necessary to think about how the frames of a film join together—does the camera move within a shot, how long are the separate shots, how are the shots cut together—and to think about how all this combines with the use of sound. In other words, how do the individual elements of the film work together to form a televisual narrative? My second example addresses some of these questions by looking at a film entitled 'The Six Billion Dollar Experiment', which was produced and broadcast by the BBC as part of the long-running science documentary strand, *Horizon*.[3] I'll focus on the opening couple of minutes of this documentary to illustrate how the analytical techniques outlined above in relation to the newspaper article can be applied to a television programme. Similar footage is repeated throughout the rest of the programme.

Once again, a good way to approach this text is by way of a commutation test—looking at how the meaning of the text changes if one element of it is altered. In this case, imagine watching the opening sequence without the sound turned on. What can be seen? What sense can be made of it? The answer to the first question is 'quite a lot' and the answer to the second question is 'very little'.

The film opens with a long shot of a cityscape (Figure 2a). The camera pans across the scene for a few seconds and begins to zoom in on the upper floors of one of the skyscrapers before cutting away to another scene. This is reminiscent of an introduction to, say, a sit-com where the cut would take us to a flat inside the skyscraper, e.g. think of the long-running US-produced sit-com *Friends*. In that case, the opening long shot would act as an establishing shot introducing us to the environment in which the action takes place. But in this case, the film cuts to the silhouetted back view of a person in a hard hat walking down a corridor (Figure 2b); the meaning isn't immediately clear. Tubes and wires run along the ceiling and walls of the corridor, which is bathed in a dull red light. This is more like an industrial space than the upper floor of a skyscraper. It doesn't seem connected to the opening shot.

As we follow the person, the light gradually changes from red to a bluish-white. The light dims red again as he approaches a bend. The changing light and low light levels create a sense of foreboding and with the camera following the man we get a sense of being led somewhere. What is it that is around the corner? The sequence then cuts to a shot looking up at what seems to be the inside of a large domed building. Perhaps this is what was around the corner. Yet the light here is entirely different, the camera is now in a fixed position rather than moving and it is angled upwards rather than being at eye-level as in the previous shot. The person we were following has vanished. Once again, there seems to be little continuity with the previous shot. The cut seems unmotivated and there are no cues that this space should be read as in any way connected with the corridor.

The camera slowly tilts down from the domed ceiling to reveal a brightly coloured machine of some sort, symmetrically positioned in the wide-angle shot (Figure 2c). This is followed by a cut to what could be a side view of the same machine as the camera slowly tilts upwards again following the machine's circular shape. Together these two shots of the machine last for 13 seconds—a long time for a single shot with no obvious action—before cutting back to the cityscape (Figure 2a). After a few seconds this blurs into streaks

---

3. This programme was first broadcast on BBC Two on 1 May 2007 at 9 p.m.

(a)

(b)

Figure 2 (a)–(f) Frames from the opening sequence of 'The Six Billion Dollar Experiment'
(© BBC: *Horizon*, 2007).

of coloured light, wipes to a black screen and cuts to a blurred shot of a dark figure walking past a shop window. The picture then distorts, seeming to bulge outwards, and fragments into grey shards flying out of a dark void (Figure 2d). It is as if space itself is disintegrating.

The flying fragments last for 5 seconds before cutting to a medium close-up of a man stepping towards the camera out of a circle of darkness that matches the void of the previous shot. This match cut makes it seem as if the man is stepping out of the

(c)

(d)

Figure 2 continued

disintegrating space. The edges of the circle of darkness are a lighter colour, framing the picture like the mouth of a tunnel. Is the fragmenting void a tunnel in space–time? Does this man come to us from another world?

The man is initially looking down but as he steps forward he looks up and faces into the camera. He wears a black coat that merges with the darkness behind him, so all that can actually be seen of him is his head and coloured scarf. He is disembodied, adding to his sense of otherness. The film then cuts to similar shots of three more men and one woman, all dressed in dark clothes and pictured against more or less dark

(e)

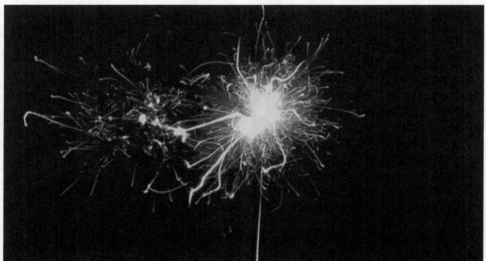

(f)

Figure 2 continued

backgrounds. One is looking sideways, smiling slightly, another man and the woman are adjusting their clothes, the other man is smoothing down his hair. They all seem to be preparing to address the camera, to talk to us. Finally another younger man appears, dressed entirely in black and pictured against an entirely black background. All that can be seen against the blackness is his face and hands as he rubs his eyes and pulls his hands down over his cheeks (Figure 2e). His head and hands seem to float in the darkness, just like Holly, the face of the computer on the popular BBC-produced sci-fi comedy, *Red Dwarf*.

More shots of the cityscape, the tunnel and the machine follow, along with some shots of streaks of light flashing across a black screen and a blue-yellow bubble like a ball of gas expanding across the screen (Figure 2f). The picture then fades to grey and the familiar *Horizon* logo appears accompanied by the usual music, followed by the programme title 'The Six Billion Dollar Experiment'. The screen freezes for a few seconds before dissolving into the next sequence.

It has taken a long time to describe this sequence of images, but it takes just 1 minute and 40 seconds to watch it. So one thing can be noted right away: the shots change very quickly, requiring viewers to speedily process an awful lot of audio-visual information. This is similar to Hollywood films, where the average shot length is often less than 5 seconds. In Hollywood films, however, films are edited for continuity. The rapid succession of different shots is cut together in such a way to help give a sense of a coherent three-dimensional space in which the action takes place. For instance, a typical sequence might start with an establishing shot, consisting of a long shot of, say, two people in a room, followed by a close-up of one person, then a close-up of the other, then a shot of what that person is looking at, and so on. The camera angles and order of the shots is such that it is clear how the subjects of each shot relate to each other. Once viewers have learnt how to read these media and genres, they understand that the different shots combine to present a single sequence of actions all occurring in the same room and interpret the shots as following each other in time just as the actions unfolded in time. In the *Horizon* documentary, by contrast, the shots seem quite disconnected from each other—they switch from one subject to another and from one place to another, from cityscape to machine, from street to person. Although later on in the film there are also short sequences of shots all clearly located in the same space, visual discontinuity predominates throughout the film. Viewers can safely assume that this has been done deliberately; but why?

Where there does seem to be some connection between two successive shots in the opening sequence—the match cut between the flying fragments and the man stepping forwards—the matching is brought about via digital effects. Far from creating a coherent space in which the man is located, the man is positioned as stepping out of a disrupted space beyond the everyday world. The shots which follow serve to introduce a number of characters who we can expect to meet again in the rest of the film. Yet their disembodied heads and the intertextual allusion to the face of the *Red Dwarf* computer, dehumanize these characters and further reinforce their otherworldliness. The industrial space and oppressive lighting of the corridor, the blurring and fragmenting of the city street, the strange lights that flash across the screen, and the rapid cutting between seemingly disconnected scenes, all contribute to a sense of alienation. All of these effects and techniques are repeated throughout the film. Together they signify that this is not our ordinary world and these are not ordinary people. The form of the film works to create a distance between viewer and subject. This is precisely the opposite of what most films strive to do.

This sense of distance is reinforced by the ways in which the film draws attention to its own construction. We see the disembodied heads prepare to address the camera and by looking straight into the camera, they remind us that the camera is there. This, along with the various special effects of blurring and exploding coloured shapes, highlights the act of making the film. Again, this is very different from a standard drama which works hard to hide its constructedness in order to draw the viewer into the story, especially through continuity editing and realistic *mise-en-scène* (the pro-filmic elements of setting, costumes, lighting and so on).

Because this documentary makes little use of continuity editing, it is very difficult to get any sense that a story is being told from the images alone. With a fiction film, even with the sound turned off, it is obvious that there are characters interacting with each other and reacting to events. The visuals provide a sense of time passing and of one action lead-ing to another—the basic elements of a story. The visuals of the *Horizon* documentary do not tell a story in the same way. Yet the film does tell a story. It does so through the sound track. In one sense, the sound track anchors the visuals in the same way as the written text of the news report anchored the photograph in the previous example. But in that case the anchorage served to confirm a reading that was already suggested by the composition of the photograph. In this case, as with many other science documentaries, the anchorage imposes on the visual text a narrative which it would not otherwise signify.

As in most science documentaries, the film has a voice-over which narrates the story. In this instance, the voice belongs to the actor Samuel West. Some viewers might recognize the voice, but even those who don't will recognize an authority in this clearly articulated, slow, almost ponderous, male voice. His narration begins with a precise date, the date on which a new particle accelerator at CERN in Geneva is due to be commissioned. This anchors the location of some of the visuals and sets up an unfolding time frame which anticipates a story that will finish on the specified date. The narrator states that this pro-mises to be 'an extraordinary day' and that turning the accelerator on could 'conceivably trigger a catastrophic event'. This possibility is firmly discounted later in the film, but by mentioning it at the start, and thus framing the whole film in terms of a possible forthcoming catastrophe, the drama of the story is heightened. The narrator's final words in this opening sequence are: 'countdown has begun'. So what we are about to be told involves the passing of time building up to a climatic event. This is the typical arc of a narrative.

The anticipation of a dramatic narrative is heightened by the music, a repetitive sawing of strings in a minor key which carries us forwards with the promise of a resolution. The music is highly reminiscent of (and is probably a repeated sample from) the music from Errol Morris's award-winning documentary *The Thin Blue Line*. That film also opened with shots of a cityscape and a haunting score by the minimalist composer Philip Glass. Viewers of 'The Six Billion Dollar Experiment' may not pick up on this intertextual reference, but in terms of production choices it is significant that the style of music used in a film about particle physics is the same as that in a film about the arrest and wrongful conviction of a man for the murder of a police officer. Once again, an aspect of this film draws on a signifier of alienation. Elsewhere in the film, different jollier music signals changes in scene and in pace and thus helps to move the narrative forwards. This switching between familiar musical motifs—what one TV critic has referred to as 'plinky-plonky popular science music and the occasional burst of portentous strings' (Brooker 2007)—is a common way for science documentaries to define the narrative trajectory in films whose visuals lack any narrative continuity.

However, the burden of the storytelling lies with the narrator. West is an external narrator. He takes no part in the action of the story. He never appears on screen. Compare this with a presenter-led documentary such as the BBC's *Life of* . . . series of natural history documentaries with David Attenborough. In these, a much-loved and knowledgeable presenter moves through the world he is talking about. He shows as well as tells. As with continuity editing, this helps draw the viewer into the story world. It is as if we are there with him, watching the animals. He is our friend and guide on an adventure into the natural world. We suspend our disbelief and don't question whether all the animal antics he shows us really took place there and then, rather than being spliced together from shots taken on different occasions and possibly even in different places. By contrast, in 'The Six Billion Dollar Experiment' we assume that the narrator has not even observed the events he is telling us about. His absence from the story and his measured speech remind us that he is a hired hand reading a script prepared by someone else. As a narrator, he exists in an abstract plane mediating between the story world and the world of the audience. As with the visual cues, there is again a sense of distance.

The narrator does not tell the story alone. He is helped by interviewees—the disembodied heads who are introduced in the opening sequence. As scientists working on the project, these interviewees signify expertise and authenticity. Although they sometimes speak in voice-over, they also appear on screen as they talk and in the course of the film some of them are shown in the space of the story—walking to work, giving a lecture, working at a computer. These sequences are cut together using continuity editing but they make up a relatively small proportion of the whole film. The sound track provides some personal touches, as when one interviewee mentions the time he met his future wife, but generally the viewer is not invited to identify with these characters. At this point it is worth noting that 'point-of-view shots', where a sequence cuts to a shot of what one of the characters sees, are common in fiction films. They invite the viewer to adopt the position of the character. In this film, by contrast, point-of-view shots are almost completely absent. Overall, then, there is little to humanize the disembodied heads.

Despite the variety of voices provided by the talking heads, the narrator's voice-over hangs heavily, almost oppressively, over the film. There is a tension between the demands of a narrative—coherent connected spaces, causally connected actions, characters who undertake the actions—and the many elements of this film which signify discontinuity and distance. Shots of an interviewee laughing or making a personal remark are presented as disembodied heads. Brief sequences giving a sense of characters placed in the story world are overwhelmed by discontinuous cuts and digital effects. The spaces portrayed are largely unpeopled. One sequence in which a scientist visits an old radio telescope starts with a shot of a mottled grey surface and an echoing voice calling 'Hello, hello. Is there anyone out there?'. In the previous sequence, a scientist lectures to an empty auditorium.

The visual montages of this film have a compelling aesthetic to them. Many of the shots are beautifully composed. But it is an aesthetic of abstraction to which these images appeal. The form of 'The Six Billion Dollar Experiment' repeatedly connotes distance and discontinuity. In its composition of shots, editing techniques and digital effects, 'The Six Billion Dollar Experiment' presents the science of particle physics as otherworldly, alienating and distant from the viewer. The narration celebrates science's ability to solve problems, but the film form positions science as something far removed from the everyday experiences of the human world.

Conclusion

The analyses presented here have shown that the form of a text shapes its meaning as much as the overt content does. Such things as picture composition, camera angles, background music, editing technique, the type of narrator and the structure of the text, all signify. To understand how science is represented in the media, it is important to pay attention to these features as well as to the explicit messages of a text. This chapter has been able to introduce only a few of the concepts which can be brought to bear on a given text. Semiotics, discourse analysis, narrative analysis and rhetorical analysis together offer a wide range of conceptual tools with which to deconstruct popular science texts (see, for instance, Myers 1990; Bastide 1992; Curtis 1994; Seguin 2001; Calsamiglia and López Ferrero 2003).

The type of textual analysis illustrated here also provides a starting point for reaching beyond the text itself, both in terms of the intertextuality of the text and in terms of its social and cultural context. In the analyses above, I have touched on issues of production processes (the use of a press release, the decision to use Philip Glass music) and issues of reception (a Muslim audience may read the photograph of the piglets differently from a secular audience). Interviews with the media producers would be necessary to probe why they covered the stories in the ways they did, but ultimately it is the finished media texts which circulate through culture and it is on these that semiotic analyses focus. Likewise, reception studies can reveal something about what actual audiences make of particular media texts (see Hansen, Chapter 3.1 this volume; Hornig Priest, Chapter 6.1 this volume; Carr *et al.*, Chapter 6.2 this volume; Holliman and Scanlon, Chapter 6.3 this volume, for further discussion). The point of semiotics, however, is for the analyst to question what often goes unquestioned in the daily round of media production and consumption.

One thing that the semiotic approach makes clear is that the analysis of media texts is a matter of interpretation. It is inevitably personal and subjective. Quantitative methods of analysis have a tendency to hide their subjectivity behind a screen of numbers, but the fact remains that each text *has to be interpreted*. (Of course, there are also social researchers that favour quantitative analytical techniques who willingly acknowledge, and aim to make visible, these layers of subjectivity in their work.) Semiotic analysis, by contrast, is open about the interpretative act. It lays before readers the textual evidence to support the interpretation offered and allows them to judge for themselves how persuasive they find it. This chapter is clear evidence of such an approach.

One criticism often made of qualitative textual analyses is that because they usually only deal with a small, and possibly unrepresentative, number of texts, the findings cannot be generalized. However, this is to miss the point of such analyses. For example, as Myers has argued about a semiotic analysis of popular science by Bastide:

I do not think it would improve the article if Bastide had written on forty articles rather than four and had somehow shown empirically that these narrative patterns were a general feature of these genres. The appeal is in the surprises and the juxtapositions one finds at this level of abstraction.

Myers (1992, p. 279)

As Myers suggests, what semiotics has to offer is a particular style of thought and a level of granularity in analysis that would be very difficult to reproduce on a large scale. Most importantly, this type of analysis makes available a critical mode of very detailed reading that can help challenge the taken-for-granteds of popular science.

■ REFERENCES

Ahlstrom, D. (15 January 2000). How Dolly brought the media flocking to an unassuming man. *Irish Times*, p. 10.

Barthes, R. (1977). *Image–Music–Text*. Fontana, London.

Bauer, M. and Bucchi, M. (eds) (2007). *Journalism, Science and Society: Science Communication Between News and Public Relations*. Routledge, London.

Branigan, T. (13 October 2001). Beacons of hope: coloured pigs could aid transplant research. *The Guardian*, p. 15.

Brooker, C. (3 November 2007). Charlie Brooker's screen burn. *The Guardian: Guide*, p. 52.

Calsamiglia, H. and López Ferrero, C. (2003). Role and position of scientific voices: reported speech in the media. *Discourse Studies*, **5**(2), 147–73.

Conrad, P. (1999). Uses of expertise: sources, quotes, and voice in the reporting of genetics in the news. *Public Understanding of Science*, **8**, 285–302.

Curtis, R. (1994). Narrative form and normative force: Baconian story-telling in popular science. *Social Studies of Science*, **24**, 419–61.

Holliman, R. (2004). Media coverage of cloning: a study of media content, production and reception. *Public Understanding of Science*, **13**, 107–30.

Jenkins, J. (9 October 2001). Animal-to-human organ transplants one step closer at MU—Genetic modification could benefit biomedical research and agricultural advancement. *Press Release*, University of Missouri News Bureau, Colombia, USA.

Myers, G. (1990). *Writing Biology: Texts in the Social Construction of Scientific Knowledge*. University of Wisconsin Press, Madison, WI.

Myers, G. (1992). Translator's note, in Bastide, F. A night with Saturn. *Science, Technology and Human Values*, **17**, 259–81.

Nelkin, D. (1995). *Selling Science: How the Press Covers Science and Technology*, 2nd revised edn. W.H. Freeman, New York.

Neresini, F. (2000). And man descended from the sheep: the public debate on cloning in the Italian press. *Public Understanding of Science*, **9**, 359–82.

Priest, S. (2001). Cloning: a study in news production. *Public Understanding of Science*, **10**, 59–69.

Seguin, E. (2001). Narration and legitimation: the case of in vitro fertilization. *Discourse and Society*, **12**(2), 195–215.

■ FURTHER READING

The following books do not focus specifically on representations of science, but they do provide good introductions to the analytical approach outlined in this chapter:

- Fiske, J. (1987). *Television Culture*. Routledge, London. Fiske's book draws on detailed textual analysis and audience studies to examine television's role in popular culture. He includes discussion of core analytical concepts, including polysemy, intertextuality and narrative structure with reference to genres ranging from news to game shows.

- Fiske, J. (1990). *Introduction to Communication Studies*. Routledge, London. This book provides a clear introduction to theories of communication with an emphasis on semiotic approaches.

- Hall, S. (ed.) (1997). *Representation: Cultural Representations and Signifying Practices*. Sage, London. This is a collection of essays looking at how meaning is produced through visual images, language and discourse. The introductory chapter by Hall on the practice of representation and his later chapter on stereotyping are particularly useful in this respect.

- Kress, G. and van Leeuwen, T. (2006). *Reading Images: the Grammar of Visual Design*, 2nd edn. Routledge, London. This text offers a systematic approach to the analysis of visual images. The authors draw on a wide range of examples, from children's drawings to textbook illustrations, to compare the grammar of visual culture with that of language.

■ USEFUL WEB SITES

The following web sites provide further information that is relevant to the two examples explored in this chapter:

- **Semiotics for Beginners: http://www.aber.ac.uk/media/Documents/S4B/semiotic.html**. This is a comprehensive web site by media studies lecturer David Chandler introducing the key ideas in semiotics. The content of this site is also published as a book.

- **The University of Missouri press release: http://web.missouri.edu/~news/releases/octnov01/ Pratherpigs.html**. The University of Missouri press release (see also Jenkins 2001), discussed in relation to the newspaper article analysed in this chapter, can be accessed at this link.

- BBC *Horizon* **'The Six Billion Dollar Experiment': http://www.bbc.co.uk/sn/tvradio/ programmes/horizon/broadband/tx/universe/**. This URL links to the BBC's pages for the *Horizon* documentary analysed in the chapter above, and includes a short trailer for the programme.

- **'Fears over factoids'**, *Physics World*: **http://physicsworld.com/cws/article/indepth/30679**. This critique of 'The Six Billion Dollar Experiment' by physicist Frank Close highlights how a concern with *accuracy* rather than *meaning* leads to a very different sort of analysis from the one offered in the chapter above.

SECTION 6

Examining audiences for popular science

. . . the complex processes of reception and consumption *mediate*, but do not necessarily *undermine*, media power. Acknowledging that audiences can be 'active' does not mean that that the media are ineffectual. Recognising the role of interpretation does not invalidate the concept of influence.

Jenny Kitzinger (1999) A sociology of media power: key issues in audience reception research (emphasis in original; in Greg Philo (ed.) *Message Received: Glasgow Media Group Research 1993–1998*).

6.1

Reinterpreting the audiences for media messages about science

Susanna Hornig Priest

Introduction

Historically, scientific researchers have thought of themselves as communicating primarily with their own colleagues, whether through the publication of traditional journal articles or through the presentation of results at seminars or academic conferences—the vehicles most commonly used for the communication of research results (see Wager 2009 for discussion). Even more modern developments such as disciplinary listserv discussions and internet-only journals have been directed primarily at other specialists (see Chalmers 2009 for discussion). And science journalists have not always been much different, thinking not so much in terms of the variety of publics that exist for science but primarily in terms of a much narrower groups of individuals who are actively interested in (and already supportive of) research activities. In other words, science journalists may tend to write for the 'fans' of science, just as scientists write for other scientists. In this postmodern world, however, both of these views are myopic.

The audiences for science consist of a variety of groups: not just scientists but non-scientist citizens, not just schoolchildren but their parents and other family members, not just the 'fans' of science but activists who may adopt a stance opposed to conventional scientific wisdom. Policy-makers and opinion leaders with no pretensions to scientific expertise can make productive use of scientific results, and this is as it should be. It is crucial that members of the scientific community understand that not everyone who will scrutinize—and in some cases make use of—their results shares their values or outlook. Scientists and non-scientists are not divided solely on the basis of education or expertise, but on the centrality of science itself to their worldview.

This chapter provides some guidance and perspective with regard to how to think about the various audiences for messages and information about science and technology, and how this picture has changed (and how it has not) with the explosion of new digital media such as the internet and interactive DVDs. Although the material is presented from an arguably US-biased perspective, given that its author is located there and is most familiar with US approaches, it is important to note that this is unambiguously a global process. All around the world, with each generation, a larger and larger proportion of individuals have access to higher levels of education. They also have access to more diverse sources of information and a greater and greater volume of information, particularly with respect to science. (Of course, this education and access is unevenly distributed, leading to the observation that increasing 'information gaps' among various social groups are creating important new inequities.) For this reason some scholars have described the most developed parts of the world as 'information societies' and worry about 'information overload' affecting our ability and willingness to keep up with the flood, a situation in which the sheer volume of information about world events available to any individual is simply overwhelming. Navigating this new digital world will require new skills, both among communication professionals and among consumers.

While librarians and publishers have been scrambling to prepare for this new world in which the information floodgates stand wide open (see Gartner 2009 for discussion), media professionals had better not get left behind. Many of the most pressing global policy issues (not just developments within the scientific research community itself) have crucial scientific and technological components: global climate change, overpopulation and the food supply, the prospect of outstripping natural resources, the development of alternative energy sources, the promotion of 'green' technologies, threats of new pandemic diseases, the appearance and impact of fundamentally new classes of technology such as bio-technology and nanotechnology and advances in preventative medicine, genetics and the treatment of disease, to name only a few obvious ones. These are areas where non-scientists can—and must—become informed and involved, multiplying many times over the size and complexity of the audience that must keep up with at least some emerging developments in science. The opinions and attitudes of these broader audiences matter. And in a 'new media' or digital media world, we are not talking about passive 'mass' audiences but about individuals making active information-seeking choices.

Political theory provides two fundamental options within democratic societies: either the management of government and politics can be left to elite groups who are able to master the information overload while most other people remain uninformed—and thus disenfranchised—on most current policy issues, since media appear inadequate to allow ordinary people to do otherwise; or means must be found to educate those ordinary people about enormously complex problems on an ongoing basis. While it may be naïve to suppose that all contemporary societies aspire to become democracies, or better democracies in some cases, this tension between expert or elite control of decisions with a scientific foundation versus some level of popular participation in such decisions exists across the spectrum of political systems, and continues a discussion dating back to at least the early 20th century debate between Walter Lippman and John Dewey. In advanced capitalistic countries such as the USA and much of western Europe, a variety of audiences have begun to demand a voice in the way such decisions are made—and are not always willing to be excluded from such debates on the basis of an alleged deficiency of under-standing of the evidence at stake.

Some new developments in science and technology—such as genetic modification of organisms, or the conduct of stem cell research—challenge traditional social values (in somewhat different ways in different cultures). Others (such as nanotechnology) are less problematic in terms of these challenges but raise other issues involving the distribution of benefits and the rights of individuals to participate in societal decision-making about technology and science, as in many other areas. Yet other areas, such as global climate change, present pressing environmental challenges that can only be met through broad public participation. And finally, even in areas as seemingly divorced from everyday life as string theory or other developments in theoretical physics, it is ordinary people who bear the burden (generally through taxation) of supporting the research, a dimension that would seem to generate both a moral and a strategic imperative to involve them as an audience for the results.

Increasingly, however, the model that assumed that the role and purpose of science communication was to present (to so-called 'mass' audiences) the results that scientists had discovered has come under question. Science itself may be seen as a social construction— that is, as a fallible collective perception inseparable from culturally based assumptions. The concept of a 'mass' audience that will react in largely uniform and predictable ways to 'mass' messages has been largely discarded by media researchers, as well as by practical professionals ranging from journalists to advertisers who now think more in terms of targeting particular audiences than explaining complexities to an undifferentiated 'mass'. In science communication, as in all kinds of communication research, this develop-ment accompanies a renewed concern with understanding the concepts of 'audience' and 'reception'. Audiences are no longer conceptualized as homogeneous or passive, and this has implications for popular responses to science as it does for other categories of information.

This chapter considers the traditional and the not-so-traditional audiences for informa-tion about science, attempts to set that discussion in the context of lofty goals for using public engagement to improve the function of democracy with respect to science, and then also sets it in the context of how theories of media and audience are evolving to encompass the digital world of interactive and internet communication. These audiences should not be conceptualized as consisting only of scientists and individuals who are already attentive to, and supportive of, science. Rather, the practice of science communica-tion must now take into account that a broad variety of people—that is, many different audiences—may be affected in different ways by the public representations of science. Further, these representations matter to questions of democratic practice and the empower-ment of non-scientists to participate in decision-making about science-related issues.

Traditional audiences for science

Most scientists and many science journalists have traditionally made the assumption that the only audiences for science are either scientists themselves, the policy elite working in science-related areas such as energy policy or health care and a limited number of indi-viduals who are in a way the 'fans' of science, that is, for one reason or another they have a special personal interest in science. Scientists, therefore, often assume that the only way to get people to agree with them on a given policy issue is to educate them on the science, and conversely, that if people are scientifically educated they will see the world in roughly the same way that scientists do. This is a naïve set of assumptions that ignores the fact that policy decisions involve priorities, values and ethical considerations that lie entirely outside (narrowly defined) science. Science journalists often make a similar set of assumptions. Their materials—whether the contents of a newspaper 'science page', a science documentary like television's *NOVA* in the USA, or a web site about the international space programme—are often targeted exclusively at a narrow subset of the general public. Of course, more specialized publications like the magazines *Nature* and *Science* are entirely aimed at elite audiences.

It may be true that, thinking in terms of the traditional newspaper, a section on business news may be read most carefully by businesspeople and should be written primarily with them in mind, while a section on sports news may be read most attentively by sports fans and should be written primarily with *them* in mind. The increasing audience specialization of news media everywhere has often been noted and is partly a result of trends in advertising practice; specialized publications on everything from aviation to zookeeping exist in part because advertisers wish to target specially interested audiences. But in today's globalized world at least some of the international news needs to be accessible to everyone, and so does some of the science.

The idea that there are specific, identifiable 'attentive' and 'interested' publics for science, just as political scientists had earlier identified 'attentive' and 'interested' publics for news of foreign affairs, is often associated with the work of Jon Miller (1983). From this view only a small percentage of the population, perhaps 10–15%, has any noticeable interest in (or knowledge of) science; the implication is that messages about science may as well be targeted only to this group. For many years, Miller led the development of the science literacy and attitudinal indicators used by the US National Science Foundation in their ongoing assessments of the US population, resulting in similar questions being incorporated in the ongoing Eurobarometer studies conducted in all EU member states and a handful of others. These studies have been implicitly, and often explicitly, quite negative in their view of the level of scientific sophistication of the so-called 'general' public, but whether this is literally true is a matter of interpretation. When individuals become interested in a scientific topic, for whatever reason, they often seem quite capable of educating themselves on it, regardless of their educational background, a conclusion that would be consistent with the tradition of 'knowledge gap' research on media effects.

For example, one of the standard questions in this survey series asked how many days it takes for the Earth to travel around the Sun. Does not knowing the answer to this question really mean that the respondent does not understand the basic configuration of the solar system? How important an indicator is it whether someone can come up with the correct answer of 365 days in a quick telephone survey? Does this reflect in a meaningful way their membership in an 'attentive' or 'interested' public? Another question asks whether human beings are descended from other primates. Europeans tend to score higher on this question than those in the USA, even though people in the USA score higher on the whole scale on average than do Europeans. This pattern probably reflects the existence of a larger subpopulation in the USA that rejects evolutionary theory in favour of a fundamentalist interpretation of the Christian Bible. But it could also reflect a problem with question wording, in that human beings are *not* believed by scientists to be descended from other *contemporary* primates but from some sort of a common ancestor. And do higher US scores on some of the other questions reflect greater 'science literacy' in the USA than in Europe—or simply a US school system that consistently favours quick recall of the answers to multiple choice questions from early childhood onward, rather than 'understanding' in any deeper sense? Further, knowledge (in the sense of being able to choose a correct answer) does not readily translate into attitudes or real understanding. Factual mastery has even been argued to be counter-productive to understanding; award-winning children's television programming in the US, such as the acclaimed educational series *Sesame Street*, presents all kinds of educational material as a rapidly changing kaleidoscope of isolated facts, which some commentators have suggested may leave some children confused and frustrated.

Recently an attempt has been made in the USA to extend the methodology used to assess science literacy to the assessment of technological literacy (National Academy of Engineering 2006). Arguably, here the waters get even murkier because policy about technology is almost always associated with value choices, whereas science has often been conceptualized as an objective body of knowledge existing outside of politics. Whether this is the way science actually operates is beyond the scope of this chapter; suffice it to say that political scientists, sociologists and others who study the way science really operates in society point to the political and social character of the way we set the scientific research agenda through the entire gamut of educational, funding, publishing and reward structure decisions that control it as counterevidence. Both engineering and science take place in a social context.

Traditional and non-traditional audiences

For many years, the largest gathering of science journalists in the USA and most likely the world could regularly be found at the annual meeting of the American Association for the Advancement of Science (AAAS), publishers of *Science* magazine and promoters of science in the USA (and now throughout the world) since the 19th century. Until quite recently, the annual meeting of the US National Association of Science Writers (NASW) took place concurrently with AAAS, giving journalist attendees two motives to turn up there and two chances to find the money to go. While employers might not fund NASW members to attend their own professional meeting, they would much more often fund them to attend the AAAS meeting because finding a good science story or two there would be almost guaranteed.

And the AAAS has gone to great length and expense to make sure journalists attending get advance information to allow them to identify the most promising of those stories and are well treated once they are there. Scientists who want publicity for their research efforts scramble (along with the public relations and public information workers who support their universities) to cooperate (and compete). It was at an AAAS meeting, for example, that the nearly completed human genome code was originally unveiled to great fanfare, accompanied by the distribution of associated special issues of both *Science* and *Nature* (the latter literally hot off the press at the time it arrived at the meeting). It was also at a more recent AAAS meeting in 2004, this time in Seattle, that the Korean researcher Woo Suk Hwang announced his unprecedented successes in cloning human stem cells, an announcement that rocked US policy circles (operating under a George W. Bush administration prohibiting federal funding for stem cell research) before it was denounced as a fraud (see Kruvand and Hwang 2007 for a comparative discussion of press treatment of these developments in the USA and Korea).

The journalist attendees at this and similar events make up the small, elite group Dunwoody (1980) has characterized as the 'inner circle' of specialized science journalism. While Schudson (1978) has traced the early evolution of the 'objectivity' ethic in US journalism generally to 19th century economic trends in the news business—a matter of maximizing profit by maximizing audiences—Nelkin (1995) did a masterful job of identifying the historical factors that led US journalists generally, and science journalists in particular, to embrace the scientific world's emphasis on 'objectivity' and facticity in the 20th century (even though scientists and journalists may mean subtly different things by the term 'objectivity') (see also Allan, Chapter 4.1 this volume). This commitment to a nominally de-politicized form of journalism is rapidly becoming a world standard. However, 'objectivity' is not an adequate standard for science journalism under many circumstances; science operates by social consensus, and no one study ever represents the absolute truth. Until recently, over-reliance on the 'objectivity' standard has allowed non-believers with respect to global warming to point to journalistic treatments that 'objectively' provide evidence both pro and con, whereas the existence of an overall scientific consensus on climate change has been apparent for many years.

A good proportion of the journalism of the 'inner circle' has been directed at the traditional 'attentive' and 'interested' audiences as described above, and it has also been coloured by the competition among both scientists and journalists to make headline news. While the culture of science tends to discourage self-promotion and publicity-seeking on behalf of individual scientists, to the extent that scientists like astronomer Carl Sagan who end up devoting large parts of their careers to media-based science education may find their reputations as scientists damaged as a result, the culture of today's universities is increasingly different. In a sharply competitive funding environment in which the tuition fees paid by students do not cover the costs of running the university, small armies of public information and public relations people seek to promote the work of 'their' researchers. Because these people are working to get their institutions noticed, and because individual studies are generally publicized one by one as they are completed and published, each new result tends to be presented as a 'breakthrough' result. No wonder more conservative scientists are sometimes reluctant to talk to the media.

The traditional system through which science news is generated at major public meetings is also supported by the embargo system utilized by the editors of major scientific and medical journals such as the *New England Journal of Medicine* (see Allan, Chapter 4.1 this volume; Trench, Chapter 4.2 this volume). These publications may send journalists press releases and information kits in advance of publication of an issue but strictly prohibit dissemination of the actual contents until the moment of publication. This makes sure that unverified science that has not passed the test of peer review does not receive undue publicity, which might confuse 'the public'. But it also maintains the position of the journal in question as a premier source of 'breakthrough' science, and may in some instances serve to protect universities' interests in associated intellectual property (see Schulze 2009 for discussion). The journal is almost certain to be cited as the source of the story. All of this feeds the idea of an imagined audience supportive of science and eager for the next advance, always just around the corner—an audience that may not exist, of course, as recent controversies over genetically modified (GM) organisms attest (Irwin, Chapter 1.1 this volume). Given that the science in question probably passed peer review many months before the journal issue in question was sent to press, the idea of withholding the information until the day of publication in order not to confuse news audiences seems less persuasive an explanation of embargoes than the desire to preserve the 'breaking news' character of the content of the current issue.

The way this traditional system has functioned has non-trivial consequences for audience response. Since the audience is initially assumed to be 'attentive' or 'interested', with a well-above-average background in science, there is little motive for journalists to seek ways to make the information relevant for other possible audiences or to challenge the implication that each new study is a 'breakthrough' by providing context in the form of reference to other studies on the same topic. If they do so, they tend to fall back on the 'objectivity' ethic and attempt to 'balance' the story with opposing views (pro- and anti-global warming, or a scientist's views on stem cell research with those of a theologian, and so on). This does little to educate the public—whether 'attentive' and 'interested' or not—about the true nature of scientific consensus and the complex and value-laden trade-offs that actually underlie policy decisions in science and engineering. The persistent uncertainty of science is ignored. Rather, scientists continue to be presented as a sort of closed priesthood and the social process through which scientific truth is produced continues to be rendered invisible.

In today's on-line world of news, journal publishers have begun to offer scientists the opportunity for their work to appear in electronic form before it appears in print (Chalmers 2009; Gartner 2009). How this will change the nature of science journalism and audience reaction remains to be seen. Newspaper science pages are becoming less common, suggesting that the whole notion of science journalism as a highly specialized 'beat' may be changing. The producers of television's *NOVA* now offer a magazine format involving a series of short stories designed to interest a less dedicated audience—and a more demographically diverse one—than the 'active' and 'interested' audiences for its traditional hour-long format.

Indirect evidence that the audiences for science are broader than the 'attentive' and the 'interested' can be found in the popularity of non-traditional forms of science, often in fictional formats ranging from science fiction and fantasy to mysteries and crime shows (see also Carr *et al.*, Chapter 6.2 this volume). Science teachers have reportedly been astonished at the increase in student interest that appears to be associated with the rise of audience attention to the original *CSI* (*Crime Scene Investigation*), set in Las Vegas, and its popular spin-offs, *CSI Miami* and *CSI New York*. Blockbuster sci-fi productions like the *Star Wars* series (not to mention the ever-popular *Star Trek* television series and the spin-offs it spawned) invite audiences of all ages to think creatively about what technological and scientific futures might await us. Movies like *Jurassic Park*, while controversial within science because the fiction is taken as a distortion of scientific reality, provide one of the few truly public forums for considering the trajectory of science in a critical way—and evidence the public is interested after all. Finally, there appears to be consistent public interest in science as it applies to everyday life in the form of health news, weather reports, stories about new consumer technologies and sports.

Along these same lines, and given that not everyone shares the worldview of scientists in which knowledge (empirically validated) is an inherent good, much of science communication will capture audiences only if it also meets other needs—if it entertains as well as informs.

Science, public policy, public engagement and democracy

Increasingly, concerns have been raised about whether (in this context) our traditional 'top-down' approach to communicating with the public about science and public policy is adequate (e.g. see Irwin, Chapter 1.1 this volume). This shift in thinking has been a gradual one. Opposition to stem cell research in the USA and to food biotechnology in Europe, for example, was widely interpreted by the scientific and policy communities as evidence of public misunderstanding of science. In fact, this opposition in both cases had very little to do with the science itself and everything to do with perceptions of ethics and due process. Scientific knowledge has only a weak link to attitudes (Sturgis and Allum 2004), and many people make up their minds on such matters based on other elements altogether—such as ethical considerations (Priest 2005; Gaskell *et al.* 2005) or trust in the spokespeople involved (Liu and Priest forthcoming). As this information gradually accumulated, it became clear that telling the public about science was not always going to change their minds about policy. This paradigm shift is well documented by Gregory and Miller (1998).

The (largely false) assumption that conveying accurate scientific information will dispel opposition to the policy options embraced by scientists and other elites is referred to as the 'deficit model' (Ziman 1991) and often represents a patronizing attitude with respect to ordinary people and their ability to hold well-reasoned opinions despite limited scientific literacy as measured by survey results. Such attitudes are unfortunate and impede meaningful communication; they also overlook the wisdom and value of what is sometimes called 'local knowledge', i.e. relevant knowledge from lived experience that is not necessarily unscientific or anti-scientific but in particular controversies may suggest different solutions from those arising from the scientific community. In this regard Wynne (1989) demonstrated that the local knowledge of sheep farmers was a necessary complement to the laboratory knowledge of radiation scientists in determining how best to engage in sheep farming after the Chernobyl disaster. In the same vein, by some accounts, native groups in northern North America reported observing subtle ecological changes attributable to climate change well before the current era of heightened controversy, but could not get government officials to take them seriously.

However, in some cases public failure to embrace science policies may have serious consequences for society as a whole, as well as for the individuals involved, and they fly in the face of the best scientific consensus. When people decline to have their children vaccinated because they fear the risk of vaccination may be greater than the risk of the disease, or when they deny the existence of global warming because they believe the evidence has been gathered to further specific political agendas that they reject, everyone's well-being is put at risk. Policy trade-offs are complex and ever-shifting; the rejection of nuclear power as an alternative energy source in the 1960s and 1970s seemed to be the 'safe' choice for the environment to some North Americans, but faced with astronomical fuel bills and limited oil supplies—not to mention the known contribution of the burning of fossil fuels to climate change—the public opinion picture looks quite different now. Of course, such shifts are not universal, and controversy over the wisdom of a return to nuclear power is currently a 'live' issue in the UK.

Genetic testing for disease risk might appear to be a universal good, but if that testing is unregulated and if a finding of being at risk (whether accurate or not) can affect the person's employability and insurability, perhaps citizens need to rethink this.

There seems to be a growing consensus that part of the solution to these difficult policy trade-offs somehow involves more 'public engagement'. The National Science Foundation in the US and equivalent bodies in Europe have invested huge sums in 'public engagement' exercises with respect to the emerging products of biotechnology and nanotechnology. With biotechnology, this seemed to be a matter of too little, too late. With nanotechnology, which poses quite a different set of issues, right now public engagement appears to be a solution in search of a problem (Irwin, Chapter 1.1 this volume). Public consultation is firmly entrenched as an element of environmental policy, especially at the local level, where it makes more sense that the opinions of local residents need to be taken into account. It is more difficult to envision as a reasonable way to engineer 'upstream' consent for huge policy decisions with very uncertain consequences, such as the gargantuan international investment in nanotechnology, predicting the end results of which would appear to require a crystal ball even for the most scientifically literate of us all.

Nevertheless, while not a panacea for the apparent crisis of public faith in science, the principle of public engagement (or public consultation) as one element of 'best practice' with respect to science communication is here to stay, as it should. This fact underscores the necessity for reconceptualizing the audiences for news and information about science—to go beyond the 'attentive' and 'interested' to everyone who might ultimately have a stake in policy decisions, which almost invariably is all of us. In fact, one of the key purposes of public engagement probably ought to be to remind all of us just what is at stake—no less than the future of humanity, in some cases, and in others no less than the character of future human society. While not everyone will participate in face-to-face activities such as public meetings, consensus conferences, science cafés, and so on, the online world will also provide new opportunities for more of us—regardless of free time or mobility—to do so. Science education and science communication are already emerging presences in the multiplayer fantasy world of *Second Life*, for example. Such efforts are intriguing, but longer term may serve to exacerbate the 'information gap' between those with and without access to advanced digital technologies.

Another dimension of the relationship between science and society is suggested by consideration of the role of the non-expert who participates in research. Historically, amateur entomologists and astronomers (to suggest just two germane examples)—for whom science might best be conceptualized as a hobby—have contributed to important discoveries. Today, individuals ranging from bird watchers equipped with checklists to backyard weather watchers equipped with rain gauges are actively engaged in producing scientific data. To the extent that the credibility of data produced by amateurs might be called into question, the dynamics at play illustrate the 'sacred' quality of data produced by 'real' scientists, whose actual practices may or may not be any more precise or less intuitive than those of others seeking equivalent information.

Media theory and audience theory

Just as the planning process for science and technology is gradually becoming more two-way through a variety of public engagement strategies, media theory has moved from the consideration of one-way media 'effects' to a concept of active audiences that exert their own influences on messages and their interpretation. Increased bandwidth available via cable and satellite television systems and via the internet is clearly associated, from an audience perspective, with increased interactivity, further 'demassification' or individualization of programming choices, and asynchronicity with respect to scheduling of programmes, as individuals can record and re-order material to their own taste (Rogers 1986; Bennett, Chapter 5.1 this volume). This has important political consequences; it is more difficult to argue that a single 'hegemonic' message ever dominates. Increasing cultural pluralism in much of the world makes it more apparent that the same information will be understood or 'decoded' differently by different people. The lines between news and entertainment are blurring, as are the lines between 'mass' and interpersonal media as the use of mobile phones, iPods, text messages, e-mail and web-based social networking sites such as *Facebook* and *Myspace* that create new electronic communities all proliferate. Individuals make active choices about what messages to consume and when. Different audience members may be seeking similar information for quite different purposes— ranging from entertainment to investigating the latest consumer gadget to making major life decisions with respect to health-care treatment.

What does all of this mean for science communication? To offer one simple and very practical example, web sites for science-related activities need to allow for multiple types of users—perhaps through the use of different 'portals' or start-out menus. A site providing information about a polar research expedition, for example, might be visited by teachers of young schoolchildren, college students writing essays, government officials wondering if their investment in the research was a good one, commercial interests seeking evidence of exploitable natural resources, citizen scientists wanting to know how they can become involved, environmentalists searching for evidence of climate change and its impact, global warming researchers, idle minds of any age seeking something interesting and unusual, and other polar scientists wondering how their colleagues are faring. One size no longer fits all; the publications manager and the web designer for even the simplest science project must consider a host of audiences and purposes. And if the project involves a broader mission to educate or communicate its results—as it is increasingly recognized that many, perhaps all, of them should—this recognition becomes an imperative. New media in today's world both splinter or fracture traditional audiences and create new communities of interest that span geographical distances. Communicating science has become much more than a translation problem.

Research on the formation of public opinion on biotechnology, to take just one controversial example, has shown that there are at least five different audience groups: those that are ready to accept science at face value and assume that it is beneficial (or at least benign); those that want experts to take a technology assessment approach, researching the risks and benefits of individual technologies before making up their minds; those who want ethics to be the primary consideration in evaluating whether to adopt particular technologies; those who want the opportunity to weigh the risks and benefits for themselves; and those who also want to make up their own minds, but on moral or ethical grounds (Priest 2005). The distribution of these groups varies from country to country (and should be considered only an illustration; no doubt any country's population could be divided up in any number of ways). The patterns of trust in different actors also varies among different countries, and is one of the best predictors of varying national climates of public opinion for biotechnology (Priest *et al.* 2003). A consequence of living in a post-modern world is that both science communicators and science policy-makers must recognize this diversity of perspectives and proceed in the face of it. While these particular ways of categorizing audiences should not be reified, they illustrate some of the many ways in which differences of perspective matter.

Data suggesting that non-expert audiences interpret information through 'heuristic' or 'peripheral' processing (rather than 'central' processing, involving somehow coming to terms with the actual scientific data) is sometimes presented in ways that seem to suggest it is inferior. 'Heuristic' processing relies on 'affective' or emotional dimensions such as trust, which (again, largely by implication) are not considered as reliable. But in reality, heuristic cues such as trust are essential elements of completely rational and legitimate strategies for deciding what science to accept and what to reject. Most of us are simply not in a position to decide which scientific conclusions (from the causes of AIDS to the validity of string theory) are best justified. We do, however, evaluate the credibility of those who bring us these conclusions. This is a highly rational—not at all irrational—strategy for making decisions in areas where we do not have specialized expertise.

All of the diverse audiences for science matter, not just because bad science makes bad policy but in a more positive sense because the involvement of multiple stakeholders and so-called local or non-scientific expertise increases the odds that good policy will result. This is a two-way communication problem, not just a 'top-down' one. From 'mad cows' to measles vaccines, from nuclear power to GM foods, would policy-makers have done better if they had listened more carefully to public concerns? Can the destructive potential of pandemic diseases or of yet more radical climate change be prevented without listening to—and further engaging—the public? Only time will tell, but it is pretty clear that the same old approaches to the audience will not be enough.

Within the scientific community, the expectation that public engagement is a good thing seems more generally accepted, but the future is uncertain. As modern science and technology that is expected to have broad social impact unfolds—for example, in the case of nanotechnology—there is an unmistakably heightened expectation that engaging the public is crucial to avoiding public opinion fiascos of the type that are believed to have characterized the introduction of biotechnology (although the assumption that better 'public engagement' will guard against public displeasure is not necessarily warranted). Newer generations of scientists may accept public engagement as a moral imperative, as well as a strategic necessity. Audiences who have instant access via education, traditional media forms and the internet to an incredible range of expert and non-expert opinions and perspectives are likely to demand further involvement in future policy decisions. But the reality of public engagement as currently conceptualized is highly uncertain. Most people will not, cannot or simply do not devote time and effort to public engagement activities; simply put, if they were that interested in science, they probably would have become scientists.

In the longer term, process may be more important; that is, it may be more important that people feel that they have the opportunity to become involved—that they are not excluded—than that large numbers actually choose to become involved, a prospect that may not be realistic.

■ REFERENCES

Chalmers, M. (2009). Communicating physics in the information age. In: *Practising Science Communication in the Information Age: Theorizing Professional Practices* (ed. R. Holliman, J. Thomas, S. Smidt, E. Scanlon and E. Whitelegg). Oxford, Oxford University Press.

Dunwoody, S. (1980). The science writing inner club: a communication link between science and the lay public. *Science, Technology, and Human Values*, 5(30), 14–22.

Gartner, R. (2009). From print to online: developments in access to scientific information. In: *Practising Science Communication in the Information Age: Theorizing Professional Practices* (ed. R. Holliman, J. Thomas, S. Smidt, E. Scanlon and E. Whitelegg). Oxford, Oxford University Press.

Gaskell, G., Einsiedel, E., Hallman, W., Priest, S., Jackson, J. and Olsthoorn, J. (2005). Social values and the governance of science. *Science*, **310**, 1908–9.

Gregory, J. and Miller, S. (1998). *Science in Public: Communication, Culture, and Credibility*. Plenum, New York.

Kruvand, M. and Hwang, S. (2007). From reviewed to reviled: a cross-cultural narrative analysis of the South Korean cloning scandal. *Science Communication*, **29**(2), 177–97.

Liu, H. and Priest, S. (forthcoming). Understanding public support for stem cell research: Media communication, interpersonal communication, and trust in key actors. *Public Understanding of Science*.

Miller, J. (1983). *The American People and Science Policy: the Role of Public Attitudes in the Policy Process*. Pergamon Press, New York.

National Academy of Engineering and National Research Council, Committee on Assessing Technological Literacy (2006). *Tech Tally: Approaches to Assessing Technological Literacy in the United States* (ed. E. Garmire and G. Pearson). National Academies Press, Washington, DC.

Nelkin, D. (1995). *Selling Science: How the Press Covers Science and Technology*, 2nd revised edn. W.H. Freeman, New York.

Priest, S. (2005). The public opinion climate for gene technologies in Canada and the United States: competing voices, contrasting frames. *Public Understanding of Science*, **15**, 55–71.

Priest, S., Bonfadelli, H. and Rusanen, M. (2003). The 'Trust Gap' hypothesis: predicting support for biotechnology across national cultures as a function of trust in actors. *Risk Analysis*, **23**(4), 751–66.

Rogers, E. (1986). *Communication Technology: the New Media in Society*. Free Press, New York.

Schudson, M. (1978). *Discovering the News: a Social History of American Newspapers*. Basic Books, New York.

Schulze, C. (2009). Patents and the dissemination of scientific knowledge. In: *Practising Science Communication in the Information Age: Theorizing Professional Practices* (ed. R. Holliman, J. Thomas, S. Smidt, E. Scanlon and E. Whitelegg). Oxford, Oxford University Press.

Sturgis, P. and Allum, N. (2004). Science in society: re-evaluating the deficit model of public attitudes. *Public Understanding of Science*, **13**, 55–74.

Wager, E. (2009). Peer review in science journals: past, present and future. In: *Practising Science Communication in the Information Age: Theorizing Professional Practices* (ed. R. Holliman, J. Thomas, S. Smidt, E. Scanlon and E. Whitelegg). Oxford, Oxford University Press.

Wynne, B. (1989). Sheepfarming after Chernobyl: a case study in communicating scientific information. *Environment*, **31**(2), 10–39.

Ziman, J. (1991). Public understanding of science. *Science, Technology, and Human Values*, **16**, 99–105.

■ FURTHER READING

- Bauer, M., Allum, N. and Miller, S. (2007). What can we learn from 25 years of PUS survey research? Liberating and expanding the agenda. *Public Understanding of Science*, **16**, 79–95. This paper examines large-scale survey approaches to the public understanding of science. In so doing, it reviews 25 years of survey research arguing that, if developed and expanded appropriately, such a tradition still has much to offer scientists and social researchers alike.

- Bryant, J. and Zillman, D. (eds) (2002). *Media Effects: Advances in Theory and Research*, 2nd edn. Lawrence Erlbaum, Mahwah, NJ. Although not specifically examining audiences of science communication, this edited collection, currently in its second edition, introduces a range of theoretical and methodological approaches to studying audiences, including those consuming and contributing to new forms of communication.

- Lewis, J., Innthorn, S. and Wahl-Jorgensen, K. (2005). *Citizens or Consumers: What the Media Tell us About Public Participation*. Open University Press, Buckingham. Although not specifically examining audiences of science communication, this book examines the relationship between news media in the (re)construction of democratic citizen-consumers, drawing on research conducted in the UK and USA.

- Rogers, E. (1986). *Communication Technology: the New Media in Society*. Free Press, New York. Rogers turns diffusion theory toward consideration of the personal computer revolution. While readers will find many technical details out of date after 20 years, Rogers was the first—and still among the most influential—writers to characterize what is unique about computers in their emerging role as a communication medium.

■ USEFUL WEB SITES

- **Eurobarometer: http://www.esds.ac.uk/international/access/I33089.asp**. Eurobarometer Surveys were introduced in 1973 by the European Commission. Since 1989, this twice-yearly survey of 1000 citizens of the European Union, has measured knowledge of and attitudes relating to science.

- **National Science Foundation: http://www.nsf.gov/statistics/seind08/pdf/c07.pdf**. This chapter is currently (2008) the most recent in a long-standing effort by the National Science Foundation (NSF) to summarize available statistics on public attitudes toward science and technology, as well as public understanding, as a regular feature of its Science and Engineering Indicators report series. The 2008 report includes data from the 2006 General Social Survey science and technology module, building on an earlier NSF-sponsored series that extended from 1979 to 2004. The report's Survey Data Sources table also provides a very useful compilation of relevant data from international, as well as other US, surveys. Earlier reports in this ongoing biennial series are also available at **http://www.nsf.gov/**.

Investigating gendered representations of scientists, technologists, engineers and mathematicians on UK children's television

Jenni Carr, Elizabeth Whitelegg, Richard Holliman, Eileen Scanlon and Barbara Hodgson

Introduction

A recent research review into the participation of girls in classroom physics (Murphy and Whitelegg 2006) synthesized the evidence from over 15 years' worth of research into the participation of girls in physics and described the multi-faceted nature of the issue of girls' participation in physics, and within science more generally. The review revealed that one of the key factors determining students' attitudes to physics is 'self-concept' —how students see themselves in relation to the subject both now and in the future. Students' self-concept, therefore, affects educational subject choices that can subsequently open up career opportunities that may lead to enhanced life chances and improved salary prospects.

Whilst compulsory education forms a major part of the lived experiences of children and young people, a much broader range of experiences gained beyond the classroom also help to influence children's lives. These provide the context within which children and young people begin to construct their own identities. It could be argued, therefore, that the development of children's and young peoples' self-concept is also shaped within this broader context.

With this broader context in mind, the *(In)visible Witnesses* project looked beyond the school environment to investigate another potential sphere of influence on the development of children's and young peoples' self-concept, that of the mass media, specifically children's television (see Whitelegg *et al.* 2008). The project was funded by the UK Resource Centre for Women in Science, Engineering and Technology (UKRC). Broadly speaking, the UKRC aims to increase the participation and position of women in science, engineering and technology, and so it is interested in examining the range of influences that may affect girls' and women's opportunities for engaging in these fields. Within this overall context the principle aims of the *(In)visible Witnesses* project were to:

1. Study the (re)construction of gendered representations of science, technology, engineering and mathematics (STEM) in order to investigate the continuing portrayal of established stereotypes of STEM and document the emergence of new images.

2. Investigate to what extent these images might have an effect on children's and young people's perceptions of STEM.

Underpinning these principal aims is the methodological premise that there are three elements of mass communication—production, content and reception—which are inextricably linked in a dynamic, organic and continuous cycle (Philo 1999). In this project we have studied two of these elements in detail—content and reception—through an analysis of media content (television) and a reception study (involving children and young people).

In this chapter we provide a more detailed overview of the project, discussing further the rationale behind the methods used in the data collection and analysis. In order to provide the context for this discussion of the project, however, this chapter begins by exploring two key issues that underpinned the rationale for this project, and thus shaped the methodology used.

Self-concept and the formation of identity

Firstly, we discuss issues relating to the notion of self-concept and the formation of identities. In 1957 Margaret Mead and Rhoda Métraux, both cultural anthropologists, reported on a pilot study they had carried out, which was commissioned by the American Association for the Advancement of Science (AAAS). In outlining the rationale for their project the authors identified a concern that underpinned much of the public understanding of science movement (PUS) that was to follow some 30 years later (see Irwin, Chapter 1.1 this volume for an overview of the PUS movement):

... there is a great disparity between the large amount of effort and money being devoted to interesting young people in careers as scientists or engineers and the small amount of information we have on the attitudes those young people hold towards science and scientists

Mead and Métraux (1957, p. 384)

In attempting to address that disparity, Mead and Métraux did not simply ask children about their attitudes towards scientists and science in general, but also explored what we have referred to as self-concept. In a more detailed discussion later in this chapter, we will use this study to highlight the important insights that can be offered by this focus on the notion of self-concept. We also outline how we took into account the weaknesses we argue are inherent in Mead and Métraux's (1957) approach, and sought to develop what we consider to be a more rigorous approach. These weaknesses relate to the second contextual issue we will discuss; how researchers should attempt to carry out research *with*, rather than simply *about*, children and young people.

Fraser (2004) outlines how theories of childhood development, particularly those within the field of developmental psychology, have helped shape ideas about what constitutes 'child-centred' or 'young person-centred' research methods. He highlights how these ideas have led to some researchers focusing solely on ensuring that their methods are 'appropriate' for use with children and young people. That is to say, the methods chosen are shaped by generalized notions of the developmental stages children are likely to have achieved in terms of their cognitive abilities. Fraser argues that this can lead to a rather simplistic interpretation of 'child-centred' research—an interpretation that fails to take into account the importance that context can make in terms of the ways in which children are able to deploy those cognitive abilities. Orientating Fraser's more general point to the approach adopted for our project, we will draw on the work carried out by Buckingham (1993) and Gauntlett (1996) in order to discuss the notion of children and young people as 'active viewers'.

Having discussed these important contextual issues we will move on to discuss the project itself in more detail, discussing each of the methods used during the different stages of the project. As part of this discussion we will draw out the methodological benefits and challenges of linking investigations of media content with reception analysis, documenting the strategies used to ensure a rigorous approach to the research and thus enhance the validity of the findings and recommendations that we made.

In the study carried out by Mead and Métraux (1957) the participants were divided into three cohorts and asked to complete one of three writing tasks. In order to gain an understanding of young people's impressions of scientists in general, what Mead and Métraux refer to as the 'official image' (Mead and Métraux 1957, p. 387), one cohort was asked to write an essay about their opinions of scientists, responding to the statement 'When I think of scientists I think of . . .'. The format of the exercise for the further two cohorts, however, was designed to elicit information relating to what we have referred to as self-concept. The second cohort was asked to imagine themselves as scientists, and then write about what kind of scientist *they would not like* to be. The final cohort was asked to imagine themselves as scientists, and then write about what kind of scientist *they would like* to be. Reflecting, and it could be argued reinforcing, gendered cultural expectations of the time, female students were given the additional option of imagining themselves as being married to a scientist, rather than becoming scientists themselves.

Discussing their findings, Mead and Métraux (1957) highlight that when students wrote about their opinions of scientists in general terms, the images of scientists they related were, in the main, positive. The researchers then contrast these images with those constructed by the cohorts of students who had been asked to imagine themselves as either being scientists or being married to a scientist. Mead and Métraux comment that these responses were 'likely to invoke a negative attitude as far as personal career or marriage choice is concerned' (Mead and Métraux 1957, p. 387). In part this was because the participants tended to describe science as a career that involved hard work, risk and responsibility, but very few rewards.

In arguing for the usefulness of their approach, Mead and Métraux point out that if the study had focused simply on the 'official image' of science and scientists, and not asked participants to imagine themselves in the role of scientist, 'it would have been possible to say that the attitude of American high-school students to science is all that might be desired' (Mead and Métraux 1957, p. 387). It follows that by extending the range of questions asked of analogous participants the researchers gained further important insights.

Self-concept and media influence

It is worth noting that in their discussion of the findings of the study the researchers outline what they believe to be the effects of media representations on young people's perceptions of science and scientists, but they do so based solely on their own analyses of these media images. That is to say, they did not attempt to explore the way in which these specific images were interpreted by the research participants themselves.

Strategies used to check the accuracy of researchers' interpretations are often referred to as 'member validation' (Lincoln and Guba 1985). Member validation can be used in different ways within a research project, but the overarching aim is to establish the credibility of the researchers' observations. Mead and Métraux do not discuss their reasons for relying solely on researchers' interpretations of media images. One aspect that we might consider is the large-scale nature of their study. With the help of seven consultants and fourteen graduate students, Mead and Métraux collected essays from approximately 35,000 students, and 1000 of the essays were analysed in detail. Introducing an element of member validation may well have been beyond even the considerable resources that these researchers could call on. In addition, it would be a considerable period of time before what we would refer to today as 'audience reception studies' moved away from a simple 'cause and effect' model of the way in which media images influence people's opinions and behaviours (see Hansen, Chapter 3.1 this volume; Hornig Priest, Chapter 6.1 this volume)—a theoretical and methodological shift that would take even longer to come about in relation to projects involving children and young people.

Media literacy and children

In his review of research projects exploring children's television viewing, Gauntlett (1996) argued that researchers have too often underestimated children's ability to be active viewers and interpreters of media representations. He states that:

Almost always 'researched on' rather than 'worked with', children are constructed—by psychologists in particular—as inadequate and uncritical of their encounters with the mass media, which itself is represented as forbiddingly powerful.

Gauntlett (1996, p. 48)

Likewise, Buckingham (1993) has argued that if 'television literacy skills' are viewed as a series of developmental stages, then children's skills will always be constructed in terms of what they are lacking:

Mainstream research on children and television has tended to define children as more or less 'incompetent' viewers. What children do with television is typically compared with adult norms, and thereby found wanting.

Buckingham (1993, p. 282)

Buckingham's approach was deceptively simple—he asked children to talk about television. To be more specific, he developed a series of activities that were carried out by groups of between four and six children (aged between 7 and 12 years) and one adult researcher. Some of the activities were 'open-ended' discussions and others included some form of stimulus, such as viewing an extract from a television programme. Underpinning Buckingham's approach was the perspective that television viewing, rather than being an individualized and isolated activity, is predominantly a social activity. He highlights how, if watching television with others, people will talk about the programmes they are watching and even, at times, talk back to the screen. Even when we watch television alone, we may talk about what we have watched to others—'talk about television is a vital element of our everyday lives' (Buckingham 1993, p. 39). Children, Buckingham argues, are no different in this respect and therefore by encouraging them to take part in an activity with which they are already familiar, and where no prior assumptions are made as to their 'competence', researchers can gain insights into the way in which children make sense of the television programmes they watch.

Gauntlett's (1996) approach was different in that rather than focusing on children talking about the television programmes they had watched, he focused on children talking about the television programmes they were creating. Such an approach is similar to those used by some sociologists to investigate how adults interpret and contextualize media reporting (see Kitzinger 1990; Philo 1990; Holliman 2005; Holliman and Scanlon, Chapter 6.3 this volume for an overview of these methods; see Holliman 2004 for a study of science reporting).

Gauntlett worked with seven small groups of children (between six and ten children in each group) from different schools to create videos that reflected the children's concerns about environmental issues. These children ranged in ages from 7 to 11 years. The researcher met with each group of children and explored what the children understood about environmental issues and what their concerns were. Following this initial meeting the children were taught the technical aspects of using the video and recording equipment. The children then chose the topic of the video, planned the content and carried out the filming. Following the plans devised by the children as closely as possible, the videos were edited by the researcher, who added effects such as music and on-screen credits. Gauntlett states that although the footage filmed by the children contained useful data, the finished product of the 'polished' video, which was produced by the researcher, served more as a gift to the children and the schools involved than as a piece of research evidence. It follows that Gauntlett's research focus was on the process of producing the video:

. . . the ideas, the planning, comments and suggestions made during filming, debates that took place between the children, the narrative style and tone favoured and so on.

Gauntlett (1996, p. 78)

In this sense, Gauntlett's approach is similar to that taken by Buckingham in that both researchers sought to develop research methods that privileged children's own perspectives.

That is to say, they sought to explore the ways in which children can, and do, interpret the television programmes they watch, and did so by including the children as active participants in the research process.

Both Buckingham and Gauntlett argue that their evidence supports their claims that, given the opportunity, children and young people demonstrate quite sophisticated 'media literacy' skills. Buckingham, for example, claims that:

> Their debates about the relationship between television and reality were complex and often extremely lucid. They knew a good deal about the way in which television is produced . . . They were well aware of the persuasive functions of advertising, and often sceptical about many of its claims. Their discussions of their favourite comedies and soap operas displayed a complex awareness of the development of narrative, and of the constructed, fictional nature of the text.
>
> Buckingham (1993, p. 283)

To summarize, Mead and Métraux's (1957) study highlights how important the notion of self-concept can be if researchers seek to go beyond examining the opinions and impressions that children and young people might have about science and scientists and explore the ways in which media representations can either encourage or constrain options for girls (and boys) to study post-compulsory science and choose to become scientists. We note, however, that there were methodological issues in relation to establishing the credibility of the researchers' interpretations of media imagery. We sought, therefore, to address these issues through designing activities, reflecting the approach taken by researchers such as Buckingham (1993) and Gauntlett (1996), which engaged children more directly in the research process; to allow them to demonstrate their media literacy skills in productive ways.

Data collection and analysis

As outlined in the introduction to this chapter, there were two principal aims that informed the *(In)visible Witnesses* project, the first of which was to explore gendered representations of STEM, and it is to this element of the project that we now turn.

Analysing media content

Our starting point was to collect a sample that could be considered representative of the types of STEM programmes broadcast using analogue signals in the UK during a specified period. Once collected, the programmes in the 2-week-long samples were analysed to confirm whether they included STEM content (see Whitelegg *et al.* 2008 for an extended discussion).

The identification of STEM content presented the first methodological issue that needed to be addressed in this sampling process—we needed to clarify what exactly constituted 'STEM content' within the context of this project. In order to enable a consistent approach to the identification of STEM content, a set of operational definitions were required that could be applied to factual and fictional representations, and to adults, children and non-humans. To achieve this we employed the following operational definitions of STEM. These definitions were adapted from earlier work (Holliman *et al.* 2002):

Science	These programmes should include significant explicit scientific content, namely a reference or references to scientific findings, scientific research, scientific procedure, science as an intellectual activity or scientists.
Technology	These programmes should include significant explicit technology content, namely a reference or references to technological design, technology research, technological procedures, technology as an intellectual activity or technologists.
Engineering	These programmes should include significant explicit engineering content, namely a reference or references to engineering design, engineering research, engineering procedures, engineering as an intellectual activity or engineers.
Mathematics	These programmes should include significant explicit mathematical content, namely a reference or references to mathematical concepts and formulae, mathematics research, mathematics as an intellectual activity or mathematicians.

When the researchers were unsure about whether to select a programme or programme extract as STEM, this was viewed and discussed with other members of the project team so that a consistent application of the operational definitions could be maintained. These kinds of discussions between project team members represent one form of what is referred to as 'triangulation'. Hammersley and Atkinson (1995) describe the concept of triangulation as deriving from 'a loose analogy with navigation and surveying' (p. 231). In the same way as taking bearings from at least two landmarks can more accurately pinpoint one's position on a map, cross-checking interpretations between researchers can help clarify and consolidate the deployment of analytical tools. The discussion of interpretations between project team members, sometimes referred to as 'investigator triangulation', is only one form of triangulation (Denzin 1978). Other examples of this strategy, specifically 'methodological triangulation' (see also Jensen and Holliman, Chapter 2.1 this volume) and 'data triangulation', will be discussed at other points in this chapter.

Once the initial selection of extracts that portrayed STEM was completed,[1] we produced a further subset of STEM content for analysis, focusing on extracts broadcast within specific schedules for children and young people. This resulted in a total sample of 154 extracts containing STEM: 74 coded as containing science; 38 coded as containing technology; 28 coded as containing engineering; and 14 coded as containing mathematics.

Transcripts of these programmes were produced, and these were checked for accuracy and coded for all recorded speech using the analytical concept of a speaking actor. For the purposes of our analysis 'speaking actors' were defined as any actor making an audible

---

1. The details of each of the extracts were entered into a data base. The data base records included the channel, the name of the programme, when it was broadcast, whether it portrayed science, technology, engineering or mathematics and the type of programme (e.g. whether it was news and current affairs, pre-school, learning/educational, animated cartoon or other).

utterance or utterances that could be recognized as belonging to a known language or dialect, whether this be on-screen or off-screen (e.g. narrators) This included humans, cartoon characters and non-humans (e.g. anthropomorphized animals). The speaking actors were coded deductively in terms of their gender, which was classified as either female, male, mixed or unidentified,[2] and in terms of their role, e.g. narrator, presenter, scientist. The number of words spoken by each actor was then quantified. Further to this, we produced qualitative narrative summaries of these extracts, conducting inductive coding to investigate emerging themes relating to the nature of STEM and representations of STEM practitioners.

Our approach to the analysis of these transcripts provides one example of how combining quantitative and qualitative methods, and underpinning the analysis with both deductive and inductive coding, can bring additional insights to the analytical process. Seale (1999) discusses the influence of methodological triangulation in relation to the development of approaches that seek to integrate quantitative methods into projects that draw mainly on qualitative data. Seale argues that there can be real benefits for qualitative research in 'counting the countable' (Seale 1999, p. 119). Illustrating the usefulness of this approach, Seale discusses how quantitative data can be used to enable readers of the research 'to judge whether the writer has relied excessively on rare events, to the exclusion of more common ones that might contradict the general line of argument' (Seale 1999, p. 128). He also discusses how the use of quantitative approaches can highlight 'deviant cases' within the data, and how a particular focus on these cases can facilitate qualitative approaches. We can illustrate this aspect of Seale's argument through an example from our project.

The quantitative analysis of our sample demonstrated a clear emphasis in the overall distribution of words spoken in favour of males (66% males to 31% females). A more complex gender distribution of spoken actors was identified, however, by investigating the differences between the types of programmes. In the category animated cartoons, for example, the distribution is clearly in favour of males (72%), but this contrasts with news and current affairs where the gender distribution is more evenly balanced (52% males to 48% females). Further analysis of the annotated transcripts, however, reveals a yet more complex pattern. The speaking roles in the programmes that were included in the news and current affairs category are almost exclusively newsreaders and reporters. Scientists appear in only three of the extracts, and all of these scientists were male. So, whilst the news and current affairs extracts might appear to represent examples of 'good practice' in that the overall distribution of speaking actors is relatively balanced between genders, whether they have a positive affect on children's self-concept in STEM required further analysis of what children think it means to present/practise STEM. We will look at this issue in more depth at a later stage in this chapter. At this point, we want to note how this example of combining quantitative and qualitative analysis produced observations that highlighted issues the researchers could explore further as part of the reception study—and it is to this reception study that we now turn.

2. The category 'mixed' includes groups of speaking actors, e.g. a group of children (mixed gender) singing. The category 'unidentified' includes speaking actors that could not be categorized as female, male or mixed, e.g. anthropomorphized 'gender neutral' creatures. Where it wasn't clear from the extract itself whether a speaking actor was female or male, we have conducted web searches to try to confirm the gender.

The audience reception study

For this phase of the project we worked with two class-sized groups of children and young people. One group consisted of 30 participants (13 males and 17 females) aged 8–10 (Key Stage 2) and the other group consisted of 15 participants (4 males and 11 females) aged 11–15 (Key Stages 3 and 4). The reception study consisted of several related activities, which are discussed below.

The questionnaire and draw-a-scientist activity

The first two activities, a questionnaire and the draw-a-scientist (DAS) activity, were designed to elicit the participants' pre-existing images of STEM. The design of both of these activities, including the analysis of the data, drew on projects carried out previously by researchers working on projects with aims similar to our own (e.g. Chambers 1983; Rodari 2007; Steinke *et al.* 2007). As such, although the size of the sample of participants involved in our study was relatively small (therefore not allowing us to generalize from our findings taken in isolation) we could compare our analysis with other larger studies and interrogate any similarities or differences.

The questionnaire was based on a previous one developed as part of the 'Science and Scientists' (SAS) project (Sjøberg 2000), the design of which was intended to 'tap into aspects relating to the interests of children, their experiences, their perceptions of science, their hopes, priorities and visions for the future' (Sjøberg 2000). (Researchers working on the SAS project also made use of the DAS activity, providing a further point of reference between our project and previous studies.)

We adapted the SAS questionnaire for our purposes, e.g. also including questions aimed at gathering background data on participants' television viewing patterns before they began interacting during the group activities. Taken as a whole, therefore, our questionnaire allowed us to gain valuable background information about our participants whilst also allowing us to compare our findings with those of other researchers.

The DAS activity is a research instrument that has proved popular amongst researchers seeking to explore children's perceptions of science and scientists. Chambers (1983) developed this activity in which children were asked simply to 'draw a scientist'. The drawings were then coded according to a protocol that comprised 'indicators', which Chambers argued represented stereotypical images of scientists. This coding protocol used a scale of 1 to 7, with 7 being considered to be the most stereotypical, and with 1 point being given for each of the following:

1. laboratory coat
2. spectacles, including protective goggles
3. facial hair (itself a gendered code)
4. symbol of research, scientific instruments and/or laboratory equipment
5. symbols of knowledge, e.g. books, laboratory notebooks and filing cabinets
6. technology, or 'products' of science, e.g. test tubes, explosions
7. relevant captions, formulae and equations.

When asked to 'draw a scientist' the majority of the participants in our study had the capacity to draw on a sufficient number of shared cultural codes for us to interpret the images they created as being 'scientists'. To ensure consistency in coding, the data were discussed by all members of the research team. As part of this process the team also noted that the Chambers protocol failed to take account of what we considered to be important and relevant symbolic codes. We therefore included the following additional codes: gender of the scientist; full figure or head only; type of scientist; if the hairstyle was 'unkempt' or 'unusual'; and noted the name the children gave their scientist.

Within this context we can say that our findings were in line with those of other studies, in that when the participants wanted to represent an image of a scientist to the researchers, the majority of them (73%) drew male characters. Six of the participants (13%), all girls, drew female characters.[3] Again, these findings are consistent with recent DAS studies that also record greater numbers of girls and young women drawing female scientists than was the case in the earlier studies (e.g. see Rodari 2007; Steinke *et al.* 2007).

In her discussion of the various critiques of the DAS activity, Schibeci (2006) outlines the arguments both for and against the usefulness of drawings as a method of eliciting children's understanding and interpretation of issues and phenomena. Those who argue for the usefulness of this method base their arguments on ideas derived from developmental psychology. Their argument is that whereas young children may not yet be fully competent in terms of their oral and written literacy skills, they are able to represent quite complex ideas in drawings. We should note, however, that this perspective is underpinned by the assumption that those viewing the drawings, usually adults, will be able to interpret them in the same way as children do. Furthermore, when the drawings are carried out as part of a research project, it is important that the assumptions of the framework used to interpret the drawings are themselves interrogated.

Relating this point to Chambers' (1983) protocol, we can see that in order for images to be interpreted as the 'most stereotypical'—i.e. to score 7 out of 7—the scientist drawn would need to have facial hair and therefore be, one assumes, male. Keeping in mind that Chambers protocol is designed to explore a composite image, any single indicator, taken out of context, would not necessarily represent a stereotypical image of a scientist. It could therefore be argued that this protocol only examines representations drawn of female scientists in terms of how closely they relate to a previously established male 'norm'. In the light of this, we question how useful this protocol is in exploring any stereotypical representations of *female* scientists, rather than of scientists in general. Given that there are some more recent indications to suggest that new stereotypes featuring women scientists are emerging in mass media representations, this is of particular significance. For example, Flicker (2003, p. 316) suggests in discussing feature films that:

The woman scientist tends to differ greatly from her male colleagues in her outer appearance: she is remarkably beautiful and compared with her qualifications unbelievably young. She has a model's body—thin, athletic, perfect, is dressed provocatively and is sometimes 'distorted' by wearing glasses.

3. Several drawings could not be definitely coded as either female or male.

It follows that if new sorts of gendered images of scientists, technologists, engineers and mathematicians are becoming available to children and young people via the mass media we need to produce coding schemes that address these new images. Furthermore, we also need to consider the impact that these images might have. Images of female STEM professionals that appear unattainable to young girls could prove to be equally constraining in terms of girl's self-concept as images that represent STEM practitioners as being mainly men.

To summarize, in adapting from previously established methods both the questionnaire and the results of the DAS activity provided us with information about our participants that meant that we could compare them with other studies that sought to explore similar topics. At the same time, in comparing our analyses of the DAS activity across the project team, another example of 'researcher triangulation', we established a number of additional codes that could be added to Chamber's (1983) protocol, and we would argue that these additions have the potential to enhance the usefulness of this protocol for examining stereotypical images of scientists. Finally, we have identified the need to develop a protocol for examining stereotypical media images of female scientists, rather than just scientists in general. Whilst the development of such a protocol would be a project that needs to be developed in the future, in terms of this study this analysis again 'sensitized' the researchers to the need to consider not just whether or not women are being represented in STEM-related television programmes, but what form those representations take. When discussing the content analysis stage of the project we also highlighted how certain issues raised by the quantitative data also served to 'sensitize' the researchers to issues that required further exploration in the reception study. With these two issues in mind, we will now turn to look at the activities we carried out during the reception study that were designed to explore directly the ways in which our participants interpreted images and extracts from television programmes that were included in our sample.

The carousel activity

For the carousel activity[4] participants were shown a range of still images from popular TV programmes and some video extracts from our STEM sample. We asked participants to identify the scientist(s), the scientific activities, give their opinions of the participants they had identified and describe the context of the 'action'. The children gave their answers on a series of worksheets but, in order to facilitate 'data triangulation', the discussions between the children whilst completing these worksheets were also recorded on digital audio recorders.

Furthermore, images from CBBC's *Blue Peter* and *Newsround,* the examples of news and current affairs programmes that formed a part of our sample, were included in the carousel activities. Participants were asked to view these short extracts and complete worksheets. Both KS2 and KS3/4 groups appeared to recognize the particular role of the

---

4. A carousel is a term familiar to schoolteachers. We have used the term carousel to define a series of data collection activities, each of which requires participants to address an issue related to children's STEM television programmes. Participants were given the same amount of time for each activity, moving in small groups from one to the other. These activities were not designed to be completed sequentially and there was no competitive motive for completing them.

presenters, newsreaders and reporters in relation to the STEM content of the programming. Questions about whether the programme contained any STEM content, for example, elicited the comments below:

It includes engineering because they show what people have invented (CBBC *Newsround*)

Yes but only if it's on the script (CBBC *Blue Peter*)

Similarly, when asked to comment on the presenters, newsreaders and reporters and whether they talked about STEM, responses included:

They talk about them if it's in the news (CBBC *Newsround*)

Talk about it if they have a guest. Go to an engineering place to see how it works (CBBC *Blue Peter*)

Recordings made of the participants discussing these programmes also illustrate how the participants sought to make the distinction between 'doing science' and 'talking about science'. This distinction was further illustrated in the responses from participants when discussing other types of television programmes. When discussing, for example, Michaela Strachan, who has (co)presented various natural history programmes on television for a number of years the participants generally recorded positive comments about her personality, they also made the distinction that she talked 'about science' rather than being a scientist herself. Similar distinctions were made in relation to an extract from a BBC education programme, *Primary Geography: the Science of Wind*. Given that our participants made this distinction between presenters and authentic scientists, and in doing so validated, at least 'provisionally', our original concerns relating to this issue, this is an area that we will revisit in planned future projects. We were also able to use these data to highlight these issues in the recommendations included in our project report (Whitelegg *et al.* 2008).

We have extracted this example from what is a substantial body of data collected during this carousel activity in order to highlight the iterative nature of our research methodology and to illustrate the usefulness of data and method triangulation and member validation. Readers wanting to examine further the other issues addressed in the findings that we presented and recommendations that we made based on these findings should consult the full project report (Whitelegg *et al.* 2008), details of which can be found at the end of this chapter.

The storyboard and reflective writing activities

The final two activities that the participants completed were designed to reorientate the focus of the methodology back towards the participants constructing images of STEM and STEM professionals themselves, as with the DAS activity, rather than (re)interpretation of images presented to them by the researchers.

The first of these activities retained the focus on children's television programmes by asking the participants to take part in a 'storyboard' activity where they discussed and then constructed a story and/or some dialogue for a STEM-related children's programme. This storyboard activity provided a very rich source of data, which we will be building on in the next stage of our work. In terms of the project we report on here, the data confirmed to us that our participants were capable of demonstrating sophisticated media literacy skills. The participants produced some potentially fruitful ideas for programmes that combined popular types of programmes such as game shows and reality TV in a STEM context. These data were used to reinforce one of the key messages of the project—that children and young people should not be viewed as passive recipients of media messages. As such, this activity provided the research team with a further example of 'data triangulation' with which to support other findings of the project that relied, in terms of their validity, on claims concerning children's and young people's media literacy.

The second activity was designed to address again the notion of self-concept in that we asked participants to undertake some reflective writing where they imagined themselves as adult STEM professionals in the future; similar to the method described by Mead and Métraux (1957). However, unlike Mead and Métraux, our analysis of the narratives that the participants created was informed by not only our own analysis of media representations but by examples provided by our participants of the way in which they made sense of examples of the same media representations.

The participants were given the choice of imagining themselves as an engineer, scientist or mathematician and they were supplied with a list of topics that they might like to use to help them plan their writing. These included: what age they were; what kind of environment they worked in; what they would wear and what equipment they would use; what they liked most/least about their work; what topics they would like to find out more about; and why they thought these topics were interesting or important. We then analysed these narratives, identifying recurring themes and exploring the similarities and differences in the ways in which these themes shape the content of the account being told (Mason 1996).

In line with Mead and Métraux's (1957) study, we could identify examples that emphasized negative aspects of working as a STEM practitioner, but we also found examples of what we referred to as 'aspirational narratives'. These narratives contained themes such as: a particular 'problem solving' role for STEM; independence within the workplace; status and financial rewards; and job satisfaction linked to helping to solve environmental problems and cure diseases. Drawing on data relating to other activities completed as part of the reception study to make comparisons, we identified that aspirational narratives that contained all of these themes were produced by participants who seemed to have a relatively sophisticated understanding of other aspects of what 'being a STEM professional' involves. It follows that such an approach illustrates one of the key strengths of data triangulation.

Conclusion

We began this chapter by highlighting how the methodology adopted for the *(In)visible Witnesses* project reflected the premise that mass communication comprises elements—production, content and reception—that interact within a 'dynamic, organic and continuous cycle' (Philo 1999). We will conclude this chapter by applying this same 'cyclical' image to the research process itself in order to summarize how the different methods and strategies deployed by the research team enhance the coherence and rigour of the methodology as a whole, and thus the validity of the findings and recommendations.

One of the first issues we discussed was the importance of recognizing that children and young people are active interpreters of media representations, not simply passive recipients. We have also argued, however, that this recognition requires researchers to do more than attempt to 'read' the interpretations that children and young people might produce as part of the research process. Researchers also need to ensure that the children and young people play an integral, and indeed 'dynamic', part in the process itself. Our discussion of the various insights offered by the participants has highlighted the efficacy of this approach.

The carousel activities provided an element of member validation, an aspect of the research that we noted was important in terms of further exploration of an issue that had arisen during the content analysis phase of the research. This process of validation was fruitful not only in providing a basis for recommendations made in the report for this project, but also in terms of highlighting an area needing further research. Whilst it is important to evaluate research in terms of the contribution it makes to knowledge about a particular topic or issue, it is also important to evaluate research in terms of its usefulness in highlighting the 'gaps' in that knowledge that still need to be explored. In this sense then, the participants contributed to the 'organic' nature of research into this issue—their insights and observations will 'feed into' the development of research methods that might be deployed on future projects.

Throughout our discussion we have emphasized various strategies of triangulation. We have highlighted how discussions between investigators of deductive coding protocols (identifying STEM and the DAS activity) can be useful in relation to both clarifying the deployment of these protocols and generating new insights. We have examined how a combination of quantitative and qualitative methods can be used to substantiate the validity of claims made and can also prove fruitful in terms of identifying and exploring 'deviant cases'. We have illustrated how we brought together data from a variety of sources—data generated by our own project and data generated by other researchers exploring similar issues. We have explained how, by adopting an iterative approach to our analysis, we believe we have produced a more nuanced, but also more comprehensive and rigorous, interpretation of these data.

The rationale that underpinned the development of this project was that understanding the ways in which children and young people develop self-concept and identity is an essential part of understanding both the barriers to participation in STEM and the ways in which these barriers might be overcome. The formation of self-concept and identities is a complex process—and this process is not just confined to one context and does not happen as a result of one single experience. It is, one might say, 'dynamic, organic and continuous'. We argue that researchers hoping to gain a better understanding of this process need to recognize this, not only in the issues that they seek to explore but also in the formulation of the methodology that they employ.

■ REFERENCES

Buckingham, D. (1993). *Children Talking Television: the Making of Television Literacy*. Falmer Press, London.

Chambers, D. (1983). Stereotypic images of the scientist: the draw a scientist test. *Science Education*, **67**, 255–65.

Denzin, N. (1978). *The Research Act: a Theoretical Introduction to Sociological Methods*, 3rd edn. Prentice Hall, Englewood Cliffs, NJ.

Flicker, E. (2003). Between brain and breasts – women scientists in fiction film: on the marginalization and sexualization of scientific competence. *Public Understanding of Science*, **12**, 307–18.

Fraser, S. (2004). Situating empirical research. In: *Doing Research with Children and Young People* (ed. S. Fraser, V. Lewis, S. Ding, M. Kellett and C. Robinson), pp. 15–26. Sage, London.

Gauntlett, D. (1996). *Video Critical: Children, the Environment and Media Power*. University of Luton Press, Luton.

Hammersley, M. and Atkinson, P. (1995). *Ethnography: Principles in Practice*, 2nd edn. Routledge, London.

Holliman, R. (2004). Media coverage of cloning: a study of media content, production and reception. *Public Understanding of Science*, **13**(2), 107–30.

Holliman, R. (2005). Reception analyses of science news: evaluating focus groups as a method. *Sociologia e Ricerca Sociale*, **76–77**, 254–64.

Holliman, R., Trench, B., Fahy, D., Basedas, I., Revuelta, G., Lederbogen, U. and Poupardin, E. (2002). Science in the news: a cross-cultural study of newspapers in five European countries. *Proceedings of the 7th International PCST Conference Science Communication in a Diverse World, Cape Town, South Africa*. Available online at: **http://www.saasta.ac.za/scicom/pcst7/holliman1.pdf**.

Kitzinger, J. (1990). Audience understandings of AIDS media messages: a discussion of methods. *Sociology of Health and Illness*, **12**(3), 319–35.

Lincoln, Y. and Guba, E. (1985) *Naturalistic Enquiry*. Sage, Beverley Hills, CA.

Mason, J. (1996). *Qualitative Researching*. Sage, London.

Mead, M. and Métraux, R. (1957). The image of the scientist among high-school students: a pilot study. *Science*, **126**, 384–90.

Murphy, P. and Whitelegg, E. (2006). *Girls in the Physics Classroom: a Review of Research on the Participation of Girls in Physics*. Institute of Physics, London. Available online at: **http://www.iop.org/activity/education/Making_a_Difference/Policy/file_22225.pdf**.

Philo, G. (1990). *Seeing and Believing*. Routledge, London.

Philo, G. (1999). Introduction – a critical media studies. In: *Message Received: Glasgow Media Group Research 1993–1998* (ed. G. Philo), pp. ix–xvii. Longman, Harlow.

Rodari, P. (2007). Science and scientists in the drawings of European children. *Journal of Science Communication*, **6**(3), 1–12.

Schibeci, R. (2006). Students' images of scientists. What are they? Do they matter? *Teaching Science*, **52**(2), 12–16.

Seale, C. (1999). *The Quality of Qualitative Research*. Sage, London.

Sjøberg, S. (2000). *Science and Scientists: the SAS-study*. University of Oslo, Oslo.

Steinke, J., Knight Lapinski, M., Crocker, N., Zietsman-Thomas, A., Williams, Y., Higdon Evergreen, S. and Kuchibhotla, S. (2007). Assessing media influences on middle school-aged children's perceptions of women in science using the Draw-A-Scientist Test (DAST). *Science Communication*, **29**(1), 35–64.

Whitelegg, E. Holliman, R., Carr, J. Scanlon, E. and Hodgson, B. (2008). *(In)visible Witnesses: Investigating Gendered Representations of Scientists, Technologists, Engineers and Mathematicians on UK Children's Television*. Research Report Series for UKRC No. 5. UKRC, Bradford.

■ FURTHER READING

- Buckingham, D. (2005). *The Media Literacy of Children and Young People: a Review of the Research Literature*. Available online from **http://www.ofcom.org.uk/advice/media_literacy/ medlitpub/medlitpubrss/ml_children.pdf**. This literature review, commissioned by the Office of Communications (Ofcom), summarizes evidence from research projects that examined the way in which children and young people 'access, understand and create communications in a variety of contexts', including television programming.

- Kitzinger, J., Haran, J., Chimba, M. and Boyce, T. (2008). *Role models in the media*. Research Report Series for UKRC No. 1. UKRC, Bradford. The aim of this research project was to explore the possible effects of media representations on the under-recruitment of, and failure to retain, women in science, engineering and technology. In this report, the first of four, the research team document their analysis of interviews with female scientists, engineers and technologists.

- Sjøberg, S. (2000). *Science and Scientists: the SAS-study*. University of Oslo, Oslo. Available online at: **http://folk.uio.no/sveinsj/SASweb.htm**. Professor Svein Sjøberg, who designed the questionnaire for the SAS project, is currently working as the project organizer on the 'Relevance of Science Education' (ROSE) project (**http://www.ils.uio.no/english/rose/**). Readers wishing to follow up with further reading on this topic may find the web sites for both the SAS and ROSE studies useful.

- Whitelegg, E. Holliman, R., Carr, J. Scanlon, E. and Hodgson, B. (2008). *(In)visible Witnesses: Investigating Gendered Representations of Scientists, Technologists, Engineers and Mathematicians on UK Children's Television*. Research Report Series for UKRC No. 5. UKRC, Bradford. This report provides further details of the data collection and analysis for the project outlined in this chapter. The report concludes with recommendations made by the research team in light of the findings of Phases 1 and 2 of the project, which includes recommendations for further research on this issue.

■ USEFUL WEB SITES

- **UK Resource Centre for Women in Science, Engineering and Technology (UKRC)**: **http://www.ukrc4setwomen.org/**. The UKRC web site outlines the work of the centre and provides links to research projects, reports (including Whitelegg *et al.* 2008 and Kitzinger *et al.* 2008), and related publications and statistics.

- **STEM Partnerships**: **http://www.stemcentres.org.uk/home**. STEM Partnerships are a series of 'hubs' that provide information, support and advice to schools about STEM activities designed to encourage children's and young people's participation in STEM. As such, they give useful background information regarding initiatives being implemented within the compulsory education sector.

- **Centre for the study of children, youth and media:**
  http://www.childrenyouthandmediacentre.co.uk/. David Buckingham is the director of
  The Centre for the Study of Children, Youth and Media, which is based at the Institute
  for Education, London. Details of the research carried out by members of the centre and
  research reports and associated publications can be found on the Centre's web site.

- **ArtLab: http://artlab.org.uk/index.htm**. David Gauntlett, who directs ArtLab, has developed a
  number of research projects that have sought to explore children's understanding of media.
  Many of his research projects have used methodologies based on children creating their own
  media 'messages'. Outlines of his work, including a more detailed account of the project
  referred to in the chapter, can be found at this web site.

6.3

Interpreting contested science: media influence and scientific citizenship

Richard Holliman and Eileen Scanlon

Media influence and scientific citizenship

Ideas about scientific citizenship are closely related to ongoing discussions about public engagement with science (for discussion see Irwin, Chapter 1.1 this volume; Stilgoe and Wilsdon, Chapter 1.2 this volume; Irwin 1995; Irwin and Wynne 1996; Irwin and Michael 2003; Schibeci and Lee 2003; Jenkins 2004). Within these debates it is often argued that media coverage and science education are significant sources that can influence public attitudes and reactions to science (e.g. Royal Society 1985; House of Lords 2000; Schibeci and Lee 2003). In support of this premise, social researchers have discussed the important role that media coverage plays in raising awareness of (particularly newly emerging) scientific and biomedical issues and then in framing the terms for public debate thereof (see Hansen, Chapter 3.1 this volume for an extended discussion). Overall, studies by these researchers suggest that media reporting of science is important for informing citizens, but that these representations provide a partial, mediated view of new developments in science (see Allan, Chapter 4.1 this volume).

We argue that media reception is a complex process. Our view of these processes is encapsulated in the following quote:

. . . people do not passively absorb everything that is beamed from their television set. Instead they interpret and contextualise. Public views are not formed from thin air. Equally, they are not simply dictated by the media or by ministerial pronouncements or by lay 'perspectives' or 'cultures'. Judgements are made according to information available from the media, education, friends and family and other sources are evaluated against previous experience and information.

Miller (1999, p. 218)

The resulting picture is one where citizens regularly engage in complex processes of audience interpretation and contextualization: a mixed picture of active/passive assimilation–avoidance–rejection of new (scientific) information from a range of sources including (digital) media, in discussion with peers, and in terms of pre-existing knowledge, expertise, attitudes and beliefs. In this sense, the indeterminate influence of media and the (re)construction of scientific citizenship are linked. As social researchers, we seek a more sophisticated understanding of these processes by adopting a methodological approach that allows participants to articulate how they perceive media reporting of science (Holliman 2005), situating this work within the 'ethnographic turn' in studies of public engagement with science (see Irwin and Michael 2003 for discussion).

This situation is further complicated because media professionals tend to favour stories about newly published research and/or controversial issues in science (Allan, Chapter 4.1 this volume; Miller 1999); in other words, they value scientific knowledge which is subject to dispute, and that is more likely to be open-ended, uncertain and contingent (Macnaghten and Urry 1998). This means that it can be open to a number of interpretations, both within the scientific community and in wider public debates (including media reporting). In this sense, contested science has the potential to create the conditions whereby it becomes the subject of media reporting and wider public debate, because it is controversial (Shapin 1992; Holliman 2004). And, having become the subject of media reporting and wider public debate, contested science has the potential to generate further controversy, therefore maintaining its profile in the public sphere. Thus at least some of the scientific knowledge that citizens will encounter in media reporting is likely to be contested because of the scientific or science-based controversies associated with the work (see Brante 1993 for discussion).[1] In addressing these issues in this chapter we report on the interactions observed during a reception analysis conducted as part of the *Reported Contested Science* project.

The *Reporting Contested Science* project

The aims and objectives of the *Reporting Contested Science* project that are relevant to this chapter are:

- To explore participants' prior knowledge, experience, attitudes and beliefs in relation to media reporting of contested science.
- To investigate the influence of media reporting on the (re)construction of scientific citizenship.

We are interested in questions such as: how is contested science—defined as open, contingent and provisional knowledge—represented in news media; and how does an examination of contested scientific research illuminate beliefs about the nature and validity of scientific knowledge? The research project deals with interviews with key informants involved in producing media reports, content analysis of news coverage of contested science (see Holliman *et al.* 2002) and reception studies investigating how

1. Scientific controversies involve contests over knowledge claims, e.g. whether experimental results are valid and reliable. Science-based controversies involve other factors that influence the controversy, e.g. ethical issues (Brante 1993).

participants interpret and contextualize media coverage of contested science. The reception study is the focus of this chapter.

Two scientific topics that generated news media reporting were chosen as the basis of the research discussed here: finger length and sexuality and genes and intelligence. We have included brief descriptions of the topics and the key findings of our content analysis below to facilitate comparison with the results of the reception study.

Finger length and sexual orientation

The finger length and sexual orientation topic involved a research team led by Professor Marc Breedlove at the University of California, Berkeley. Published in the high-profile peer-reviewed journal *Nature* (Williams *et al.* 2000), the researchers produced a short research paper that claimed a correlation between finger length and sexual orientation. The research involved a survey of 720 adults at street fairs in San Francisco, in which the length of the second and fourth fingers was measured and a questionnaire administered requesting information on sexual orientation and birth order.

Breedlove and his colleagues argued that on average homosexual women were more likely to have been affected by higher levels of male hormones (androgens) in the womb, when compared to heterosexual women, as indicated by finger length. They argued that this interpretation was less clear-cut for homosexual men, at least until birth order was accounted for. In this instance, Breedlove and his colleagues argued that men with more than one older brother were more likely to be homosexual and that this also correlated with higher exposure to male hormones in the womb, again indicated by finger length. Taken together they argued that these findings suggested that, among other factors including genetics, female and male homosexuality was partly determined by prenatal exposure to male hormones.

We identified 12 UK newspaper articles that reported this research, of which five were 'elite' (broadsheet) and seven were 'popular' (tabloid). Of these 12, seven included photographs or 'info-graphics' inviting readers to 'check their own hands' against the findings of this research study. Four of these articles also included pictures of well-known celebrities and public figures, often showing their hands and using captions to describe their sexuality.[2] Of these, two popular articles used multiple pictures, inviting readers to 'test' the validity of the research findings. The following example illustrates these points:

How your hands can reveal if you're gay – JUST TRY THE FINGER TESTS

Here's a handy way to see if a pal is gay—look at the fingers on their right hand. . . .

[The article included two pictures of hands, one captioned 'GAY HANDS The index finger is much shorter than the ring finger—a pointer that the person could be lesbian or gay', the other 'STRAIGHT HANDS The length of index and ring fingers are very nearly equal']

In the interests of research of course, we have taken the matter in hand. Here we put some well-known celebrities to the test.

The Sun (31 March 2000, emphasis in original)

---

2. Captions are used to 'anchor' the producer's preferred reading of an image (see Mellor, Chapter 5.2 this volume; also Barthes 1977; Hall 1981 for discussion).

The article went on to discuss eight further pictures of celebrities and public figures who were either known to be gay or lesbian or who were assumed to be heterosexual. Of these, only one image challenged the interpretations of Breedlove and colleagues; the remainder reinforced pre-existing ideas about these individuals' sexuality, therefore implicitly supporting the research findings.

A further two articles illustrate how newspapers can frame scientific research as contested, both in terms of scientific and science-based controversy. In contrast to an article published in the same newspaper the previous day, which reported the findings without criticism, these items were more sceptical. For example, a front-page banner advertisement asked the question:

Can your index finger REALLY determine your sexuality?

Daily Mail (31 March 2000, emphasis in original)

This item linked to an article, published on pages 32 and 33 of the same edition, which answered this question, both through the headline, article text and use of pictures and captions:

That's handy – The pictures that prove your index finger doesn't determine your sexuality

. . . Despite being derided as far-fetched by many, some in the scientific community rushed to support Dr Breedlove. Dr Nick Neave, a biological psychologist from Northumbria University, said: 'Dr Breedlove's findings mirror research we have been doing.' . . .

So, in the interests of scientific debate – and public curiosity – the Mail took a series of well-known figures whose sexuality – present and past – is, almost without exception, known and put their hands to the Breedlove test. You can judge what the pictures prove for yourself.

Daily Mail (31 March 2000)

Twelve pictures were included, each with a caption, e.g. 'Self-confessed lesbian, tennis champion Martina Navratilova's index finger is longer than her ring finger' and 'Even notorious womaniser Rod Stewart fails the Dr Breedlove test his ring finger is plainly longer.' Of the 12 pictures and captions, eight clearly challenge the proposed theory about finger length and sexuality and a further three are ambiguous. Only one picture and caption, of the actor Stephen Fry, clearly supports the interpretation of Breedlove and colleagues.

All but one of the elite articles was written by a science journalist; the exception critiqued genetic determinism and simplistic conceptualizations of sexuality and gender (*Guardian: G2*, 31 March 2000). The articles written by science journalists did not provide detailed comment on the reliability or validity of the research, or its implications. As such, it could be argued that they do not discuss issues of scientific or science-based controversy. Rather, they report the announcement of the publication of this research, discussing how it was conducted, and noting the broad conclusions. This emphasis on reporting the methods and findings, particularly in elite newspapers, was reflected in the analysis of citations. Of these, the vast majority were from scientists (seven from Dr Breedlove). A further three were quotes from non-scientists—all in popular newspapers —including two from the *Pink Paper*,[3] both of which challenged the research findings.

---

3. The *Pink Paper* is a free weekly UK-based newspaper for lesbians and gay men.

Genes and intelligence

The genes and intelligence topic was based on two research projects. The first consisted of a number of US and UK-based scientists, including Professor Robert Plomin of the Institute of Psychiatry, London. This team published a paper in the journal *Psychological Science* which argued that the gene variant *IGF2R* on chromosome 6 was more common among children with high IQ scores (Chorney *et al.* 1998). However, they also argued that this association was based on an average effect within the studied population accounting for about 2% of the variance in intelligence, and that many genes may be responsible for the inheritance of intelligence.

The second project was conducted by a team of researchers from Princeton University, MIT and Washington University. Published in *Nature* (Tang *et al.* 1999), this team produced a paper which suggested that mice which had been given extra copies of the gene *NR2B* performed better in experiments to test memory and learning. In effect, the researchers argued that the extra copies of the gene helped the mice to learn faster, taking in large amounts of information, and to remember this for longer. The researchers named these mice 'Doogies' after the 16-year-old child prodigy Doogie Howser MD from the fictional television sitcom of the same name. (This is an example of 'intertextuality'. In this US-produced show the main character was a graduate of Princeton University and a medical doctor who worked as a second-year resident at the Eastman Medical Centre.)

Nine newspaper articles reported the findings from the two projects, emphasizing the scientific research findings and the implications of the work for human intelligence. Several aspects of the finger length coverage also featured in the coverage of genetics and intelligence. For example, of the three articles reporting the study published in *Psychological Science* two included images of celebrities, in this instance those associated with being extremely intelligent (e.g. Albert Einstein), or whose intelligence was questioned. The following illustration shows how one popular newspaper article introduced a scientific controversy by extending the purely academic conceptualizations of intelligence discussed in the original study, using the example of the now retired footballer (soccer player) Paul Gascoigne (also known as Gazza):

PURE GENE-IUS – Being brainless isn't your fault, Gazza . . . you just lack that vital chromosome 6

It is a question which has troubled the greatest minds through the ages: What makes you clever?

Whatever it is, poor old Gazza – yesterday voted the thickest person in Britain – needs more of it. . . .

Even daft as a brush Gazza emerged from a tough upbringing to find fame and fortune. And though he may be lacking in conventional wisdom, he is incredibly gifted in the bodily kinaesthetic department – the instinctive movements which make you good at sport.

Daily Mirror (18 September 1998, emphasis in original)

Of the 23 individuals directly cited in these articles, 20 are scientists, most of whom are supportive of the researchers' interpretations. Notably, however, one of the reports of the *Nature* article also included aspects of scientific controversy through criticism of the research teams' conceptualization of intelligence:

Mice given extra gene become smarter

'This is a real piece of vulgar hype from Princeton. I'm rather shocked', said Steven Rose, head of the brain and behaviour research group at the Open University. 'The work is interesting. It uses novel genetic techniques. It doesn't tell us anything much that we didn't know before, but it is neatly done.'

He added, 'They shouldn't do this stuff, it really is irresponsible. Intelligence doesn't reside in a gene, or in a cell, or even in a brain. Human intelligence is something that develops as part of the interaction between children and the social and natural world, as they grow up. It's not something locked inside a little molecule in the head.'

The Guardian (2 September 1999)

Furthermore, all three of the non-scientists quoted were critical of the implications of the study involving mice, introducing issues of science-based controversy, in particular raising ethical concerns about the prospects of producing 'designer children'. In contrast, consistent references were also made to the prospect of developing therapeutic treatments, particularly those related to cognitive disorders such as Alzheimer's disease.

In summary, we argue that much of the reporting of the two topics is uncontested, uncritically describing the research findings. However, elements of scientific and science-based controversy were also apparent in the reporting of both topics and these aspects were used to reinforce or challenge the validity, reliability and significance of the research findings. Media professionals used a range of journalistic techniques to facilitate reader understanding, including headlines, images of celebrities and public figures with captions to anchor the preferred meaning, info-graphics, and direct quotations (see Mellor, Chapter 5.2 this volume for a discussion of these techniques). Furthermore, we have found evidence that scientists used similar techniques in the genes and intelligence topic, e.g. by naming the 'intelligent' mice after the TV character Doogie Howser MD. These examples suggest that both media professionals and scientists were aware of the importance of identifying shared concepts to aid reader comprehension.

Methods

This chapter deals with an aspect of the reception analyses from the *Reporting Contested Science* project involving focus group interviews. During these interviews participants participated in an activity called the 'newsgame' (see Philo 1990), where they were asked to produce a newspaper article about one of the scientific topics, *either* finger length and sexuality *or* genetics and intelligence. In adopting this method, we aimed to investigate group *process* accounts and *products* as a way of assessing participants' views about contested science. In effect, we argue that the products themselves (the news stories) and the interactions engaged in by subjects *en route* to completing the products reveal much about participants' knowledge, conflicts, attitudes and beliefs about complex contested science topics and media reporting thereof.

Focus group interviews were chosen as the method for this analysis because they provide a forum to debate specific issues within a relaxed interactive environment, producing a wealth of data which facilitate understanding of the similarities and diversity of opinions from a variety of participants. Furthermore, in situating this work within ethnographic studies of public engagement with science, they provide opportunities for researchers to investigate what participants *know* as scientific citizens (see Irwin and Michael 2003 for discussion), allowing participants to articulate their views and then discuss them using their own language and terminology (Kitzinger 1994). As Kitzinger and Barbour (1999, p. 4) argue:

. . . instead of asking questions of each person in turn, focus group researchers encourage participants to talk to one another: asking questions, exchanging anecdotes, and commenting on each others' experiences and points of view.

Focus group interviews therefore hold a unique position when compared with other reception analysis methods (such as questionnaire-based studies and surveys, individual interviews or participant observation), because they allow a variety of participants to discuss technically detailed scientific topics in a focused way (Holliman 2005).

The structure of the focus groups

Beginning with Philo's (1990) work on the 1984 to 1985 UK miners' strike, reception studies have used the 'newsgame' method as a way of understanding how groups interpret and contextualize media coverage (e.g. Kitzinger 1990). Principally, although not exclusively, these studies have asked participants to construct television news bulletins using a range of images taken from coverage of these issues.[4] The research presented in this chapter is informed by these earlier studies, for example in terms of the structure of the groups.

The group sessions were organized into seven stages, taking an average of 2 hours to complete. Initially (Stage 1) participants were briefed on the structure and purpose of the activity. Importantly for our discussions here, participants were informed that the groups were not designed to test their knowledge of contested science, but rather to investigate their views of media coverage of the topics.

In Stage 2 participants completed an initial questionnaire which documented their educational qualifications, prior knowledge and experience of media coverage of science in general and media coverage of *either* genetic explanations for intelligence *or* sexuality in particular. These questionnaires were completed individually, allowing participants to comment outside of the group interaction (see also Stages 6 and 7).

In Stage 3 participants were introduced to the newsgame, where they were asked to work together as a group to produce a newspaper article on one of the two scientific topics. To this end we provided paper and pens and groups were given between 20 and 30 minutes to complete the activity. The groups were also provided with stimulus materials drawn from actual media coverage of genetics and the topics. Each group was provided with 14 images, eight of which were the same for all the groups. Examples included: scientists working in laboratory conditions, a pregnant woman undergoing

---

4. We note that previous researchers have used exercises where participants have been asked to generate the text, but not the headline, layout or use of pictures, of newspaper articles (Philo 1996).

an ultrasound scan and scientific diagrams, such as an illustration of the DNA helix. A further six images were topic specific. For example, the finger length and sexuality groups had access to pictures of lesbians and gay men, a high-profile member of parliament (MP), Anna Nolan, a contestant from the first series of the reality show *Big Brother*, Dean Hamer, author of the 1993 'gay gene' study, and an info-graphic showing images of male and female hands. In contrast, the genes and intelligence groups had access to pictures of young children reading and playing chess, Albert Einstein, Carol Vorderman (a presenter on the UK's Channel 4 game show *Countdown*, well-known for her mathematical prowess) and a graphic illustrating a baby's brain.

Participants were asked to decide which newspaper they were writing for and to provide the lay-out, headline, correspondent and captions for any pictures they used. In addition, they were asked to generate 'copy' for the article. No further details were provided by the moderator to encourage participants to generate their own language and terminology. On completion of this task (Stage 4) the group members read the report, demonstrating the lay-out of the article.

In Stage 5, a semi-structured discussion followed, led by the moderator. Here participants were asked about a range of issues including: how they had made their decisions about the lay-out and content of their article, how accurate and authentic they considered it to be, what they remembered about the actual coverage and whether they felt media coverage influenced their views about science.

In Stage 6 participants were asked to complete a final questionnaire that asked them to document their media consumption and whether the focus group had changed their opinions about media reporting of science. As a final activity (Stage 7) participants were invited to ask questions about the case study under consideration. This also provided the moderator with an opportunity to address any problematic issues that were discussed, either with the group as a whole or with individuals.

Sampling

Overall, we collected data from 14 focus groups, involving 73 participants (36 female and 37 male participants) from a number of geographical locations (Table 1). The size of group ranged from three to eight participants, with seven groups examining biological explanations for sexuality and seven investigating genetic explanations for intelligence (Table 1).

We employed a structured sampling technique (see Kitzinger and Barbour 1999 for discussion) that included those expected to have an interest in science (e.g. postgraduate science students) and those with no perceived interest (e.g. office workers) (Table 1). Participants were drawn from pre-existing groups in that they consisted of participants who were already known to each other, e.g. in a work- or study-related capacity, or as friends. This facilitates a more relaxed atmosphere between participants (Kitzinger 1994). Further to this, participants were guaranteed anonymity and the group interviews were conducted in locations where participants would normally expect to meet (e.g. in the workplace in the case of the office workers, or the home of one of the participants in the case of the group of friends). We also included both single and mixed gender groups and the groups were moderated by both male and female researchers.

Table 1 Illustrating the group ID, description, gender distribution, date, geographical location and topic

ID	Description	Gender	Date (dd/mm/yy)	Place	Topic
1	Office workers	3 females	16/11/00	Buckinghamshire	Sexuality
2	Postgraduates – social science	3 females, 3 males	30/01/01	Buckinghamshire	Intelligence
3	Postgraduates – science	2 females, 3 males	01/02/01	Buckinghamshire	Sexuality
4	Local journalists	3 females, 2 males	05/02/01	Buckinghamshire	Sexuality
5	Neighbours	3 females, 3 males	08/02/01	Buckinghamshire	Intelligence
6	Group of friends	4 females, 2 males	21/02/01	Bedfordshire	Sexuality
7	Group of friends	4 males	03/03/01	Brighton	Intelligence
8	A-level students	6 females	04/03/01	Brighton	Intelligence
9	Group of friends	5 females	04/03/01	Brighton	Intelligence
10	Undergraduates – science	8 males	06/03/01	Liverpool	Sexuality
11	Postgraduates – science	3 females, 2 males	27/03/01	North Wales	Intelligence
12	Postgraduates – science	6 males	27/03/01	North Wales	Intelligence
13	Group of friends	4 females	02/04/01	Bedfordshire	Sexuality
14	Gay men	4 males	04/04/01	Buckinghamshire	Sexuality

Data collection and analysis

The data collected were of several different types. The questionnaires from Stages 1 and 6 were completed individually by participants, in hard copy, and collated using a spreadsheet. Primarily, however, focus groups generate data through interaction. Accurately recording these dynamic interactions requires some form of permanent record. To this end, field notes were generated by the moderator during Stage 3—itself a theoretically saturated activity (Silverman 2000)—to record the use of stimulus material, group interaction and non-verbal communication, and participants were also asked to leave the plans for their newspaper article. (Following the completion of the focus group interviews, computer-generated copies of the articles were produced for analysis.) On their own, these approaches are generally deemed to be insufficient to provide an accurate record of the group interaction, however (Silverman 1993). Stages 3, 4 and 5 were therefore audio-taped and transcribed.

The difficulties of generating reliable coding categories from dynamic group interactions are particularly relevant to this study (for discussion, see Mercer and Wegerif 1999). For example, choosing an appropriate unit of analysis can be challenging when working with 'language in use' because the phenomena of interest may be spread over several utterances (Mercer and Wegerif 1999). To address this issue we worked iteratively and inductively with the transcripts.

Furthermore, it may not always be clear what the meaning of a single utterance is, particularly when working from a transcript alone; saying 'yes', could mean agreement, or it could mean 'please continue with what you are saying' (Dillenbourg *et al.* 1996). Moreover, utterances may have more than one meaning and be renegotiated during the interactions (Mercer and Wegerif 1999). These concerns about over-reliance on transcribed recordings of group interactions can be addressed, in part, by analysing the data in combination; for example, we compared the full transcript, audio-tape recording and field notes with the final product. Overall, the resulting analysis of the group interactions can therefore be characterized as a descriptive account of language in use.

We analysed the group products, both in terms of the lay-out and structure of the articles, the headlines and the article copy, also noting any aspects of scientific or science-based controversy that emerged. Furthermore, we compared the selection of images across the groups with those provided as stimulus materials. The results of these investigations were then compared with the results of our analysis of media content (as discussed earlier).

Results and discussion

In this section we focus on the results from Stages 3 and 4 of the focus groups: the process and products from the newsgame activity.

Investigating process: newsgame preparation

The analysis of the process accounts illustrates the decision-making processes that groups underwent in producing the newsgame artefacts. As such, they explain far more about what the groups knew about media reporting and the two scientific topics than an analysis of just the product can provide. These process accounts also demonstrate that, apart from occasional requests for additional guidance or clarification about the task, participants worked effectively without direct moderation in all 14 groups. Indeed, the groups were far more free-flowing in terms of dynamic interaction between participants during the production phase (Stage 3), particularly when compared with Stages 4 and 5.

We found many features in the recorded dialogue which suggested collaborative working between participants during Stage 3, such as sentence completions, conflicts and repair sequences and task negotiation. We also found evidence of shared understandings, in particular when discussing which pictures to include as part of the newsgame artefact, discussion of references to popular culture deemed relevant to the production of the article and media production processes more generally. To illustrate these points we have included descriptive accounts of two of the groups, one producing a report on the finger length and sexuality topic, the other genes and intelligence.[5]

Finger length and sexuality
Group 10 included eight males studying for undergraduate geology degrees at Liverpool University. The group began Stage 3 by discussing whether to write a popular or an elite article, a feature that was common in all the groups. Having agreed to write a popular article one of the participants briefly outlined what they remembered of the scientific story

5. These groups were moderated by Simeon Yates and Claire Donovan, respectively.

behind the finger length topic. This participant became central to the production of the article, writing much of the text and expanding on their understanding of the science at various points. In this respect, there was very little collective discussion of the science in the early exchanges, apart from brief mentions of human cloning. Instead, the early discussions between participants, which became more collaborative, centred on the choice of headline, pictures and captions, and the overall layout of the article. However, discussion was not limited to the production of the article and the scientific information related to the story. The group also discussed issues related to the finger length and sexuality topic more generally, such as Clause 28,[6] effectively framing this issue as a science-based controversy. However, this issue was not mentioned in the final article and there was little discussion specifically related to the finger length and sexuality research at this time.

The group appeared to be confident about how to structure the layout of the article, also making reference to specific columnists and other similar storylines that might be covered by a popular newspaper. Further to this, the production of the article featured a number of conflicts that were resolved through negotiation, for example although no vote was actually taken the choice of headline was resolved when one participant was 'outvoted'. The chosen headline (see Figure 1 overleaf) itself also illustrates evidence of intertextuality, by using a catch phrase from the popular UK television comedy quiz show *Shooting Stars*, hosted by Vic Reeves and Bob Mortimer.[7]

Once the headline, layout and pictures were agreed participants discussed the copy for the article. During these exchanges there was more detailed discussion of the science related to genetic explanations for sexuality (e.g. issues of nature/nurture), but also the finger length and sexuality research and media reporting thereof (see the earlier discussion). In these exchanges participants considered whether surveys investigating sexuality and the statistics they generate could be informative about distributions among the wider population, issues that do feature in the group's final article (Figure 1). To this end, the group considered how statistical data might be represented in media reporting, for example:

M1:[8] talk about the link between finger length and gender first

M3: then go on to talk about the possible links of sexual orientation, if any

M1: the testosterone thing

M2: I just made that up! So if we start with that, and then copy that, then we'll describe how this supposed like survey took gay men and straight men and then looked at their fingers and there was an overwhelming

M3: positive correlation

M2: positive correlation

M4: think of a number, 0.7

M2: statistics

M1: stick a graph in at the bottom

M3: yeah, no axis [laughter and cross-talking] I love that

Group 10 (undergraduate scientists) discussing finger length and sexuality

WE REALLY WANT TO SEE THOSE FINGERS

NEW SCIENTIFIC RESEARCH FINDS LINK BETWEEN FINGER LENGTH AND SEXUALITY!!!

Recent research featured in New Scientist links finger length with sexual orientation. In women the ring finger and index finger tend to be about the same length. But in men the index finger is usually the shorter of the two digits.

A poll conducted by New Scientist surveyed 500 men from mixed social and ethnic backgrounds, out of these 500, 30 were gay and 27 or 90% of these had the same length index and ring finger.

Our scientific researcher A N Other can exclusively reveal that this is no fluke occurrence and a link definitely exists.

[Info-graphic showing male and female hands]

ADVERT

IS HE OR ISN'T HE?
Prominent MP XXXXX XXXXX (pictured here at a party conference) shows the characteristics mentioned in the report.

[Photograph of the high-profile MP with his hands in the air]

READERS, YOU DECIDE!!
YES 0800 ☐☐☐☐☐☐
NO 0800 ☐☐☐☐☐☐
Calls cost 10p

Figure 1 The newspaper article produced by Group 10 (undergraduate scientists)

6. Amongst other things Section 28 of the UK Local Government Act 1988 prohibited local authorities from promoting the teaching in state schools of the acceptability of homosexuality. Clause 28 was repealed in Scotland in 2000 and in England and Wales in 2003.

7. In each example we have used the group's plans to illustrate the proposed layout of the article, including the newspaper, date and headlines and verbatim article text. Text in square brackets has been added by the authors for illustration.

8. Where multiple participants have been quoted they are distinguished as F (female) or (M) male. Numbers also appear alongside these letters, e.g. M1 and M2, illustrating that different participants were speaking.

This exchange illustrates two important features in the production of the article: evidence of effective collaborative working, not least in terms of sentence completions, and knowledge of how to represent newly published science in a newspaper article, features that were also apparent in the other groups.

Genes and intelligence

Group 8 included six female A-level students studying in Brighton, one of whom had aspirations to become a journalist. This group is slightly atypical in terms of the length of the finished product; Group 8 produced a two-page spread (Figure 2) and even considered adding a third page during the preparation of the article. (All the others groups produced articles of one page or less.) This may be partly explained by the extensive discussion about the suggested scientific content of the article, genes and intelligence. Indeed, despite the fact that 74% of the participants who worked on the genes and intelligence topic stating that they had knowledge of media reporting of this issue, none remembered either of the specific research projects discussed earlier. They did, however, have a good working knowledge of issues that they considered to be linked to this topic. Group 8 discussed three in detail: human cloning, in particular ethical issues related to this issue; the announcement of the completion of the first working draft of the human genome; and the possible link between genes and intelligence. In the following exchange, the participants were choosing which pictures to include in the article:

F2: not Einstein

F1: no, Einstein was a good scientist, like if you clone Einstein

F2: you will get genius

F4: but how could you clone Einstein? He's dead

F2: you need his blood cells

F3: you need his blood, or even a hair

F5: no

F3: maybe his corpse is still around

F2: his?

F3: his corpse

Group 8 (A-level students) discussing genes and intelligence

In the end the group decided not to include a picture of Einstein. They did choose to address all three issues in their article—human cloning, the first working draft of the human genome, and genes and intelligence—linking them through ideas about genetic manipulation and 'designer children' (Figure 2).

As per Group 10 one participant became central to the production of the article, in particular organizing the task distribution. The group began by briefly discussing whether to write a popular or an elite article, agreeing to write for the former because they did not 'have enough scientific knowledge to do a broadsheet one'. The production of the article took longer than for other groups with extensive discussion of the use of images and the layout of the article. The groups decided to produce two pages, one that focused on the science, the other on the public reaction and the pros and cons of conducting these kinds

THE NEW AGE OF INTELLIGENCE?

[Graphic of a DNA strand]

Change you [sic] whole identity by just changing part of your DNA

DNA gives instructions as to which type of protein should be manufactured. Changing part of your DNA can actually change your whole identity because each element gives different instructions to your body, such as your eye colour.

By manufacturing the optimum amount of proteins the level of intelligence can be maximised and hence more intelligent human beings can be cloned or produced.

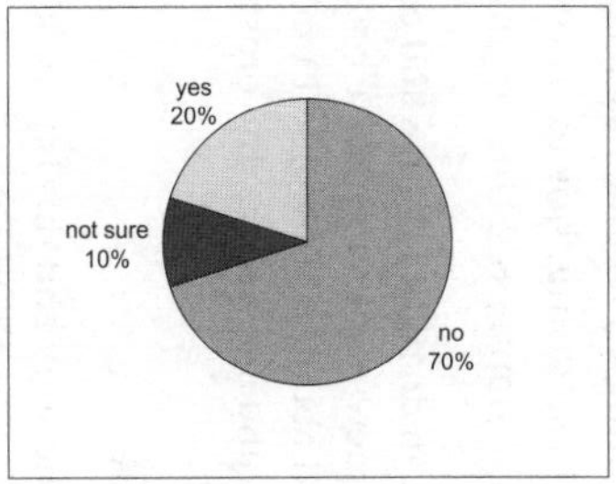

[Photograph of a laboratory-based scientist]

Research taken [sic] place in USA

Bill Clinton and the scientists who are involved in the Human Genome Project have released on 3 March 2001 to the press at the White House the latest research results by the Human Genomics Company. They have mapped out the code of human DNA.

[Graphic of designer baby]

You can change your baby into a genius

Activists outside the White House yesterday were violently arrested and beaten by police. Sarah Howard from the University of Connecticut was released from jail this morning and said that she was disgusted by claims that unethical research was being practised.

Pros & Cons

Pros
* Medical breakthrough - production of human organs which can save lives.
* Cloning of geniuses (e.g. Einstein)
* Agricultural breakthrough ("perfect" seeds, crops, poultry, animals)
* Boost economy (More educated/intelligent workforce and GDP.)
* Perfect race

Cons
* "Playing God"
* Degrading human lives (foetuses are cloned simply for human organs - no different from abortion)
* Disruption of circle of life (intrusion of nature's way of life)
* Disgusting images of mouse with human ear
* Cloning of Hitler/Mussolini

Elvis Presley born again?

We surveyed 3,000 individuals across the UK and asked their opinions on cloning, etc. The results are as follows....

Who do you want to be cloned?

[Two bar charts were included – male and female - showing voting as percentages. Those listed were actors or musicians, e.g. Julia Roberts, Marilyn Monroe, Jennifer Lopez, Eminem, Brad Pitt and River Phoenix.]

[Photograph of child reading]

Children's IQ can be increased by 40% in the year 2010

yes 20%
not sure 10%
no 70%

[Photograph of Carol Vorderman]

Carol Vorderman was voted the most unlikely female 2 be cloned

Figure 2 The newspaper article produced by Group 8 (A level students)

of research; in effect, commenting on science-based controversies associated with the science, but not the scientific ones. In so doing, however, they demonstrated sophisticated media literacy skills (see Carr *et al.*, Chapter 6.2 this volume).

F2: okay, well this is what we can do right . . . pros and cons, yes, so, if we like have like the article about it like here and then somewhere in a separate column we can have the pros and cons, this is blatant, what is being talked about, if they didn't understand the article [Page 1 reporting the science] then at least they can understand this, the pros and cons

F3: do newspapers list?

F1: they do, sometimes they do

F2: okay, exactly, this is like the story and this is part of it, the summary of it, whatever, in some cases you'd have like a separate column, like that

Group 8 (A-level students) discussing genes and intelligence

On reflection the group considered the first page to be more representative of an elite article, examining the science, and the second to be more like a popular report, which is consistent with the findings from the analysis of newspaper content discussed earlier.

The requirement for and position of the headline was discussed repeatedly. Initially, the group discussed the possibility of using the character 'Mini Me' from the second *Austin Powers* film[9] as the basis of the headline, showing evidence of intertextuality. In the end one participant chose a headline in the form of a question (Figure 2) and one participant supported that choice. (Interestingly, the group discussed the speculative nature of the scientific research they reported in their article in Stage 5 by emphasizing the role of the question mark in this headline.)

Again, the process accounts from the groups working on the genes and intelligence topic suggest that participants worked together effectively and were aware of aspects of media reporting of genetics.

Investigating products: the newsgame artefacts

All 14 groups completed the activity by producing a newspaper article, regardless of whether they were discussing media reporting of genetic explanations for sexuality or intelligence. Further analysis suggests that participants made similar decisions about which newspaper they were writing for and the perceived importance of the story—as indicated by the type of newspaper and page number—with many shared features, such as heavy use of images and captions, quotations from 'experts' and intertextual puns in the headlines (Table 2), demonstrating evidence of media literacy.

When questioned participants gave a number of reasons for their choice of newspaper, for example several reasoned that they could use a greater number of pictures and less copy, also requiring less detailed scientific knowledge. Others argued that they associated the controversial nature of the topics with 'a scandal – that's what I associate a tabloid with' (Group 10, undergraduate scientists). However, not everyone agreed. In Group 8,

---

9. The character Austin Powers is a parody of the fictional spy James Bond. In the second film in the series of four, 'Mini Me' is introduced as the clone of Austin Powers.

Table 2 Illustrating the group ID, topic, choice of newspaper, headline, pictures and page number

ID	Topic	Newspaper	Headline	Pictures	Page
1	Sexuality	The Daily Planet	HUMAN PERFECTION	Parliamentary debate, chromosomes	5
2	Intelligence	The Sun	Scientists Say S'eyes Matters!	Carol Vorderman, designer baby, baby brain	–
3	Sexuality	The Sun	BOFFINS PROBE GAY GENES	DNA strand, high-profile MP, pregnant women, female/male hands	2
4	Sexuality	The Sun	Hands up if you're gay!	Anna Nolan, high-profile MP, female/male hands	5
5	Intelligence	The Sun	BEAUTY OR . . . . THE BEAST	Carol Vorderman, Albert Einstein	5
6	Sexuality	The Sun	IT'S ALL IN THE HANDS!	Pregnant women, female/male hands, high-profile MP	4
7	Intelligence	The Sun	WHAT'S IN CAROL'S GENES REVEALED	Carol Vorderman, Albert Einstein	1
8	Intelligence	A tabloid	THE NEW AGE OF INTELLIGENCE?	Carol Vorderman, child reading, DNA strand, designer baby, laboratory scientist	1
9	Intelligence	The Tabloid	CAROL'S BRAIN CHILD	Carol Vorderman, Albert Einstein, baby brain	1
10	Sexuality	The Sun	WE REALLY WANNA SEE THOSE FINGERS	High-profile MP, female/male hands	7
11	Intelligence	The Trash	CK BABIES	Albert Einstein, child playing chess	1
12	Intelligence	The Tamper News	BOFFINS FIND SMART GENE!	Carol Vorderman, baby brain, DNA strand	1
13	Sexuality	The Sun	CHROMOSOME 6 DISCOVERY	Pregnant women, scientist and DNA, gay men and lesbians	1
14	Sexuality	Daily Mail	GAY Population Revealed!	High-profile MP, scientist and DNA	1

for example, we note that the issue of whether to produce a popular or an elite article was contested, with one participant arguing for the latter because 'it would be more interesting'. On reflection the Group 8 participants argued that, although they had chosen to produce a popular article, the final product resembled perceived characteristics of elite reporting. One participant from this group even went on to argue that popular reporters were more skilful than their elite counterparts because they had the ability to mediate complex information for a popular audience.

The terminology used in the articles is similar in many of the stories, often framing the articles in terms of science-based controversies that featured in actual reporting of these topics. For example, even though none of the groups remembered the specific report-ing of genetic explanations of intelligence discussed earlier, four of the articles which examined this topic reported that developments of this technology might lead to 'designer babies'; an issue that did feature in reporting of this topic. However, in contrast to actual reporting of this topic none of the group products discussed possible therapeutic benefits for cognitive disorders.

Moreover, several of the groups examining the topic of genetics and sexuality also discussed science-based controversies. For example, one of the groups reported the idea that embryos could be screened for sexuality so that: 'By 2012 it may be possible for unborn babies to be genetically screened, raising the possibility of homosexuality being eradicated' (Group 3, postgraduate scientists), whilst another considered the implications of the findings for the insurance industry (Group 14, gay men). By contrast, two of the genes and sexuality groups implicitly discussed aspects of scientific controversy, explicitly mentioning the nature/nurture dichotomy, and accepting a gene-centred view of behaviour over environmental explanations.

The use of images and captions suggests that participants were aware of reporting of the topics and/or analogous issues, such as genomics. In the first instance the groups used a similar number of pictures; typically two to three, and many included captions to anchor the preferred meaning of the message. This suggests that participants were aware of how pictures are used in newspaper reporting. Further analysis shows that the choice of images was also fairly consistent for the stories. For example, all seven of the groups reporting genes and intelligence included pictures of at least one celebrity or public figure (six groups used the picture of Carol Vorderman, four used Albert Einstein). This is consistent with actual reporting of this topic. Furthermore, four of the articles examining genes and sexuality included the info-graphic of female and male hands, three of which invited readers to 'test yourself', elements that were also apparent in the actual reporting of this topic. In addition, five of these groups included the image of the high-profile MP; three invited readers to decide whether he was gay; one confirmed that he was.

In contrast, Group 14 (gay men) used the picture of the high-profile MP in his capacity as a government minister, supporting the introduction of a genetic data base to be used by the insurance industry. Further analysis of this group's transcript shows that this group had considered whether to use this image in a similar way to the other groups.

In summary, we note that aspects of contested science were apparent in several of the group products, often reflecting aspects of science-based controversy that featured in actual reporting of the topics or analogous issues. We argue that this is evidence that participants were aware of media reporting of these issues, combining media literacy skills with evidence of prior knowledge, experience, attitudes and beliefs in relation to these topics and science more generally.

Conclusions

News media reporting of science is a key source of newly published scientific information for citizens, particularly those who have completed their formal education. This reporting provides partially mediated representations of the social construction of scientific knowledge. These filtered representations are delivered within an increasingly diverse and competitive media marketplace, resulting in interpretation and contextualization by media-literate scientific citizens. It is within this overall context that we have explored participants' views of media reporting of contested science. To do so we have employed methods that allowed participants to discuss *their* views within a supportive atmosphere working with pre-existing groups. In reporting the results of the focus groups, we have documented evidence to show that participants arrived at the groups with well-formed opinions about genetic explanations for sexuality and intelligence.

Overall, it is argued that the stories themselves are best understood as displaying participants' collective understanding of what constitutes a science story alongside a demonstration of what they comprehend news values to be. Two factors were key to this: first, participants were informed that the exercise was not a test; second, participants were aware that to produce the artefact they were required to reach a 'consensus', but that they would be given an opportunity in the general discussion to debate the authenticity of the artefact. Taken together, these premises facilitated useful collaborative discussion and debate between participants.

The relationship between news media coverage of contested science and the (re)construction of scientific citizenship is complex. Our research has provided us with some opportunities for examining both the development of participants' understanding of science topics and their views of scientists and media professionals because they allow us to investigate what participants already understand about these processes. We believe that such an approach could also inform those interested in the teaching and learning of controversial scientific and science-based issues.

Acknowledgements

This research was funded by an Open University Research Development Fund Grant (Number BR60751). In addition to the authors the project team included Claire Donovan, Hilary MacQueen and Simeon Yates.

■ REFERENCES

Barthes, R. (1977). *Image–Music–Text*. Fontana, London.

Brante, T. (1993). Reasons for studying scientific and science-based controversies. In: *Controversial Science: From Content to Convention* (ed. T. Brante, S. Fuller and W. Lynch), pp. 177–91. State of New York University, New York.

Chorney, M.J., Chorney, K. Seese, N., Owen, M.J., Daniels, J., McGuffin, P., Thompson, L.A., Detterman, D.K., Benbow, C., Lubiski, D., Eley, T. and Plomin, R. (1998). A quantitative trait locus associated with cognitive ability in children. *Psychological Science*, **9**(3), 159–66.

Dillenbourg, P., Baker, M., Blaye A. and O'Malley, C. (1996). The evolution of research on collaborative learning. In: *Learning in Humans and Machines: Towards an Interdisciplinary Learning Science* (ed. P. Reimann and H. Spada), pp. 189–211. Pergamon, Oxford.

Hall, S. (1981). The determinants of news photographs. In: *The Manufacture of News: Social Problems, Deviance and the Mass Media* (ed. S. Cohen and J. Young), pp. 226–43. Constable, London.

Holliman, R. (2004). Media coverage of cloning: a study of media content, production and reception. *Public Understanding of Science*, **13**(2), 107–30.

Holliman, R. (2005). Reception analyses of science news: evaluating focus groups as a method. *Sociologia e Ricerca Sociale*, **76–77**, 254–64.

Holliman, R., Scanlon, E. and Vidler, E. (2002). Reporting contested science: comparing media coverage of genetic explanations for sexuality and intelligence. *Proceedings of the 7th International Public Communication of Science and Technology Conference: Science Communication in a Diverse World*. University of Cape Town, Cape Town, South Africa. Available at **http://www.saasta.ac.za/scicom/pcst7/holliman2.pdf**.

House of Lords, Select Committee on Science and Technology (2000). *Science and Society*, Third Report. HMSO, London.

Irwin, A. (1995). *Citizen Science*. Routledge, London.

Irwin, A. and Michael, M. (2003). *Science, Social Theory and Public Knowledge*. Open University Press, Buckingham.

Irwin, A. and Wynne, B. (eds) (1996). *Misunderstanding Science: The Public Reconstruction of Science and Technology*. Cambridge University Press, Cambridge.

Jenkins, E. (2004). School science, citizenship and the public understanding of science. In: *Reconsidering Science Learning* (ed. E. Scanlon, P. Murphy, J. Thomas and E. Whitelegg), pp. 13–20. Routledge Falmer, London.

Kitzinger, J. (1990). Audience understandings of AIDS media messages: a discussion of methods. *Sociology of Health and Illness*, **12**(3), 319–35.

Kitzinger, J. (1994). The methodology of focus groups: the importance of interaction between research participants. *Sociology of Health and Illness*, **16**(1), 103–21.

Kitzinger, J. and Barbour, R. (1999). Introduction: the challenge and promise of focus groups. In: *Developing Focus Group Research: Politics, Theory and Practice* (ed. R. Barbour and J. Kitzinger), pp. 1–20. Sage, London.

Macnaghten, P. and Urry, J. (1998). *Contested Natures*. Sage, London.

Mercer, N. and Wegerif, R. (1999). Is exploratory talk productive talk? In: *Learning With Computers: Analysing Productive Interactions* (ed. K. Littleton and P. Light), pp. 79–101. Routledge, London.

Miller, D. (1999). Mediating science: promotional strategies, media coverage, public belief and decision making. In: *Communicating Science: Contexts and Channels* (ed. E. Scanlon, E. Whitelegg and S. Yates), pp. 206–26. Routledge, London.

Philo, G. (1990). *Seeing and Believing*. Routledge, London.

Philo, G. (1996). *Media and Mental Distress*. Longman, London.

Royal Society (1985). *The Public Understanding of Science*. Royal Society, London.

Schibeci, R. and Lee, L. (2003) Portrayals of science and scientists, and 'science for citizenship'. *Research in Science and Technological Education*, **21**(2), 177–92.

Shapin, S. (1992). Why the public ought to understand science-in-the-making. *Public Understanding of Science*, **1**(1), 27–30.

Silverman, D. (1993). *Interpreting Qualitative Data: Methods for Analysing Talk, Text and Interaction*. Sage, London.

Silverman, D. (2000). *Doing Qualitative Research: a Practical Handbook*. Sage, London.

Tang, Y.P., Shimizu, E., Dube, G., Rampon, C., Zhuo, M., Liu, G. and Tsien, J.Z. (1999). Genetic enhancement of learning and memory in mice. *Nature*, **401**(6748), 63–9.

Williams, T.J., Pepitone, M.E., Christensen, S.E., Cooke, B.M., Huberman, A.D., Breedlove, N.J., Breedlove, T.J., Jordan, C.L. and Breedlove, S.M. (2000). Finger-length ratios and sexual orientation. *Nature*, **404**(6777), 455–6.

■ FURTHER READING

- Irwin, A. (1995). *Citizen Science*. Routledge, London. Irwin's classic text explores the concept of scientific citizenship in relation to a number of environmental risks, in so doing constructing a convincing argument for an extension of traditional conceptualizations of expertise to include contextual forms of knowledge.

- Kitzinger, J. (1999). A sociology of media power: key issues in audience reception research. In: *Message Received: Glasgow Media Group Research 1993–1998* (ed. G. Philo), pp. 3–20. Longman, Harlow. Based on the findings of several research projects conducted at the Glasgow University Media Group, including one on HIV-AIDS, this chapter provides a helpful overview of key issues relevant to the study of audience reception.

- Macnaghten, P. and Urry, J. (1998). *Contested Natures*. Sage, London. Focusing on conceptions of 'the environment', this book examines how the natural and social worlds are inextricably enmeshed, emphasizing the importance of cultural understandings of the physical world.

- McQuail, D. (1997). *Audience Analysis*. Sage, London. Although not specifically examining audiences of science communication, McQuail's introductory textbook examines a number of theoretical perspectives of media audiences, drawing on a wide range of examples from relevant research.

Final reflections . . .

Richard Holliman, Elizabeth Whitelegg, Eileen Scanlon,
Sam Smidt and Jeff Thomas

Systematic, critically informed research into the wide range of traditional and emerging forms of science communication should be a core resource for those addressing science–society relations. If conducted effectively—and potentially including stakeholders contributing to the co-production of the research design (e.g. see Jensen and Holliman, Chapter 2.1 this volume)—such research can yield important findings that are useful to those participating in science communication, including citizens, scientists and scientific institutions, media professionals, museum designers and visitors, policy-makers and so on (e.g. see Holliman and Jensen, Chapter 1.3 this volume; Davies, Chapter 2.2 this volume; Meisner and Osborne, Chapter 2.3 this volume).

It should be apparent from the selection of chapters in this volume, which we do not claim to be in any way comprehensive, that science communication research can be conducted in a number of different ways, following different disciplinary traditions, theoretical perspectives and methodological and analytical approaches (see Hansen, Chapter 3.1 this volume; Leach *et al.*, Chapter 3.2 this volume). There is no one 'science communication research method', just as there is no single 'scientific method'. As a result, we acknowledge that research findings may sometimes be contested and contradictory; or complementary and even consensual, as is the case with scientific findings (see Holliman and Scanlon, Chapter 6.3 this volume). We argue that this mixture of contest, complement and consensus, which is the feature of contemporary academic study more broadly, should be welcomed. There is no one way to communicate science and no single solution to its study. Making sense of this complex landscape requires additional *critical* skills in assessing the quality of research and the subsequent findings. One of the aims of this volume has been to encourage the development of these important skills.

In contributing to this volume the various authors have made a number of observations about the current context for science communication research, not least the increasingly blurred relationship between deskbound journalists, proactive public relations professionals and 'media savvy' scientists (see Allan, Chapter 4.1 this volume; Trench, Chapter 4.2 this volume); issues that have been addressed elsewhere in some detail (e.g. see Nelkin 1995; Bauer and Bucchi 2007). As editors, however, we consider the most important of these to be the (ongoing) theoretical and methodological realignment in reconceptualizing audiences—with an emphasis on the plural form—for science communication. Such a realignment can be identified in relation to the two broad themes adopted in this volume: public engagement and popular media. The notions of a single undifferentiated 'public' who should understand science or the 'mass audience' waiting to passively consume popular media look conceptually naïve in the current context for science communication (see Hornig Priest, Chapter 6.1 this volume). Identifying a complete picture of why such a change has happened is beyond the scope of this brief set of reflections. It is possible, however, to begin to delineate several fundamental factors that have made important contributions.

In part, these changes have been driven by emerging forms of science communication that facilitate more dynamic, interactive and participatory forms of exchange, whether face-to-face in a novel public engagement activity or online with the ability to challenge spatial and temporal boundaries. With the development of science communication that is increasingly available 'any place, any time, anywhere, through any (compatible) device', the subsequent introduction of on-demand services (e.g. see Thompson 2006 for a media industry perspective), and the emergence of user-generated content and online networks of engaged participants (e.g. see Allan 2006 for a discussion of citizen media) have led to a reconfiguration and extension of the concept of audiencehood, not least for those early adopter citizen-consumers who can afford to seek out, and have the desire to obtain, a 'future proof' communications device.

These developments place additional demands on audiences, who are expected to extend (and continually revisit) their media literacy skills beyond those required for traditional forms of science communication. As Carr *et al.* (Chapter 6.2 this volume) show, this process starts at an early age; children and young people have sophisticated media literacy skills in relation to science communication. By the time audiences reach adulthood, however, the picture can become very mixed in terms of the levels of media literacy skills that a given individual might have in relation to science communication.

Changes in the audiences for science communication have also been the result of (market-driven) technological development in relation to recent policies for popular media, for example in terms of the digital switchover in the UK, the extension of established brands (e.g. newspapers) to communicate across a range of media platforms and ongoing decisions about the funding of public sector broadcasting (see Bennett, Chapter 5.1 this volume). They have also resulted from policy changes in relation to science and society more generally, in part as the consequence of previous high-profile science–society issues including BSE and genetically modified crops, and not least in relation to the calls for greater public engagement with science (e.g. see Irwin, Chapter 1.1 this volume; Stilgoe and Wilsdon, Chapter 1.2 this volume). These changes have also been driven by the work of 'pro-am' citizens (Leadbetter and Miller 2004) and science communication (and science–society) researchers (Wynne 2006). Given that traditional forms of science communication still exist, and in some cases continue to thrive, this results in a landscape for science communication that is more complex and extended, but only sometimes participatory. In conceptualizing audiences, science communication researchers therefore need to consider the types of expertise with which participants arrive and leave acts of science communication, whether they are both producer and consumer, active or passive, participating in deficit-led, dialogic or deliberative activities, and so on; audiences who choose to communicate can now be *viewers*, *viewsers* and *users* (see Bennett, Chapter 5.1 this volume).

It is possible to argue that audiences for science can now be global, as well as local, regional or national. Online networks of engaged 'pro-am' scientific citizens are more visible and extend across spatial and temporal boundaries—sometimes fleetingly—to consider and campaign about a single science-based issue; sometimes for more extended periods of time to discuss, debate, even to contribute scientific data (Holliman 2008). At the same time, access to emerging forms of science communication is unevenly distributed. Imagining audiences as 'information rich' and 'information poor' has profound consequences for the continuing structural inequalities that exist in our techno-science saturated world, and how innovation in science might be used to counter these context-dependent situations on a case-by-case basis. The luxuries afforded by the internet and various forms of digitally enabled media are in stark contrast to local citizens seeking a voice in determining how they gain access to clean water in Zimbabwe (Stilgoe and Wilsdon, Chapter 1.2 this volume).

Such examples show that not everyone will easily be able to participate in the 'Web 2.0 information superhighway' for science communication. Others still may choose to prioritize areas of their lives other than science for more prosaic reasons: 'I'm not interested in string theory, but I'm fascinated by succulents and cacti' or 'I hated science at school, but love football. Ask me about broken metatarsals and I'll provide a detailed diagram'. The emerging context for how audiences consume is one where *choice* is paramount. We now *navigate* through scientific and other forms of information, choosing pathways that we consider to be relevant, useful, fascinating, and so on. At the same time, we may choose to ignore or reject scientific accounts (Kitzinger 1998), seek out alternative perspectives or actively avoid certain aspects of science, while focusing on others, all of which can be rationally justified (Michael 1996).

The same pattern can be seen in relation to public engagement and popular media in science. Whilst the shift towards dialogic forms of participatory engagement is welcome, the deficit model is still alive and likely to retain its zombie-like ability to endure (see Irwin, Chapter 1.1 this volume). Partly this is down to those ideologically wedded to producing these forms of science communication whilst being allied to scientism's limitations, but we argue that an enthusiasm for engagement need not imply an all-consuming abandonment of 'one-way' delivery. As others have argued (e.g. Dickson 2005) properly fashioned science information can and should continue to be delivered to those eager to know more about the ever-shifting boundaries of frontier (and ready-made) science—via screen, lecture hall, newsprint and so on—and taking full advantage of the enthusiasms and unique understandings that scientists can offer; informed democratic citizenship requires some level of informing to happen. And of course, future provision is crucially down to audiences. It is possible to imagine 'deficit desire'—audiences who actively seek the linear lecture from a trustworthy expert with little expectation or interest to challenge the scientific perspective presented to them—as well as 'dialogue fatigue'—those who actively avoid forms of engaged democratic citizenship in relation to science. And they could be the same person. To an extent this is exemplified by the long-running BBC flagship science documentary *Horizon* (see Mellor, Chapter 5.2 this volume)—analogue television being a classic example of very successful and enduring linear media, even for those who now have the option to 'press the red button'. This show been sustained for over 40 years on a relatively small audience—at least in relation to soap operas broadcast in primetime slots on mainstream television channels—of 'deficit consumers'; the science 'fans' to whom Hornig Priest (Chapter 6.1 this volume) refers.

As a final point we note our own position as science communication teachers and researchers within these debates. The Open University has successfully pioneered what might be described as 'engaged pedagogies' since its inception in the late 1960s; student-centred approaches that value and empower the learner's perspective through (peer) facilitation, for example through the use of interactive learning technologies (see Laurillard 2004) and computer-mediated conferences (see Holliman and Scanlon 2006 for discussion). Like some mainstream pedagogies for teaching science communication, for example the lecture (or even this book), such approaches sometimes involve linear media and are still very popular and effective for it. We argue that each have their place, and also their strengths and weaknesses. In short, we conclude this volume by arguing that the 'dialogic turn', in particular in relation to science policy-making and ideas about scientific citizenship, is indeed very welcome. But it is not a panacea for science communication. If produced appropriately and sensitively to address particular contexts and forms of expertise, and by trustworthy sources, one-way, information-giving forms of science communication have a place in the overall scheme of things, and can be valued by audiences.

■ REFERENCES

Allan, S. (2006). *Online News—Journalism and the Internet*. Open University Press, Maidenhead.

Bauer, M. and Bucchi, M. (eds) (2007). *Journalism, Science and Society: Science Communication Between News and Public Relations*. Routledge, London.

Dickson, D. (2005). The case for a 'deficit model' of science communication [Editorial]. *SciDev.Net*. Available online at: **http://www.scidev.net/Editorials/index.cfm?fuseaction= readEditorials&itemid=162&language=1**.

Holliman, R. (2008). Communicating science in the digital age—issues and prospects for public engagement. In: *Readings for Technical Writers* (ed. J. MacLennan), pp. 68–76. Oxford University Press, Toronto.

Holliman, R. and Scanlon, E. (2006). Investigating co-operation and collaboration in near synchronous computer mediated conferences. *Computers and Education*, **46**(3), 322–35.

Kitzinger, J. (1998). Resisting the message: the extent and limits of media influence. In: *The Circuit of Mass Communication—Media Strategies, Representation and Audience Reception in the AIDS Crisis* (ed. D. Miller, J. Kitzinger, K. Williams and P. Beharrell), pp. 192–212. Sage, London.

Laurillard, D. (2004). Rethinking the teaching of science. In: *Mediating Science Learning Through Information and Communication Technology* (ed. R. Holliman and E. Scanlon), pp. 27–50. Routledge Falmer, London.

Leadbetter, C. and Miller, P. (2004). *The Pro-am Revolution: How Enthusiasts are Changing our Economy and Society*. Demos, London.

Michael, M. (1996). Ignoring science: discourses of ignorance in the public understanding of science. In: *Misunderstanding Science* (ed. A. Irwin and B. Wynne), pp. 107–25. Cambridge University Press, Cambridge.

Nelkin, D. (1995). *Selling Science: How the Press Covers Science and Technology*. 2nd revised edn. W.H. Freeman, New York.

Thompson, M. (22 March 2006). BBC 2.0: why on demand changes everything. *Royal Television Society Baird Lecture*. Available online at: **http://www.bbc.co.uk/print/pressoffice/speeches/stories/thompson_baird.shtml**.

Wynne, B. (2006). Afterword. In: *Governing at the Nanoscale: People, Policies and Emerging Technologies* (ed. M. Kearnes, P. Macnaghten and J. Wilsdon), pp. 70–8. Demos, London.

■ INDEX

Note: 'n.' after a page number indicates the number of a note on that page.

Plate 1 *Walking with Cavemen* title sequence. The interactive features are laid out on screen for the viewser to select between using the appropriately coded colour keys on their remote control: red for 'on now'; green for 'connect' and so on. © BBC/Discovery 2003.

Plate 2 The original photograph (see Jenkins 2001) that was cropped for use in Branigan (2001) (see Chapter 5.2). Photo courtesy of MU Extension and Agricultural Information.